Research Ethics Forum

Volume 6

This Series, Research Ethics Forum, aims to encourage discussion in the field of research ethics and the ethics of research. Volumes included can range from foundational issues to practical issues in research ethics. No disciplinary lines or borders are drawn and submissions are welcome from all disciplines as well as scholars from around the world. We are particularly interested in texts addressing neglected topics in research ethics, as well as those which challenge common practices and beliefs about research ethics. By means of this Series we aim to contribute to the ever important dialogue concerning the ethics of how research is conducted nationally and internationally. Possible topics include: Research Ethics Committees, Clinical trials, International research ethics regulations, Informed consent, Risk-benefit calculations, Conflicts of interest, Industry-funded research, Exploitation, Qualitative research ethics, Social science research ethics, Ghostwriting, Bias, Animal research, Research participants.

More information about this series at http://www.springer.com/series/10602

M. V. Dougherty

Correcting the Scholarly Record for Research Integrity

In the Aftermath of Plagiarism

Springer

M. V. Dougherty
Department of Philosophy
Ohio Dominican University
Columbus, OH, USA

ISSN 2212-9529 ISSN 2212-9537 (electronic)
Research Ethics Forum
ISBN 978-3-319-99434-5 ISBN 978-3-319-99435-2 (eBook)
https://doi.org/10.1007/978-3-319-99435-2

Library of Congress Control Number: 2018953157

This Springer imprint is published by the registered company Springer Nature Switzerland AG
The registered company address is: Gewerbestrasse 11, 6330 Cham, Switzerland

Acknowledgments

It is a pleasant task to acknowledge those who have assisted me during the writing of this book. My colleagues at Ohio Dominican University continue to provide an outstanding working environment. Bruce Gartner, Theresa Holleran, Bob Gervasi, and Peter Cimbolic have offered tremendous support over the years. Many colleagues here and at other institutions have given comments on all or parts of this book. ODU Philosophy Department faculty members Lawrence Masek and Brian Besong have provided welcome encouragement and critiques. I am pleased to thank Pernille Harsting for helpful criticisms, for assistance with translations of several texts, and especially for fruitful collaboration and conversations over the last decade. Matthew Ponesse, Carl Still, Michael Barnwell, and Perry Cahall offered detailed suggestions and criticisms. Leo Madden and Kate Riley have given helpful advice. The library staff at Ohio Dominican University provided valuable assistance, and I would like to thank Tim Sandusky for tracking down many materials for me. The Galvin Family Foundation, which established the Sr. Ruth Caspar Chair in Philosophy, made work on this book possible.

At Springer, I am grateful to editors Floor Oosting and Christopher Wilby for their support for this book. Two anonymous reviewers provided helpful suggestions. I would also like to thank series editors David Hunter, John McMillan, and Charles Weijer for including this book in the *Research Ethics Forum* series.

The Spring 2017 and Spring 2018 undergraduate honors students in my Critical Research and Writing course at Ohio Dominican University assisted me in thinking through many of the ideas in this book. It has been a privilege to discuss the issue of the scholarly record with them as they begin their own research careers. For a class project for the Spring 2017 version of the course, we worked together to request a retraction of an article in philosophy on the basis of suspected plagiarism, and we were pleased to see our work come to fruition when the publisher issued a retraction confirming the evidence we provided.

I am especially grateful to the dedicated editors, publishers, and colleagues who have assisted me in supporting published corrections of the scholarly record. I understand that examining evidence of research misconduct is time-consuming and not always easy, and I have been privileged to work with many people committed to research integrity. Over the years I have also corresponded with journalists who write on research integrity, including Victoria Stern, Alison McCook, Shannon Palus, Ivan Oransky, Adam Marcus, Davide Illarietti, Justin Weinberg, Nikolas Vanhecke, and Colleen Flaherty. I am appreciative of their careful articles on many of the cases discussed in this book. Chapter 4 previously appeared in *Metaphilosophy* 48.3 (2017): 258–283, and I remain thankful to the editor and reviewers for supporting its initial publication.

My three young children, Thomas, Benedict, and Cecilia, are a continual source of joy, and I have enjoyed talking with them about authorship. My wife and colleague, Michelle Dougherty, has provided excellent editorial advice, wise counsel, and good cheer.

Contents

About the Author

M. V. Dougherty holds the Sr. Ruth Caspar Chair in Philosophy at Ohio Dominican University (USA). He is author of *Moral Dilemmas in Medieval Thought: From Gratian to Aquinas* (Cambridge University Press, 2011) and has edited *Aquinas's Disputed Questions on Evil: A Critical Guide* (Cambridge University Press, 2016) and *Pico della Mirandola: New Essays* (Cambridge University Press, 2008). He has authored and coauthored articles on academic plagiarism, and his research interests include the history of ethics and research ethics. Since 2009, he has been involved in securing dozens of retractions, errata, and corrigenda for published articles in the discipline of philosophy and in related fields. His work in generating corrections for academic plagiarism and other authorship violations has been featured on *Retraction Watch* and on other academic news outlets.

List of Figures

List of Tables

Chapter 1
Introduction

Abstract This chapter introduces a book-length study on post-publication responses to academic plagiarism in humanities disciplines. Academic plagiarism damages the integrity of the scholarly record, corrupts the surrounding academic enterprise, and creates inefficiencies across all levels of knowledge production. The correction of the scholarly record for plagiarism is not a task for editors and publishers alone; each member of the research community has an indispensable role in maintaining the integrity of the published literature in the aftermath of plagiarism. The chapter identifies different senses of the expression 'correcting the scholarly record,' discusses the integrity of authorship, outlines the purpose of academic publishing, and distinguishes post-publication from pre-publication responses to plagiarism.

Keywords Scholarly record · Plagiarism · Science · Post-publication review · Meta-science

Great rewards exist in the world of learning for those who plagiarize without being discovered. An academic plagiarist benefits from the illusion of research productivity, and this façade can generate unwarranted promotions, grants, and new offers of employment. The harm that academic plagiarism inflicts upon others is quite devastating. Plagiarized articles take up valuable space in well-reputed journals that should have gone to the work of genuine authors. Promotions, grants, and new offers of employment are no longer available to authentic researchers when they have been expended upon plagiarists. Plagiarized articles also introduce systemic harm into the world of learning that is enduring and difficult to remedy. Genuine authors are stripped of credit for their original research findings, creating a situation that distorts the history of discovery and falsifies the genealogy of ideas. The scholarly record is thereby contaminated by this duplication of research, and subsequent meta-analyses performed on the existing literature are skewed. Plagiarized articles also pollute the downstream literature when researchers unwittingly incorporate the defective work in their publications. If untreated, academic plagiarism damages the integrity of the scholarly record, corrupts the surrounding academic enterprise, and creates inefficiencies across all levels of knowledge production.

M. V. Dougherty, *Correcting the Scholarly Record for Research Integrity*,
Research Ethics Forum 6, https://doi.org/10.1007/978-3-319-99435-2_1

Academic plagiarism will likely never be eliminated, but its deleterious aftereffects can be mitigated. This book examines post-publication solutions for repairing the body of published literature in the aftermath of plagiarism. It provides clarity about how to conceptualize the scholarly record, surveys the traditional methods for correcting it, and argues for new interventions to improve the reliability of the body of published research. Recent high-profile cases of academic plagiarism have demonstrated that the problem can no longer be regarded as minor in the world of learning.

In this book, I argue that the responsibility for maintaining a reliable scholarly record is not the domain of editors and publishers alone. The many other members of the research community each serve essential roles for insuring the integrity of the published literature. A range of post-publication practices promotes a trustworthy scholarly record. The precise moment that a plagiarized article is published can be conceived as the beginning of a toxic oil spill in a pristine environment. Various containment measures exist for minimizing the expansion of pollution in both academic and natural landscapes, but their most effective use requires that they be deployed quickly with the least delay. When a plagiarized article is not initially recognized as plagiarized, and instead is cited as genuine scholarship in the downstream literature, it surreptitiously expands the range of its pollution in the months and years after publication. Like a swelling oil spill, the contamination brought about by the plagiarized article will spread unless clean-up efforts are made to restore the academic environment.

In the chapters that follow, I draw on my experiences over the last decade in seeking published corrections of the scholarly record for violations of authorship norms in the discipline of philosophy and in related fields. Most of my work in this regard has been in the form of requesting published retractions for cases of academic plagiarism. I have also sought published corrections for cases where authors discuss (and commend) their own works in the scholarly literature under undisclosed pseudonyms, and for cases involving duplicate or redundant publication (sometimes called *self-plagiarism*). Some readers may be disappointed that the names of confirmed plagiarists are absent from these pages, but this book is not about plagiarists and it is not an exposé. All the cases referenced here have previously received varying degrees of press coverage in the United States or in Europe. I consider only publicly discussed cases and the publicly available evidence without demonstrating any new claims of plagiarism in the following chapters. The book does reveal, however, the little-known background information on some of philosophy's highest-profile retractions and corrections that have appeared in the last decade. The analysis of these cases is the basis of the policy recommendations that I propose here.

Scholarly publishing is not the only venue in which accusations of plagiarism arise. Sometimes they are alleged concerning the work of artists, journalists, musicians, speechwriters, and novelists, among others. Even social media postings have been subject to accusations of plagiarism. Furthermore, student plagiarism is a pressing matter for instructors in most educational settings, and much of the current scholarship on plagiarism focuses on student plagiarism and plagiarism detection.

This book, however, focuses on academic plagiarism, which is understood as plagiarism as it is found in published items of the research literature.

1.1 Correcting the Scholarly Record

The expression 'correcting the scholarly record' is used throughout this book, and it can be understood in two main senses. First, it describes the normal path of progress within a discipline, insofar as published works seek to improve upon or 'correct' earlier ones through clarification, augmentation, or refutation. When Sir John Maddox, the British biologist and longtime editor of *Nature*, was once asked, "How much of what you print is wrong?" he replied with the quip, "All of it. That's what science is about—new knowledge constantly arriving to correct the old" (Maddox 2017). This first sense of 'correcting the scholarly record' conveys that every discipline seeks to refine its collective body of truth-claims through the work of its practitioners. Advances in the discipline are marked with additions to the published literature that update or 'correct' what came before. The research of thriving disciplines is conducted by practitioners who are aware of the defects in the research methods of earlier periods. In the celebrated article, "Why Most Published Research Findings Are False," medical meta-researcher John Ioannidis catalogued the prevalent practices that decrease the likelihood that a given scientific finding is true (Ioannidis 2005). Newer research that avoids the methodological pitfalls of older studies will likely supplant the earlier ones. On this first sense of 'correcting the scholarly record,' therefore, science is often characterized as 'self-correcting.' These internal or *ad intra* corrections of the scholarly record occur with the normal progression of a discipline, as practitioners advance a discipline through the day-to-day activities of research and then display their findings through publication.

The second sense of 'correcting the scholarly record,' however, refers to an activity that is largely external to the practice of ordinary scientific inquiry, and it involves a disruptive intervention upon the scholarly record on the part of researchers, editors, and publishers. This kind of correction consists in post-publication intrusions that most often take the form of retractions, errata, corrigenda, and expressions of concern that are issued to change the status of an already-published work. In some cases, corrections involve the expurgation of works from the body of published literature. All these intrusions are external or *ad extra* corrections, different in kind from the ordinary self-correcting progression of research.

Internal corrections are quotidian in every discipline, but external corrections are occasional disruptive interventions. Corrections of the first type are expected and frequent in any established research field, and the discipline of philosophy offers typical examples. Researchers publish new articles that engage and refute earlier ones. Books are frequently issued in revised editions that incorporate improvements that were suggested in published reviews of the earlier editions. The second version of a major philosopher's *magnum opus* might even depart in significant respects from the claims made in the first version, as has been seen in the revised works of philosophers as far-ranging as John Rawls, Immanuel Kant, and even Plato, who was reported to have revised his dialogues late in life. A journal might invite an author to respond in print to a recently published critique of that author's work. Even somewhat drastic changes can be found among internal corrections. If a published translation of a major philosopher's work is found to be deficient, the text can be entirely retranslated and published, sometimes by the same publisher in the same book series only a few years later (e.g., Heidegger 1999 and Heidegger 2012). Well-established philosophy journals sometimes modify or change entirely their titles and subtitles to reflect their aims and subject matter more clearly (Van Herck 2013; Jacobs 2012; Scharff 1997; Moran 1993; Schrenk 1989; Rodes and Rice 1970; Anderson 1946; La Direction 1997). All such internal corrections are routine in the history of a discipline.

In contrast, external corrections are far from routine. In a perfect research world, they would not exist. These corrections typically rectify some substantive problem caused by error, fraud, or negligence on the part of a researcher or a publisher. These external corrections are deployed when internal corrections would be insufficient to repair or mitigate the harm of a deficient publication in an adequate or timely way. As newspapers, magazines, and media outlets occasionally need to issue corrections of the journalistic record, scholarly editors and publishers occasionally need to issue corrections of the scholarly record.

Separating internal from external corrections is important, and knowing when a situation is sufficiently serious to justify the latter is crucial for editors, publishers, and members of a research community. Fairly recent controversies in the discipline of philosophy show that there can be profound disagreements over which of the two kinds of correction is called for in a given case, even among the practitioners of a specialized research field. In one such episode, most of the advisory board members of the prominent feminist philosophy journal *Hypatia* issued a lengthy statement in 2017 condemning the publication of an article that had appeared in the journal, and the statement favorably raised the possibility of retracting the article in question. A petition requesting such a retraction was started, and the hundreds of signatories included members of the editorial board of the journal. In response to this call for retraction, the editor-in-chief of *Hypatia* issued an opposing statement. She urged the philosophical community instead to "engage in dialogue within the pages of the journal," thereby attempting to shift the calls for an *ad extra* correction to an *ad intra* one (Scholz 2017; see also Anderson et al. 2017). The editor-in-chief of *Hypatia* also noted that the traditional peer-review process had been followed in the article's original publication, and no claims of research misconduct (such as fabrication or plagiarism) had ever been alleged. The president of the board of directors of the journal later expressed support for this view of the editor-in-chief (McKenzie 2017; McKenzie et al. 2017). The fault line of the controversy was disagreement concerning how best to address the controversy-generating article. In this situation, readers as well as various editors of the journal themselves publicly disagreed about whether an *ad intra* or an *ad extra* correction of the scholarly record was warranted. The controversy resulted in a widespread reorganization of the journal, with the forced resignation of the advisory board and the voluntary resignation of the editor-in-chief (Anonymous 2017; Solomon et al. 2017; Zamudio-Suaréz 2017; Anonymous 2018). Incidents in the philosophical community involving public calls for retractions are no longer uncommon, and these situations almost always involve disagreements over the fittingness of deploying *ad extra* corrections.

Another relatively recent episode in philosophy featuring a petition for retraction targeted *Synthese*, a prominent philosophy of science journal. The three editors-in-chief of *Synthese* had appended an editorial disclaimer to the print version of a 2011 special issue titled "Evolution and its Rivals," and they did so without the consent of the two guest editors. Titled "Statement from the Editors-in-Chief of *Synthese*," the disclaimer appeared to impugn the integrity of the guest editors and the contributors to the special issue, as it stated that "the usual academic standards of politeness and respect in phrasing [...] are not consistently met in this particular issue" (van Benthem et al. 2011: A7).[1] Without specifying which subset of the ten articles of the special issue were guilty of such violations, the disclaimer also stated: "some of the papers in this issue employ a tone that may make it hard to distinguish between dispassionate intellectual discussion of other views and disqualification of a targeted author or group" (A7). Despite calls for retraction, the status of the published disclaimer remained unchanged, but one of the parties who judged his work had been misrepresented in the special issue published a response in a 2015 issue of the journal (Beckwith 2015).[2]

Among humanities disciplines, philosophy is not anomalous in experiencing controversies and confusions over choices between internal and external corrections. The author of a volume on legal studies made the unusual request that the editor of *The European Journal of International Law* retract a somewhat negative book review that had just appeared for her newly published volume. The editor demurred, instead extending to the offended book author the opportunity to write a response to the review that would appear online next to the original review. In an exchange of letters, which were published online, the editor encouraged the book author to publish her views within the journal so that they would be "seen by anyone who reads the review" (Weiler 2010). The author spurned this offer of pursuing what would have been an attempt at an internal correction, and instead she filed a suit for criminal defamation in a French court. The suit was unsuccessful, and the author was ordered to pay fees to the journal editor (Spencer 2011). The first-person accounts by both the plaintiff and the defendant that were published shortly after the trial revealed a fundamental disagreement over which kind of correction could best solve the controversy (Howard 2011; Weiler 2011).

The calls for retraction made to *Hypatia*, *Synthese*, and *The European Journal of International Law* were ultimately unsuccessful. Those with final editorial oversight in the three cases judged that internal corrections of the scholarly record were

[1] Although the disclaimer is absent from the electronic version of the journal on the publisher's website, it is present in JSTOR's electronic version that duplicates the print version (http://www.jstor.org/stable/41477269).

[2] For an unrelated call for retraction involving another article published in *Synthese*, see Sher et al. (2016).

sufficient to remedy any harm that might have occurred. As the three were covered in separate articles in *The New York Times*, they received significant attention far outside of academia (Schuessler 2017; Oppenheimer 2011; Liptak 2011). An outside observer might have concluded from them that external corrections of the published literature in humanities disciplines must be reserved for most serious violations of research integrity, such as those involving outright fraud. Such a general conclusion about the humanities is partly right. Retractions in humanities disciplines typically involve publications discovered to be the result of substantive research misconduct, such as academic plagiarism. Yet, published retractions are extremely uncommon in humanities disciplines. Even for relatively well-known cases of demonstrated research misconduct, the needed retractions to correct the scholarly record for posterity are often never issued. Humanities disciplines, including philosophy, suffer from a twofold problem: a lack of unanimity about what kind of situations justify external corrections (as the cases above illustrate), and a failure to issue external corrections for the demonstrated cases of research misconduct.

According to one study, there are 0.14 retractions per 1000 papers in biology and medicine, but only 0.01 retractions per 1000 papers in humanities (Lu et al. 2013: 1). Retractions in humanities have been characterized as "incredibly rare" and are regarded as "rare occurrences" (Oransky 2013; McKenzie et al. 2017). One commentator has warned that "we should be highly critical and suspicious of those journals and fields in which papers are retracted very rarely, if at all" (Fanelli 2013: 6). The integrity of any discipline requires that it have the mechanisms for insuring a reliable scholarly record. Whether the lower rate of retraction in humanities is because of a lower incidence of fraud and error, or because of a lower rate of detection and correction, remains difficult to prove.

My work with colleagues over the last decade in securing dozens of external corrections of the scholarly record for plagiarized articles in philosophy and in related fields has been restricted to cases involving verbatim and near-verbatim plagiarism of various kinds and degrees. I have not focused on situations involving what one might call a "plagiarism of ideas." Allegations that an author of record has misappropriated another's ideas are notoriously difficult to adjudicate; often claims over the priority of the discovery of an idea are discussed and resolved by internal rather than external corrections of the scholarly record. One might think that word-for-word or near-word-for-word plagiarism is easy to prove, and that it must be relatively easy to obtain external corrections (e.g., retractions) for plagiarism cases of this kind. In my experience, however, such is not the case.

1.2 The Integrity of Authorship

A basic expectation in the world of learning is that an identity exists between the *author of record* and the *author of origin* for any published scholarly work. That is, the person whose name appears on the published work is assumed to be the same person who wrote or created it. Some theorists have spoken of this expectation as an "implicit contract" between author and readers (Roig 2015: 2). In acts of academic plagiarism, this fundamental expectation is violated. The plagiarist forgoes the use of conventional signals for indicating to a reader what is original and what is not. By failing to employ quotation marks, or omitting extract or block quoting, or leaving out footnotes and in-text citations, the plagiarist generates an appearance to the reader that the author of record and the author of origin are one and the same. In all acts of plagiarism, a subtle bifurcation is created between the author of record and the author of origin, and the reader has no way to determine—at least from the text itself—that the author of record is different from the genuine author. The basic expectation of this identity between the author of record and the author of origin is so deeply held in the world of learning that acts other than plagiarism that violate that expectation are also held in contempt. Ghostwriting, the use of an undisclosed *nom de plume*, and guest or gift authorship, are each considered to be substantive violations of the present-day ethical norms of academic publishing. In short, there is a strong presumption against "nominal authorship" in academic or scholarly writing.

In acts of plagiarism, the reader's trust is exploited. Readers of plagiarized articles are manipulated into assuming that the author of record is the author of origin. Of course, a plagiarized article may appropriate texts from the most reliable of sources and thereby present much information that benefits the reader. In such cases, the reader is still misled about the history of discovery and the genealogy of ideas, even though the reader may have learned much while unwittingly reading a plagiarized article.

When publishers and editors fail to correct the scholarly record for books and articles that are demonstrably fraudulent, their omissions further damage the trust that researchers have in the body of published literature. Recent significant cases of serial research misconduct in fields as diverse as psychology, economics, and leadership studies have prompted painful and detailed reflections about the research methods, publication requirements, and manner of correction in those disciplines. One can hope such soul-searching will result in an improvement in those fields (Nelson et al. 2018; Chambers 2017; Horbach and Halffman 2017; Spoelstra et al. 2016). Academic scandals can prompt critical self-reflection, and one can be optimistic when a discipline owns up to its failures. Humanities disciplines, including philosophy, will benefit from an unflinching self-assessment of the work by its practitioners in dealing with the problem of academic plagiarism.

Philosophy has provided several high-profile instances of serial plagiarism in recent years, and the plagiarism cases involve some of the most-respected and best-known scholarly publishers and academic journals of the discipline. These instances of serial plagiarism provide an unpleasant confirmation of a startling suggestion that Mark Fox and Jeffrey Beall have put forward in an article titled, "Advice for Plagiarism Whistleblowers." They state that "those who engage in plagiarism tend to do so in multiple publications; plagiarism tends not to be an isolated, one-off event" (Fox and Beall 2014: 346). Even journal editors now warn that what might initially appear as a one-time "moment of madness" can quickly develop into a "case of serial plagiarism" once one starts to examine other articles after the index case has been discovered (Martin 2007: 907). The existence of plagiarism among philosophers today has received some increased attention. A recent article calling for the establishment of a professional code of conduct for philosophy argues that such a code should begin with a condemnation of plagiarism. The authors write: "A prescription against plagiarism [...] is also of such central importance to the practice of academic philosophy that we believe it should head any discussion of a philosophers' code of practice" (Davies and Felappi 2017: 759).

1.3 Pre-publication and Post-publication Responses to Plagiarism

Important recent monographs about the problem of plagiarism today focus primarily on improving measures for identifying plagiarized manuscripts prior to publication. Debora Weber-Wulff has analyzed the problem of plagiarism in European doctoral dissertations, and she considers a culture in higher education that has enabled an acceptance of plagiarism to take hold in some universities (Weber-Wulff 2014). Other volumes have considered the tools and practices available to journal editors tasked with determining whether manuscripts submitted to journals are plagiarized (Gipp 2014; Zhang 2016). If defective manuscripts can be recognized as such in the pre-publication process, they can be rejected outright. Since publication is thereby avoided, no direct harm to the scholarly record occurs. Editors and publishers rely on peer-reviewers, text-matching software, and signed author agreements to reduce the chances that a plagiarized manuscript will ever reach print. These traditional methods of plagiarism detection are fallible, however, and peer review has been identified as a particularly weak mechanism for detecting plagiarized manuscripts (Martin 2012, 2013).

Some editors go further by creating deterrents for the submission of plagiarized manuscripts. One journal in applied mathematics publishes within its pages the names of those who have submitted less-than-original manuscripts that have been discovered, at some point during the review process, to be plagiarized (Bouyssou et al. 2006, 2009). The hope is that such public notice will minimize future deficient submissions both there and at other journals. Still further, certain journals maintain a list of banned "authors," again attempting to stem the flow of submissions from plagiarists across journals. Some of these lists of banned individuals are only available to editors (Hagen and VenGraitis 2007), but some journals maintain online public lists (Crama et al. 2006; RePEc Plagiarism Committee 2011).

Although one finds general agreement about using well-established and highly refined processes for dealing with plagiarized manuscripts prior to publication, the same cannot be said for correcting the scholarly record in the aftermath of the publication of plagiarized work. The level of theorizing about the practices for dealing with the repercussions of a published plagiarized article is much lower. Even the best plagiarism-prevention measures, which include high-quality peer reviewers, the latest technologies, and various deterrents, still have a non-negligible failure rate. After-the-fact or post-publication remedies for plagiarism are still required to address situations where plagiarized work has been published and thereby added to the scholarly record.

1.4 The Purpose of Academic Publishing

To inflict upon unsuspecting readers scholarly work that is substantively deficient is to harm them. To do so knowingly to a vulnerable readership is a failure of authorial ethics. Unfortunately, individuals might engage in scholarship for less noble ideals than sharing with others what they have learned or discovered. Some unscrupulous academics might be unconcerned with whether readers are ever injured by their deficient work; there may be some individuals, too, who do not care whether their academic publications are ever read at all. On such a view, an institution's mission-driven goal of advancing knowledge might be misinterpreted as a merely arbitrary requirement that universities have for initial hiring or for later tenure and promotion. One commentator puts the matter this way:

> Many academic fraudsters aren't aiming for a string of high-profile publications. That's too risky. They want to produce—by plagiarism and rigging the peer-review system—publications that are near invisible, but can give them the kind of curriculum vitae that matches the performance metrics used by their academic institutions. They aim high, but not too high. (Biagioli 2016: 201)

In other words, it may be the case that an academic "who plagiarizes a published article does not want to show off the article to the world as if it was his or her own. He or she wants to show off a longer publication *list* to the dean" (Biagioli 2012: 463).

An inauthentic researcher might view scholarly publishing as only instrumental, as something wholly removed from the purpose of contributing to knowledge. Publications might be seen as personal trophies primarily acquired to enhance a research profile, with little thought given to how one's publications will affect readers. Such a view of research conflates the accidental benefits of scholarship with its purpose. The *curriculum vitae* of an inauthentic researcher becomes an academic Potemkin village, where specious works masquerade as contributions to knowledge. In an essay titled "The Diffusion of Knowledge through Publishing," the historian Jarislov Pelikan proposes that the act of publishing one's research is epitomized by the medieval motto of the Order of Preachers, which is often rendered as, *To contemplate the truth and to share with others the fruits of that contemplation* ([*Contemplare et*] *contemplata tradere aliis*) (Pelikan 1992: 123). This motto is customarily attributed to the thirteenth-century philosopher and theologian Thomas Aquinas, who argued that to share with others what one has contemplated is more perfect than simply to stop at contemplation (*Summa theologiae*, III, q. 40, a. 1, ad 2). Published scholarship has as its final goal the transmission of knowledge; the purpose of scholarship is to present knowledge that will be helpful and used by others.

If members of the world of learning—including researchers, editors, publishers, and even victims of plagiarism, to name a few—remain unmotivated to respond with best practices for correcting the scholarly record in cases of proven plagiarism, the risk/reward ratio will remain unchanged and some researchers will succumb to the temptations of plagiarism with the view that the potential benefits offset the dangers. Using a game-theory model, one economist has argued that it is "rational" for researchers to plagiarize, given current structures in academic publishing and the present-day incentives that govern the academic life (Hoover 2006: 449). Similarly, a recent study using rational crime theory shows that the basic deterrents for plagiarism and other forms of research fraud are lacking (Cox et al. 2018). In a disclaimer appended to a largely tongue-in-check essay titled, "Ten Simple Rules for Scientific Fraud & Misconduct," authors Nicolas Rougier and John Timmer offer the surprisingly weak conclusion that the benefits of wrongdoing are only "probably not worth the risks" (2017: 1). The likelihood of negative repercussions for academic plagiarism may be seen to be slight, as victims of plagiarism rarely report it, and some publishers view any acknowledgment of plagiarism to be contrary to their financial and reputational interests. In the view of one theorist, "the goals of publishers and those of academia with regard to intellectual dishonesty have diverged considerably" (Harms 2006: 1; see also Lewis et al. 2011). Other researchers have argued further that academic plagiarism cases "share basic characteristics with the reasons cited for white-collar crime" and point out the lack of substantive deterrents; plagiarists "do not serve jail time or receive harsh prison sentences" (Elliott et al. 2013: 92).

The rise in notable plagiarism cases suggests that, at least in the minds of some researchers, the benefits of professional advancement through academic plagiarism outweigh the risks of disclosure. These risks are shared unequally, however. Research misconduct when performed by established scholars is generally viewed more leniently than when the same kind of misconduct is performed by early career researchers. Senior scholars are often given a pass in proven cases of academic misconduct, while early career researchers are often exiled from research tracks or academia entirely. In a discussion of academic plagiarism and other scholarly failures, theorist Brian Martin notes that "established academics have more tools to avoid or resist challenges to their abuses and prerogatives" (2016: 925). Only a change in academic culture, based on a strengthened commitment by researchers, editors, publishers, and institutional leaders to correct the scholarly record will disincentivize academic plagiarism.

Perhaps the competitive environment of academia, the pressures to receive grants and publish articles, and the perceived rewards of academic prestige are the major contributing causes generating the rise in notable plagiarism cases. A recent book on academic publishing suggests that plagiarism "can be attributed to several motivating factors, the most important of which is pressure to publish under the POP [Perish or Publish] culture" (Moosa 2018: 61–62). Plagiarism is just one of the three so-called 'cardinal sins' of academic research, however. Alongside data fabrication and data falsification, plagiarism may appear as the least serious of the three.[3] The widespread use of euphemisms for academic plagiarism suggests a lessened concern for its ill-effects. Where one would have expected to hear the word *plagiarism*, at times one encounters instead obscure expressions such as 'unauthorized collaboration.' The founding journalists of *Retraction Watch* have noted the presence of "euphemistic language and notices that border on the preposterous" in many published corrections, observing that "some of the best examples of this tortured wordplay involve retractions for plagiarism" (Marcus and Oransky 2017: 120). The failure of some researchers, editors, and publishers in humanities disciplines to use precise language to address the issue of academic plagiarism counts as evidence that the current methods for addressing the problem are either inadequate or poorly implemented.

To be sure, accusations of plagiarism within humanities disciplines are sometimes unfounded. Severe accusations have been leveled against highly successful present-day contributors to the field in philosophy, and yet over time the accusations are judged to be without merit or at least they gradually fade from view (Anonymous 1990; Neale 2001; Tudvad 2007; Hansson 2008; Anderson 2011; Timmins 2013; Wolters 2014; Hansson 2017; cf. Taylor 1916; Quintanam 2017). Even though a false allegation of plagiarism, found to be unwarranted, can be extremely damaging to a researcher's career and reputation, good-faith academic whistleblowing remains essential to maintaining a reliable scholarly record.

1.5 A Précis of Chapters

This book is divided into eight chapters. The next chapter provides a conceptualization of the scholarly record. A clear notion of the scholarly record as it exists today is a prerequisite for a successful analysis of the various instruments for correcting the published literature in the aftermath of plagiarism. I propose that items that belong indisputably to the scholarly record meet six hallmarks, which are set forth as the *Knowledge*, *Authorship*, *Publication*, *Library*, *Database*, and *Discipline* conditions. Books issued by scholarly presses and articles appearing in established journals have been the traditional formats for presenting research findings, and such items clearly meet these six conditions. Advances in technology, however, have occasioned new modes for recording and disseminating knowledge, and they create challenges to the long-standing conceptualizations of the scholarly record. Online post-publication review venues, open-access initiatives, interactive scholarly websites, online document repositories, and other venues invite a reconsideration of the precise boundary of the published literature. I distinguish between synchronic and

[3] Some disagree and tout plagiarism as the most significant academic failure. Ben Rosamond describes plagiarism as "perhaps the most grievous academic crime" and claims that there is an "almost universal consensus [...] that it amounts to just about the most serious offense that can be committed in academic life" (2002: 167, 167–168).

diachronic approaches to the scholarly record. Separating these two allows one to isolate the scholarly record as a present system from past versions that have operated under different parameters and also from future versions that are only anticipated.

In the third chapter I defend a fourfold heuristic for determining when academic plagiarism has occurred. Drawing from contemporary literature on research integrity, I propose that academic plagiarism has been committed when there is: (1) a non-trivial appropriation of words, images, or formulas, (2) with inadequate credit, (3) that generates an appearance of original authorship, (4) in a discrete item belonging to the scholarly record. This approach is sufficiently general to include a wide range of text manipulations, and yet it is sufficiently narrow to express to the particularities of plagiarism in the context of published research findings. In defending this heuristic, I argue that intent is not required for academic plagiarism, and I propose that academic plagiarism should be treated as a strict-liability offense. The presence or absence of a guilty mind (*mens rea*) is irrelevant to the need to correct the scholarly record when publications themselves are deficient. Intent may be an important element to be considered by institutions that have the role of investigating and punishing wrongdoers for scientific misconduct, but intent should be considered immaterial by members of the research community who have the privilege and responsibility of maintaining the reliability of publications for the world of learning. Too often the role of correcting the scholarly record is conflated with the role of investigating and issuing punishments for research misconduct, but the two are quite different. The chapter also considers the topic of duplicate or redundant publication (often called "self-plagiarism"), and it distinguishes academic plagiarism from copyright violation.

The fourth chapter explores the extent to which the discipline of philosophy currently falls short of maintaining a reliable scholarly record in cases of demonstrated plagiarism. Individuals discovered to have engaged in wide-scale serial plagiarism in philosophy are relatively few, but the academic publishers falling victim to them are many. Some of the most respected publishing houses in philosophy have recently faced the issue of having published plagiarized material. The chapter uses a specific context of serial plagiarism involving 43 articles and book chapters by one author of record as a test case. The various responses by these publishers to this instance of serial plagiarism provide a real-time snapshot of the practices for correcting the scholarly record in the discipline of philosophy. I propose a new rubric for evaluating published corrections of the scholarly record for cases of demonstrated plagiarism. On this rubric, the highest-scoring corrections are those that: (1) unambiguously declare that a plagiarized work is plagiarized, (2) clearly credit the original source material misappropriated in the act of plagiarism, and (3) are easily accessible to the scholarly community without registration barriers or paywalls. This analysis yields a twofold conclusion: first, relatively little uniformity exists among publishers in philosophy for responding to plagiarism; and second, the discipline of philosophy often falls short of the accepted practices for correcting the scholarly record in contrast to the natural sciences.

Academic whistleblowing is the subject of the fifth chapter. The process of issuing corrections of the scholarly record for plagiarism usually begins when a third-party researcher discovers undocumented textual parallels of sufficient import to justify notifying others of the evidence of suspected plagiarism. Despite the essential role that academic whistleblowers serve in initiating the oftentimes lengthy process of correcting the scholarly record, individuals who disclose evidence of suspected plagiarism are often subject to considerable backlash. To be sure, the evidence they provide, even when impeccable, can create a significant workload of verification for editors and publishers, as well as for research integrity officers at the institutional homes of the suspected plagiarists. I examine the benefits and hazards of multi-targeted whistleblowing and discuss the harassment and witness intimidation typically experienced by those who blow the academic whistle in good faith. The increasing awareness among researchers and institutional authorities that to harass whistleblowers is itself a form of misconduct reflects an important recent shift in academic culture. On the other hand, academic whistleblowers in recent times have been described as post-publication vigilantes for their efforts in securing corrections of the scholarly record, so the professional dangers of academic whistleblowing should not be understated.

The sixth chapter examines the responses of editors and publishers who were presented with evidence of suspected plagiarism in a series of 14 articles and book chapters for one author of record. As the publications of this author of record divide into those in philosophy and those in health communication, a comparison is possible between the manner in which different disciplines respond to evidence of suspected plagiarism. Using news reports as well as publicly issued statements from the home institution of the author of record as a backdrop, I examine various textual parallels in light of the 12 published retractions, errata, and corrigenda that have been issued in these cases.

The seventh chapter considers recent cases of contested authorship in two articles involving an American philosophy professor. The examination of these unusual publishing circumstances, which include allegations of self-plagiarism, places in high relief the requirements of genuine authorship, the complexity of some self-plagiarism claims, and the challenges of requesting corrections of the scholarly record. Differing standards of evidence are often proposed in considering whether to retract a published article for authorship violations, and the chapter evaluates these standards in light of potential benefits to the larger research community. Researchers, editors, and publishers sometimes conflate standards from legal, moral, and scientific domains, and a lack of clarity about the applicable standard can significantly delay a proper correction of the scholarly record.

The last chapter of this volume considers ways in which published corrections of the scholarly record are disseminated throughout the research community. Even when editors and publishers issue retractions, these notifications can still be minimized to the point of irrelevance if they are not reflected in the research practices of other members of the scholarly community. In some humanities disciplines, including philosophy, entries for articles in specialized research databases are not updated to reflect when the status of an article has been changed by a publisher through a retraction, erratum, corrigendum, or expression of concern. The major databases for biomedical disciplines fare much better in solving what can be called the *Database Problem*. Other problems plague the promulgation of published corrections. At times an article is retracted by a publisher, but then the article is reprinted in another venue without acknowledgment of the retracted status of the original publication (the *Anthology Problem*). Some publishers will correct the electronic version that it hosts in one system but will leave other electronic versions unchanged (the *Platform Problem*). Quite frequently the original uncorrected versions of articles are uploaded on secondary websites outside the control of a publisher. These copies are more accessible to researchers than the corrected version on the proprietary platforms of publishers, and so researchers download and use the more easily accessible but uncorrected versions (the *Repository Problem*). Furthermore, the authors of record and their institutions at times keep quiet about the changes in status of articles and do so in breach of authorial ethics. It is not uncommon for such an author of record to fail to update an online *curriculum vitae* to reflect the changes. Such an author of record often continues to cite retracted articles in later publications and to maintain the retracted articles on academic social networks or pre-print servers without offering a warning to colleagues in the general academic community about the status changes (the *Pretend-It-Didn't-Happen-Problem*). I propose clear solutions to these problems and note some positive developments.

The main purpose of this book is to argue that the correction of the scholarly literature for plagiarism is not a task for editors and publishers alone; each member of the research community has an indispensable role in maintaining the integrity of the published literature in the aftermath of plagiarism. Drawing from examples from the field of philosophy and related disciplines, I have sought to prove that current post-publication responses to academic plagiarism are insufficient. Humanities disciplines, and philosophy in particular, fall short in contrast to the natural and biomedical sciences when it comes to ensuring the integrity of the body of published research. I hope this book will be valuable not only to those in the field of philosophy and other humanities disciplines, but also to those interested in research ethics, meta-science, and the sociology of research.

References

Anderson, John. 1946. News and notes: Title of journal. *Australasian Journal of Psychology and Philosophy* 24 (3): 192.

Anderson, Mark. 2011. Telling the same story of Nietzsche's life. *Journal of Nietzsche Studies* 42 (1): 105–120.

Anderson, Elizabeth, Leslie Francis, Heidi Grasswick, Miriam Solomon, and Lisa Tessman. 2017. Board of directors' statement (May 18, 2017). *Hypatia*. http://hypatiaphilosophy.org/np-board/non-profit-board-statement-may-18-2017. Accessed 6 July 2018.

Anonymous. 1990. Bad manners? *Quadrant* 34 (10): 65–69.

———. 2017. Editors' statement (July 20, 2017). *Hypatia*. http://hypatiaphilosophy.org/editor/editor-statement-july-20-2017. Accessed 6 July 2018.

———. 2018. Announcement from *Hypatia*'s board of directors and task force co-chairs. *Hypatia*, February 23. http://hypatiaphilosophy.org/announcements/announcement-hypatias-board-directors-task-force-co-chairs. Accessed 6 July 2018.

Beckwith, Francis J. 2015. Or we can be philosophers: A response to Barbara Forrest. *Synthese* 192 (Supplement 1): 3–25.

Biagioli, Mario. 2012. Recycling texts or stealing time? Plagiarism, authorship, and credit in science. *International Journal of Cultural Property* 19 (3): 453–476.

———. 2016. Watch out for cheats in citation game. *Nature* 535: 201. https://doi.org/10.1038/535201a. Accessed 6 July 2018.

Bouyssou, Denis, Silvano Martello, and Fran Plastria. 2006. A case of plagiarism. *4OR: A Quarterly Journal of Operations Research* 4 (1): 11–13.

———. 2009. Plagiarism again. *4OR: A Quarterly Journal of Operations Research* 7 (1): 17–20.

Chambers, Chris. 2017. *The seven deadly sins of psychology: A manifesto for reforming the culture of scientific practice*. Princeton: Princeton University Press.

Cox, Adam, Russell Craig, and Dennis Tourish. 2018. Retraction statements and research malpractice in economics. *Research Policy* 47 (5): 924–935.

Crama, Yves et al. 2006. Plagiarism. *4OR: Quarterly Journal of the Belgian, French and Italian Operations Research Societies*. http://www.4or.be/Plagiarism.html. Accessed 6 July 2018.

Davies, Benjamin, and Giulia Felappi. 2017. Publish or perish. *Metaphilosophy* 48 (5): 745–761.

Direction, La. 1997. Avis aux lecteurs. *Recherches de Théologie et Philosophie médiévales* 64 (1): 1–2.

Elliott, Teressa L., Linda M. Marquis, and Catherine S. Neal. 2013. Business ethics perspectives: Faculty plagiarism and fraud. *Journal of Business Ethics* 112 (1): 91–99.

Fanelli, Daniele. 2013. Why growing retractions are (mostly) a good sign. *PLoS Medicine* 10 (12): e1001563. https://doi.org/10.1371/journal.pmed.1001563. Accessed 6 July 2018.

Fox, Mark, and Jeffrey Beall. 2014. Advice for plagiarism whistleblowers. *Ethics and Behavior* 24 (5): 341–349.

Gipp, Bela. 2014. *Citation-based plagiarism detection: Detecting disguised and cross-language plagiarism using citation pattern analysis*. Wiesbaden: Springer Vieweg.

Hagen, Bill, and Tony VenGraitis. 2007. *Appropriate use of the IEEE prohibited authors list*. IEEE. https://www.ieee.org/content/dam/ieee-org/ieee/web/org/pubs/appropriate_use_of_pal.pdf. Accessed 6 July 2018.

Hansson, Sven Ove. 2008. Philosophical plagiarism. *Theoria* 74 (2): 97–101.

———. 2017. The ethics of doing ethics. *Science and Engineering Ethics* 23 (1): 105–120.

Harms, Dan. 2006. Plagiarism, publishing, and the academy. *Journal of Scholarly Publishing* 38 (1): 1–13.

Heidegger, Martin. 1999. *Contributions to philosophy (from enowning)*. Trans. Parvis Emad and Kenneth Maly. Bloomington: Indiana University Press.

———. 2012. *Contributions to philosophy (of the event)*. Trans. Richard Rojcewicz and Daniela Vallega-Neu. Bloomington: Indiana University Press.

Hoover, Gary A. 2006. A game-theoretic model of plagiarism. *Atlantic Economic Journal* 34 (4): 449–454.

Horbach, S.P.J.M., and W. Halffman. 2017. The extent and causes of academic text recycling or 'self-plagiarism'. *Research policy*. https://doi.org/10.1016/j.respol.2017.09.004. Accessed 6 July 2018.

Howard, Jennifer. 2011. After loss in court, scholar defends libel lawsuit over negative book review. *The Chronicle of Higher Education*, March 11. http://www.chronicle.com/article/After-Loss-in-Court-Scholar/126720. Accessed 6 July 2018.

Ioannidis, John P.A. 2005. Why most published research findings are false. *PLoS Medicine* 2 (8): e124. https://doi.org/10.1371/journal.pmed.0020124. Accessed 6 July 2018.

Jacobs, Jonathan D. 2012. History. *Res Philosophica*. http://www.resphilosophica.org/history. Accessed 6 July 2018.

Lewis, Bruce R., Jonathan E. Duchac, and S. Douglas Beets. 2011. An academic publisher's response to plagiarism. *Journal of Business Ethics* 102 (3): 489–506.

Liptak, Adam. 2011. From a book review to a criminal trial in France. *The New York Times*, February 21. http://www.nytimes.com/2011/02/22/us/22bar.html. Accessed 6 July 2018.

Lu, Susan Feng, Ginger Zhe Jin, Brian Uzzi, and Benjamin Jones. 2013. The retraction penalty: Evidence from the web of science. *Scientific Reports* 3 (3146): 1–5. https://doi.org/10.1038/srep03146. Accessed 6 July 2018.

Maddox, John. 2017. John Maddox prize. *Sense About Science*. http://senseaboutscience.org/activities/2017-john-maddox-prize. Accessed 6 July 2018.

Marcus, Adam, and Ivan Oransky. 2017. Is there a retraction problem? And, if so, what can we do about it? In *Oxford handbook of the science of science communication*, ed. Kathleen Hall Jamieson, Dan M. Kahan, and Dietram A. Scheufele, 119–126. New York: Oxford University Press.

Martin, Ben R. 2007. Keeping plagiarism at bay—A salutary tale. *Research Policy* 36 (7): 905–911.

———. 2012. Does peer review work as a self-policing mechanism in preventing misconduct: a case study of a serial plagiarist. In *Promoting Research Integrity in a Global Environment*, ed. Tony Mayer and Nicholas Steneck, 97–114. Toh Tuck Link: World Scientific Publishing.

———. 2013. Whither research integrity? Plagiarism, self-plagiarism and coercive citation in an age of research assessment. *Research Policy* 42 (5): 1005–1014.

Martin, Brian. 2016. Plagiarism, misrepresentation, and exploitation by established professionals: Power and tactics. In *Handbook of academic integrity*, ed. Tracey Bretag, 913–927. Singapore: Springer.

McKenzie, Lindsay. 2017. Journal's board disavows apology for 'transracialism' article, making retraction unlikely. *The Chronicle of Higher Education*, May 18. http://www.chronicle.com/article/Journal-s-Board-Disavows/240116. Accessed 6 July 2018.

McKenzie, Lindsay, Adam Harris, and Fernanda Zamudio-Suaréz. 2017. A journal article provoked a schism in philosophy: Now the rifts are deepening. *The Chronicle of Higher Education*, May 6. http://www.chronicle.com/article/A-Journal-Article-Provoked-a/240021. Accessed 6 July 2018.

Moosa, Imad A. 2018. *Publish or perish: Perceived benefits versus unintended consequences*. Cheltenham: Edward Elgar.

Moran, Dermot. 1993. Editorial. *International Journal of Philosophical Studies* 1 (1): vii–vix.

Neale, Stephen. 2001. No plagiarism here. *Times Literary Supplement*, February 8.

Nelson, Leif D., Joseph Simmons, and Uri Simonsohn. 2018. Psychology's renaissance. *Annual Review of Psychology* 69: 511–534.

Oppenheimer, Mark. 2011. Debate over intelligent design ensnares a journal. *The New York Times*, May 13. http://www.nytimes.com/2011/05/14/us/14beliefs.html. Accessed 6 July 2018.

Oransky, Ivan. 2013. Doing the right thing: Scientists reward authors who report their own errors, says study. *Retraction Watch*. November 7. http://retractionwatch.com/2013/11/07/doing-the-right-thing-scientists-reward-authors-who-report-their-own-errors-says-study. Accessed 6 July 2018.

Pelikan, Jaroslov. 1992. *The idea of the university: A reexamination*. New Haven: Yale University Press.

Quintanam, Chris. 2017. Here's what academics are saying. *The Chronicle of Higher Education*, April 5. http://www.chronicle.com/article/Heres-What-Academics-Are/239699. Accessed 6 July 2018.

RePEc Plagiarism Committee. 2011. RePEc plagiarism offenders. Research Papers in Economics. https://plagiarism.repec.org/offenders.html. Accessed 6 July 2018.

Rodes, Robert E., and Charles E. Rice. 1970. Foreward. *American Journal of Jurisprudence* 15 (1): vii.

Roig, Miguel. 2015. *Avoiding plagiarism, self-plagiarism, and other questionable writing practices: A guide to ethical writing*. 2nd ed. U.S. Department of Health and Human Services,

Office of Research Integrity. https://ori.hhs.gov/sites/default/files/plagiarism.pdf. Accessed 6 July 2018.

Rosamond, Ben. 2002. Plagiarism, academic norms and the governance of the profession. *Politics* 22 (3): 167–174.

Rougier, Nicolas, and John Timmer. 2017. Ten simple rules for scientific fraud & misconduct. *HAL-Inria Archive Ouverte*. https://hal.inria.fr/hal-01562601. Accessed 6 July 2018.

Scharff, Robert C. 1997. Announcement. *Man and World* 30 (4): iii.

Scholz, S. J. 2017. Editor's statement (May 6, 2017). *Hypatia*. http://hypatiaphilosophy.org/editor/statement-editor-may-6-2017. Accessed 6 July 2018.

Schrenk, Lawrence P. 1989. Minutes of the 1989 executive council. *Proceedings of the ACPA* 63: 277.

Schuessler, Jennifer. 2017. A defense of 'transracial' identity roils philosophy world. *The New York Times*, May 19. https://www.nytimes.com/2017/05/19/arts/a-defense-of-transracial-identity-roils-philosophy-world.html. Accessed 6 July 2018.

Sher, Gila, Otávio Bueno, and Wiebe van der Hoek. 2016. Statement from Synthese editors. *Daily Nous: News for and About the Philosophy Profession*, January 20. http://dailynous.com/2016/01/27/statement-from-synthese-editors-moratorium-on-special-issues. Accessed 6 July 2018.

Solomon, Miriam, Lisa Tessman, Leslie Francis, Heidi Grasswick, and Elizabeth Anderson. 2017. Board of directors' statement (July 20, 2017). *Hypatia*. http://hypatiaphilosophy.org/statements/board-directors-statement-july-20-2017. Accessed 6 July 2018.

Spencer, J.R. 2011. Libel tourist ordered to pay 8,000 Euros. *The Cambridge Law Journal* 70 (2): 317–319.

Spoelstra, Sverre, Nick Butler, and Helen Delaney. 2016. Never let an academic crisis go to waste: Leadership studies in the wake of journal retractions. *Leadership* 12 (4): 383–397.

Taylor, A.E. 1916. IX.—Correspondence. *Mind* 25 (4): 550–551.

Timmins, Adam. 2013. Why was Kuhn's *Structure* more successful than Polanyi's *Personal Knowledge*? *HOPOS: The Journal of the International Society for the History of Philosophy of Science* 3 (2): 306–317.

Tudvad, Peter. 2007. SAK – an unscholarly biography of Kierkegaard. (Trans: M. G. Piety). *The Torch*. January. http://www.faklen.dk/english/engtudvad07-01.php. Accessed 6 July 2018.

van Benthem, Johan, Vincent F. Hendricks, and John Symons. 2011. Statement from the editors-in-chief of *Synthese*. *Synthese* 178 (2): A7. http://www.jstor.org/stable/41477269. Accessed 6 July 2018.

Van Herck, Walter. 2013. Editorial note. *International Journal of Philosophy and Theology* 74 (1): 1.

Weber-Wulff, Debora. 2014. *False feathers: A perspective on academic plagiarism*. Heidelberg: Springer.

Weiler, Joseph. 2010. Book reviewing and academic freedom. *The European Journal of International Law* 20 (4): 967–976.

———. 2011. In the dock, in Paris – The judgment. *EJIL: Talk! Blog of the European Journal of International Law*. https://www.ejiltalk.org/in-the-dock-in-paris-—the-judgment-by-joseph-weiler-2. Accessed 6 July 2018.

Wolters, Eugene. 2014. I nonetheless deeply regret the incident. *Critical Theory*, July 12. https://goo.gl/pGS9jg. Accessed 6 July 2018.

Zamudio-Suaréz, Fernanda. 2017. Months after 'transracialism' flap, controversy still rages at feminist philosophy journal. *The Chronicle of Higher Education*, July 21. http://www.chronicle.com/article/Months-After/240722. Accessed 6 July 2018.

Zhang, Yuehong (Helen). 2016. *Against plagiarism: A guide for editors and authors*. Cham: Springer.

Chapter 2
Defining the Scholarly Record

Abstract This chapter provides a conceptualization of the scholarly record. I propose that items that belong indisputably to the scholarly record meet six hallmarks: the *Knowledge*, *Authorship*, *Publication*, *Library*, *Database*, and *Discipline* conditions. Books issued by scholarly presses and articles appearing in established journals have been the traditional formats for presenting research findings, and such items clearly meet these six conditions. Advances in technology, however, have occasioned new modes for recording and disseminating knowledge, and they create challenges to the long-standing conceptualizations of the scholarly record. Online post-publication review venues, open-access initiatives, interactive scholarly websites, online document repositories, and other venues invite a reconsideration of the precise boundary of the published literature. I distinguish between synchronic and diachronic approaches to the scholarly record. Separating these two allows one to isolate the scholarly record as a present system from past versions that have operated under different parameters and also from future versions that are only anticipated.

Keywords Scholarly record · Published literature · Pseudonyms · Publishing · Authorship

The activities of all academic disciplines presuppose a body of published material known as the *scholarly record*. Professional researchers as well as students depend on it, and the success of their academic endeavors largely requires that it be reliable. Despite its centrality in the world of learning, the scholarly record is rarely subject to detailed consideration. Its structure, chief characteristics, and evolution over time are most often tacitly assumed rather than explicitly examined. The scholarly record functions much like the support beam of a house: both of them provide a stable foundation for day-to-day activities, but both are rarely considered by those who depend on them the most.

A clear conception of the scholarly record is a prerequisite for any successful analysis of the various instruments employed today for correcting it. Standard instruments include published retractions, errata, corrigenda, expressions of concern, and the outright expurgation of an offending item from the scholarly record.

M. V. Dougherty, *Correcting the Scholarly Record for Research Integrity*, Research Ethics Forum 6, https://doi.org/10.1007/978-3-319-99435-2_2

Products of research misconduct, such as an article based on fabricated data or a plagiarized book chapter, contaminate the scholarly record in various ways. Corrections of the scholarly record with the right instruments mitigate that harm and restore integrity to the published literature.

Misconceptions about the precise boundary of the scholarly record generate confusion about the use of tools for correcting it. Some researchers conflate the scholarly record with a set of articles published in elite venues; other researchers mistake it for a set of recently published works. Such approaches are too narrow; uncorrected fraudulent items in non-elite scholarly venues as well as fraudulent items published years ago can still harm students and researchers. Conversely, a conception that is too expansive will include items that fall outside of the scholarly record. An account of the scholarly record should be able to distinguish clearly whether items fall within or outside of it, while also providing an accurate and compelling explanation of any disputed items that are marginal or at the boundary.

Some theorists have resisted attempts to identify the boundary of the scholarly record. An extensive study by librarians asserts that "precise definition of the scholarly record is a difficult if not impossible task" (Lavoie et al. 2014: 8). Various reasons are often put forward as insurmountable impediments to circumscribing it clearly: one hears about the multiplicity of objects, the various venues in which items appear, controversies about the priority of print or electronic forms, and changing notions of scholarship. In some quarters, there has been a tendency to inflate the scholarly record to encompass all features of research inquiry, rather than to restrict it to those items that present finalized research findings. Unsurprisingly, such an enlargement yields the conclusion that the "boundaries of the scholarly record are in flux, as they stretch to extend over an ever-expanding range of materials" (Lavoie and Malpas 2015: 7). Unfortunately, such a move conflates the history of discovery with the scholarly record, and it threatens the very utility of the notion of a defined body of published literature. Despite the abovementioned concerns, the boundary of the scholarly record need not be treated as something fundamentally elusive or mysterious.

2.1 The Limits of the Scholarly Record

I propose that the scholarly record can be understood as the total collection of items (typically articles or books) where each item possesses all or nearly all of a set of essential characteristics. The undisputed items of the scholarly record will:

1. Putatively advance or summarize knowledge (*Knowledge Condition*);
2. Appear under identifiable authorship (*Authorship Condition*);
3. Be issued through an academic publisher (*Publication Condition*);
4. Be catalogued in university libraries (*Library Condition*);
5. Be catalogued in curated research databases (*Database Condition*); and
6. Belong to at least one recognized academic discipline (*Discipline Condition*).

Conditions 1–6 are hallmarks possessed by all or at least most items that are generally recognized by practitioners in a discipline as belonging to the scholarly record. As shall be seen, items that fail to fulfill some of these conditions have a marginal or questionable relationship to the scholarly record. They can still be useful to researchers and students, but utility alone does not suffice for admittance to the scholarly record. Some items that fall short of the scholarly record are sometimes characterized as belonging to the "grey literature," a term indicating they are outside the domain of the academic published literature (Lawrence et al. 2014, 2015).

2.1.1 The Knowledge Condition

The *Knowledge Condition* indicates that an item of the scholarly record makes a putative contribution to a discipline's collective body of truth claims. Items of the scholarly record typically fulfill this condition by reporting a discovery or by summarizing the current state of what is known. The *Knowledge Condition* is so central to the body of published literature that items that are later revealed to be in violation of it are ordinarily subject to some kind of correction. In the case of error, corrections by later articles typically suffice, but published corrections by editors and publishers (e.g., retractions) are justified in cases involving extreme negligence, outright fraud, or other forms of research misconduct, especially when the uncorrected items present some immediate danger to the public.

The centrality of the *Knowledge Condition* can be approached indirectly by considering published articles that are later unmasked as hoaxes. By design, hoax articles fail to provide a straightforward contribution to knowledge. In the publication of a hoax article, a host of individuals—journal editors, reviewers, and some readers—have been deceived into believing that a published work is making an explicit truth claim. Hoax articles are customarily retracted or removed because their unqualified presence in the scholarly record can impede the dissemination of knowledge and harm practitioners who are unaware that such items are inauthentic products of research. They pollute the downstream literature when taken to be genuine expressions of research by trusting researchers who cite them. Some recent work on publishing ethics proposes that articles revealed to be hoaxes should be retracted, even when their authors claim to have perpetrated hoaxes with the best of intentions. On such a view, published hoax articles that attempt to demonstrate the alleged weakness of peer review at a particular journal or to show an alleged lack of rigor of a particular subfield should still be retracted (Ronagh and Souder 2015).

Many journals have issued retractions for hoax articles in a variety of disciplines. In philosophy, the editors of the online journal *Badiou Studies* retracted a parody article featuring pseudo-academic gibberish in 2016 after the authors revealed their intention of exposing a perceived lack of argumentative integrity in both the journal and in the subfield covered by it. The article contained unintelligible sentences such as, "The void central to the theory turns out to be the essence of the manifold and the fullness that is axiomatically conceived of in a theory of multiplicities" (Tripodi

2016: 78). The editors issued a retraction, and shortly thereafter the journal apparently ceased publication (McCook 2016). More recently, philosopher Peter Boghossian had an article he co-authored retracted after he revealed that it was a hoax meant to prove problems in publishing integrity and a lack of rigor in gender studies. The article had asserted that male genitalia should be principally conceived as a social construct responsible for climate change and oppression rather than as a biological organ (Anonymous 2017a).

Such sting operations have attempted to prove something about the quality of the targeted journals, and in many instances they have succeeded in unmasking vanity outlets. Nevertheless, when occurring at more reputable journals, sting operations have created situations that damage the reliability of the scholarly record, thereby requiring published corrections by editors and publishers. Sting operations are becoming increasingly common: an anthology of especially noteworthy hoax papers has been issued as *Stinging the Predators: A Collection of Papers that Should Never have been Published* (Faulkes 2018).

In 2002, the literary journal *The Dickensian* published an article that appeared to translate a riveting letter in which Fyodor Dostoyevsky recounted a meeting with Charles Dickens in London in 1862 (Harvey 2002). On the basis of this article, Dickens scholars in several subsequent publications accepted as factual the alleged meeting between the two luminaries of Russian and English literature. A spectacular exposé a decade later revealed that the letter was a hoax and that the author of record ("Stephanie Harvey") was really the pseudonym of a disgruntled male researcher (Naiman 2013; Moss 2013). A published correction by *The Dickensian* in 2013 was added in full to the top and bottom of each page of the electronic version of the original 2002 article, as well as to new electronic pages inserted before and after the original article. It stated that "the letter extract is a hoax" and apologized for giving credence to the purported meeting (Anonymous 2002: 232A–235A). Since the article fabricated a meeting that never occurred and misled specialists about the life of Charles Dickens, the hoax article was in violation of the *Knowledge Condition*, and the correction was warranted.

For another striking violation of the *Knowledge Condition*, an erratum was issued after A. N. Wilson was duped by an academic rival to include the text of a fake love letter in his biography of British Poet Laureate John Betjeman. The elaborate hoax was indisputably confirmed when the first letter of each sentence in the forged love letter was discovered to form a profane acrostic that read "A. N. Wilson is a [expletive]." The subsequently published erratum characterized the situation as "a spoof" (Axelrod 2006). Also, a political science journal was forced to retract a 2016 hoax article that claimed that guard dogs used at the Berlin Wall were the descendants from German Shepherds used at Nazi concentration camps (Oltermann 2016). The published statement by the apparently displeased journal editors indicated that they were unaware that the preposterous thesis was a joke when they accepted the article (*Die Redaktionsleitung* 2016). Notable hoax situations have generated published corrections in many other disciplines (Van Noorden 2014; Oransky 2015). At times, the published retractions give little information about the genesis of the hoax in question; in 2012, a retraction appeared in a mathematics journal with the austere explanation that a hoax article "contains no scientific content" (Anonymous 2012).

A hoax article first deceives peer reviewers and editors about its truth claims *en route* to publication. In contrast, parody or satirical articles are often knowingly published with the imprimatur of editors. Such items are sometimes designated as articles of "ironic science" (Ronagh and Souder 2015: 1538–1539). Meant to be humorous, they are recognized by many readers as setting forth no genuine scientific claims. Most often the irony is obvious, as in the case of a five-page 2001 review by "Darcy Chopwittle" in the journal *Social Science History* of the nonexistent book *When the Cows Come Home: Barn Architecture and Changes in Bovine Public Space*. The book in question, allegedly authored by one Philinda Blank, reportedly discussed how changes in barn design over the last four centuries have violated the social, cultural, and religious practices of cows (Chopwhittle 2001). In some instances, however, the irony of such articles may be too subtle for some readers, who may lack the background knowledge to pick up on the tell-tale signs that an article is not a serious one. In the pluralistic, multilingual, and international world of learning, it should be unsurprising that a non-negligible portion of readers are unprepared to discern the ironic features of such articles. Having been misled, such readers sometimes cite these articles as genuine works of scholarship.

As well-meaning citations to ironic articles grow in the published literature, some researchers have argued that articles of ironic science should be retracted. Calls for retractions of such articles are particularly acute in the medical sciences, where the ironic published literature—at least potentially—could be misused to treat patients. Using the example of a parody article in *The British Medical Journal* by Leonard Leibovici that purported to summarize a randomized controlled trial on petitionary prayer and bloodstream infection, Maryam Ronagh and Lawrence Souder have argued for retraction, stating:

> Though it will be disappointing for the reader and tedious for the author to issue a retraction (much like the anticlimax that results from explaining a joke), Leibovici's community will not be secure in trusting the research record otherwise. (1546)

The article's inclusion in later-published meta-analyses and its ongoing citation by researchers (who are apparently unaware of the attempt at humor) provide evidence that its presence in the scholarly literature requires a remedy.

In another case in *The British Medical Journal*, 35 years passed between the original publication of a seemingly obvious parody article and its retraction (Anonymous 2009). The article had described an injury to male anatomy purportedly experienced by cellists. A significant passage of time and widespread notoriety need not be reasons to forego a correction of the scholarly record, particularly when the jesting articles continue to receive citations as they are referenced as genuine research articles in subsequent literature.

Both hoax articles (that are meant to fool readers) and parody articles (that are meant to be humorous) create difficulties for researchers when they appear within the scholarly record. When these articles are taken to be genuine expressions of research, published corrections by editors and publishers limit the damage done to the scholarly record and forewarn future readers who might otherwise be misled. The need for published corrections of hoax and parody articles underscores the centrality of the *Knowledge Condition* as a hallmark of items belonging to the scholarly record. This view does not rule out humorous or witty parodies of scholarly work; it simply requires that such items be clearly labeled as such. A good example is the collection of papers on medieval studies published as the *Proceedings of the Pseudo Society*, which presents satirical historical papers originating from "a preposterous organization" (Ring and Kay 2003). Since the papers are introduced as fictional and satirical, readers are unlikely to be confused about the nature of them.

2.1.2 The Authorship Condition

The *Authorship Condition* requires that a work is explicitly tied to an author, a set of authors, or a research group. In addition to the implication that the stated authors are the origin of what is presented, the display of authorship provides a warrant of the truth of what is set forth in a work. A stated institutional affiliation offers an additional layer of credibility. Such credibility is especially valuable when the work reports activities that others cannot easily verify, such as lab experiments, data collection, or archival work. The *Authorship Condition* implies that authors are (in principle) willing to defend what appears in print and will entertain serious queries about it. It ensures that (at least in principle) some identifiable party can be held responsible if a work is discovered to be the result of research misconduct. The *Authorship Condition* also provides a referent for any conflict-of-interest statements that accompany publications in some fields. For these several reasons, only very rarely today do publishers and editors allow research articles to appear in the scholarly record anonymously or with an undisclosed pseudonym (Neuroskeptic 2013; cf. Teixeira da Silva 2017a).[1] Clear and unambiguous authorship for products of research facilitates the trust that is required among members of a research community who depend on the work of others in contributing the development of a discipline.

Given these present-day expectations of authorship for published research, a post-publication revelation that an article has appeared under an undisclosed pseudonym can bring about strong reactions in a research community (Bohannon 2016). In a sternly worded letter appearing in *Journal of the Association for Information*

[1]Articles that do not present new research, such as published editorials, do however occasionally appear in academic journals anonymously (see, for example, Neuroskeptic 2012). Furthermore, corrections issued by editors and publishers—such as retractions, errata, corrigenda, and expressions of concern—are typically published within the pages of journals without an explicit author of record. Some journals that cover controversial research areas occasionally allow authors to use disclosed pseudonyms. The editors of the journal *Archives of Sexual Behavior* have recently allowed two articles on pedophilia to be published under the pseudonym "Max Geradt." The articles display the note: "Geradt is sexually interested in children and is co-authoring this article under a pseudonym" (Geradt et al. 2018: 375; Jahnke et al. 2015: 2173).

Science and Technology with the title, "A New Type of Misconduct in the Field of the Physical Sciences: The Case of the Pseudonyms," a physicist accused a colleague of "fraudulent scientific conduct" for using two undisclosed pseudonyms in a pair of critical articles uploaded to the *arXiv* preprint repository to attack the work of other scholars. In addition to being authored pseudonymously, the articles in question were complete with fabricated institutional affiliations and email addresses. The letter containing the condemnation of the pseudonymous attacks was later withdrawn by the publisher because of the as-yet unsubstantiated assertion of research misconduct (Anonymous 2014). Two climate-studies authors had their work withdrawn after the revelation of their use of pseudonyms, which consisted of their first and last names spelled backwards (Guarino 2016). The published retraction in this case stated that the withdrawal of the paper was "not related to the scientific merit of the study," and the publisher Elsevier later confirmed that there was an "authorship issue" justifying the change in status to the paper (Chawla 2016). More recently, the use of an undisclosed pseudonym and a false institutional affiliation led to four retractions of papers on the topic of vaccine risks (Oransky 2018).

Fraudulently stated institutional affiliations alone can also be sufficient to generate retractions. Some researchers have been discovered to have "invented academic affiliations going back decades," either by creating fictitious entities or falsely claiming association with existing ones (Abbott 2007: 632). Institutional truth-stretching has taken on new forms, as some established European researchers have reported receiving offers of over £35,000 "to list publications under the name of a lesser-known" institution in Asia, in an apparent attempt to increase university rankings of a lower-tiered institution (Grove 2017). Fraudulent representation of academic affiliations impedes accountability in research, particularly when a fabricated institutional affiliation is combined with other forms of research misconduct. In commenting on one case involving serial plagiarism in combination with fictitious institutional affiliations, a journal editor noted:

> This unfortunately means that there is no organisational body to which one can now hand responsibility for carrying out a full investigation of the wider body of the author's published work. (Martin 2007: 908)

To strengthen the *Authorship Condition*, some journals now require that correspondence be conducted via an institutional email address throughout the publication process, and sometimes the email address of the corresponding author appears in the published article.

Even though the use of undisclosed pseudonyms in scientific fields is generally disallowed in scholarly publishing, explicitly disclosed or patently obvious pseudonyms are often acceptable, provided that the genuine authors are in some way identifiable, even though the pseudonym is preserved as the official author of record. In such cases, the *Authorship Condition* is still on the whole fulfilled. The five creators of a successful text on logic that was published under the name of "L. T. F. Gamut" explained in an introductory note to the English edition that "to underline their common vision" they had "merged their identities into that of L. T. F. Gamut" (Gamut 1991: xii). This disclosed pseudonym is acronym, with the last name reflecting the institutional university affiliations of the authors (Groningen, Amsterdam, and Utrecht), and the initials standing for the first three words of the title in Dutch (logic/*logica*, language/*taal*, and philosophy/*filosofie*). Additionally, the copyright page of the volume states that "L. T. F. Gamut is a collective pseudonym" (ibid., iv). Since the use of the pseudonym and the names of the individual authors of origin were disclosed in the pages of the publication, the *Authorship Condition* is still substantially fulfilled.

In some sectors of the humanities, there is vigilance over the use of undisclosed pseudonyms in published works, but with lesser intensity than what is found in the sciences. Garland Publishing withdrew from sale a two-volume historical encyclopedia after the revelation that the editors had written a substantial number of the entries themselves using pseudonyms in addition to their real names, thereby giving the illusion of a more diversely authored work. After being out of print for two decades, the 2016 reprinting of the encyclopedia by a different press now includes an extremely brief "Clarification" on the copyright page for each volume that states, "Please note that the names Evelyn Gettone and Ward Houser are pseudonyms. These entries were written by male authors" (Dynes et al. 2016: I.ii, II.i; see McMillen 1995). Routledge similarly withdrew a 168-page monograph on performance theory that was revealed to consist almost entirely of chapters from another Routledge volume on religious studies, and no reference had been given to the original authors of the earlier volume. One critic asked rhetorically, "Can a book that describes itself as *Theory for Religious Studies* simply slip off its cover and put on a new one saying *Theory for Performance Studies*, and still work?" (Goldingay 2009: 5). The critical response to the work's publication by scholars in a variety of humanities disciplines likely assisted in reversing an original decision by the publishers to leave the scholarly record uncorrected (Schechner et al. 2009).

In the discipline of philosophy, however, there has traditionally been little discernible opposition by editors and publishers to the use of undisclosed pseudonyms, and in this respect the discipline of philosophy falls short of the publication practices of the sciences as well as many other humanities disciplines. Philosophy's acceptance of a practice that is considered in other disciplines to be outright in violation of scholarly publishing norms is remarkable, especially when editors are complicit with the practice.

One might argue that for well-known instances of pseudonymous writing, the harm to the scholarly record is minimized, since a significant portion of the relevant practitioners in a subfield would be aware of the authorial deception. Michel Foucault's use of the thinly veiled pseudonym "Maurice Florence" in publishing a detailed encyclopedia entry on himself in 1984, shortly before he died, does not appear to have deceived many, and the entry is now considered to be part of his published *oeuvre* (Florence 1994). Alain Badiou used the pseudonym "Georges Peyrol" in writing some book reviews, and they are now included in an English collection of his writings published under his real name (Badiou 2012: 191–201).

The discipline of philosophy provides distinctive examples of undisclosed pseudonyms in violation of the *Authorship Condition*. In 1982, 2 years after publishing an article in *Australasian Journal of Philosophy*, David Lewis published a critique of it under the name "Bruce Le Catt," using a pseudonym apparently crafted after his pet to discuss his prior work in the third person (Le Catt 1982). No institutional affiliation was given for Le Catt, but the location "Princeton, NJ" was printed at the end of the article. While discussing the deception shortly after Lewis's death, J. J. C. Smart observed that "Readers no doubt thought it to be by an emerging young philosopher" (Smart 2001: 3). In his published collected papers, Lewis still referenced Le Catt in the third person, indicating that he endorsed some of Le Catt's positions ("Le Catt suggests, and I agree, that [...]") but rejected others ("[Le Catt] further claims [...] but there I do not agree") (Lewis 1986: 290). The identity of Lewis and Le Catt was apparently known in some circles; in the subsequent literature Le Catt is euphemistically referenced as "a 'friendly' critic" or described as Lewis's "alter ego" (Vision 1997: 190; Sorensen 2008: 58). The Le Catt article has also been included in select published bibliographies of Lewis's work (Lewis 2015: 565). In early July 2017, I wrote to the editor of *Australasian Journal of Philosophy* to request a correction of the scholarly record, noting that some philosophers are unaware of the identity between Lewis and Le Catt, and it is conceivable that many younger members of the profession could read the 1982 article without knowing that Lewis is engaging in a dialogue with himself in the 1980 and 1982 articles of the journal. In a very efficient response, *Australasian Journal of Philosophy* published an erratum a few weeks later in late July 2017 explaining that'Bruce Le Catt' "was a pseudonym used by the author David Lewis, to discuss some work published under his own name" (Anonymous 2017d). Press coverage of the erratum noted the lack of detail regarding the circumstances of the original article in the brief erratum, and *Retraction Watch* referred to it as a "joyless notice" (Marcus 2017). Perhaps due in part to the stature of Lewis as a significant twentieth-century philosopher and to the unusual role of his pet, the erratum generated some interest (Czepel 2017; Maisonneuve 2017; Anonymous 2017c). This correction, 35 years after the original publication of the article, provides an external correction the scholarly record and brings the 1982 article into compliance with the *Authorship Condition*.

In October 2017, I was contacted by David Shatz of Yeshiva University, who had seen an online discussion of the published erratum for the Le Catt article. Shatz recounted that four decades earlier he had observed that a 1976 article in *Philosophical Studies* published under the name "M. Lisagor" seemed to be written in a style similar to the writings of Joseph Margolis, a philosopher at Temple University. He subsequently noticed that "M. Lisagor" was an anagram for "Margolis," and that the city of Philadelphia (where Temple University is located) was given in the place of an institutional affiliation at the end of the published article. Shatz sent a letter to Margolis asking about his suspected authorship of the article in *Philosophical Studies*, but received no reply. A year or so later, he approached Margolis at a meeting of the American Philosophical Association, and Margolis confirmed that he had indeed authored the article.

I wrote to Margolis in late October 2017, explaining that in some circles, there is a belief that he is "M. Lisagor," and I asked whether he would confirm that he authored the article in *Philosophical Studies*. He responded with a brief email in early November, stating, "Yes, I'm Lisagor. There's a good story there." We spoke by phone within a week. Margolis explained that Shatz was quite perceptive and the only one ever to contact him about the pseudonym. He was not surprised, however, that someone had figured him out because of the anagram name. Margolis had submitted the manuscript to *Philosophical Studies* using letterhead from a hospital were his wife worked as a clinical psychologist. He said he couldn't be traced and he could still receive letters sent there addressed to the pseudonym. The editor of the journal responded to his submission with a request for revisions, which he found agreeable, so he submitted a revised version. When the editor contacted him again with a second round of revisions, Margolis did not respond, deciding to give up on the paper. To his surprise, the article then appeared in print in the journal.

Margolis explained that he had more than one reason for using the pseudonym. He mentioned that he had already acquired some kind of reputation at the time he submitted the article, and that he had occasionally experienced some obstacles in publishing his work in some venues. Margolis believed that had he submitted the manuscript under his real name to *Philosophical Studies*, it would not have been accepted for publication. The experiment of publishing under a pseudonym was also meant to provide a counterexample for a certain view of the conditions of scientific knowledge. After the article appeared, and he had been discovered by Shatz, Margolis revealed the use of the pseudonym to the editor of the journal, but the editor at the time never came back to him with any possibility of an annotation for revealing his identity in the pages of the journal. I asked Margolis whether he would support a published clarification of authorship for the article in the journal today, so that it could be acknowledged as part of his lengthy philosophical output. He agreed and stated I could mention my conversation with him in forming a request to the journal. Margolis added that he had never used a pseudonym for any other writings. Within a week I composed a request for a published clarification of authorship and sent it to the editor of *Philosophical Studies*, copying Shatz and Margolis. Early following year, the journal published a correction note by Margolis that states, "In the original publication of the article, the corresponding author used the pseudonym 'M. Lisagor'. The correct name is given in this correction" (Margolis 2018: 1827).

Some journals are very efficient in responding positively to requests for published clarifications of authorship. In 2008, Neven Sesardić published an article in *The Journal of Philosophy* under the female pseudonym "Carmen de Macedo" (de Macedo 2008). He later explained his motivations in a 2011 essay that appeared in *Inside Higher Ed*, revealing that he feared an earlier quibble with one of the editors might have resulted in a rejection of his manuscript had he used his real name, since at the time the journal did not practice blind refereeing of manuscripts (Sesardić 2011). As David Lewis and Joseph Margolis had done, Sesardić offered his home location (Hong Kong) in place of an institutional affiliation to accompany the pseudonym. In July 2017, I wrote to the editors and requested a correction of the scholarly record. After some correspondence, I received a positive response, and the journal published a corrigendum later in the year that states, "*The Journal of Philosophy* would like to inform its readers that the following article was published under a pseudonym [...]. The author is Prof. Neven Sesardić (formerly at Lingnan University, now retired)" (Anonymous 2017e: 392).

One way of indirectly disclosing the use of a pseudonym is through the use of a comically charged name. In cases such as these, no deception is likely to occur, even though the true author of the article remains hidden. A published rebuttal to the work of Garrett Hardin appeared in *Southwest Philosophy Review* with authorial attribution to "Harrett Gardin." (Gardin 2005).[2] A satirical article purporting to examine the significance of the distinction between express trains and local trains appeared in *The Journal of Philosophy*, where the author of record is the name of a train station on a New York subway line (Myrtle Willoughby 2012). In 2010, a chapter modeled after the style and punctuation conventions of eighteenth-century English was published in a collection of new essays on the is/ought distinction in David Hume's philosophy, with authorial attribution to "A Gentlemen" and titled "A Letter from a Gentleman in Dunedin to a Lady in the Country" (Gentlemen 2010).

Early last century, Dickinson S. Miller published five articles in the journal *Mind*, but he only used his real name for only two of them. For the remaining three, he published under the undisclosed pseudonym of "R. E. Hobart" (Hobart 1930a, b, 1934). Miller later took ownership of his pseudonym in a footnote in an article he published in *The Journal of Philosophy* (Miller 1936: 314, n. 3). This brief acknowledgment, however, failed to remedy the situation that in *Mind* there were two authors of record, but only one genuine author, for the five articles written by Miller. When one considers that one of Miller's pseudonymously published articles has become quite influential since its original appearance in 1934, and that the revelation of the pseudonym has been disclosed in some, but not all, of the republications of it in other venues, the case to correct the scholarly record is quite strong. In commenting on the use of the name "R. E. Hobart," Miller's former assistant and literary executor noted that even though "the specific motives for adopting the name of Hobart are obscure," Miller had considered legally changing his name to "Hobart" and had wished to downplay his known relationship with the philosopher William James (Easton 1975: 25; see also 1, 104, 184). In 2018, my 2017 request for published clarifications of authorship for the three pseudonymous articles was denied by the publisher, Oxford University Press, on the grounds that to do so would fail to respect the author's wish to identify himself as "R. E. Hobart," and, furthermore, that permission from Miller would be required for any published clarification of authorship. Such a view is curious; Oxford University Press currently publishes the works of "Constantin Constantius" and "Johannes Climacus" under the real name of Søren Kierkegaard, presumably without having obtained permission from the Danish philosopher (see Kierkegaard 2009).

Some contemporary authors in philosophy have used a variety of pseudonyms only to assume authorship of the dispersed articles later in their careers. This re-appropriation of one's work can occur when authors acknowledge the pseudonymously published works in later publications, as did Miller, or when authors later list them on a publicly available *curriculum vitae* or republish them under their real names in other venues. Notable examples of this practice demonstrate how far

[2] This author name was supplemented in print with the qualification "as told to Michael F. Patton, Jr." Some electronic copies circulating online lack this qualification.

removed the discipline of philosophy is from the authorial publication norms of other disciplines. For two collections of essays she edited for University of California Press, Amélie Oksenberg Rorty contributed chapters under her own name and others under the pseudonym "Leila Tov-Ruach," even providing independent, detailed biographical entries for each name in the published list of contributors found at the end of each volume. Almost every reader not already in the know will be unable to discern without help that the separately listed biographical entries (1) "Leila Tov-Ruach is an Israeli psychiatrist, who writes and lectures on philosophic psychology" and (2) "Amélie Rorty is a professor of philosophy at Livingston College, Rutgers University" refer to the same person (Rorty 1980: 544; McLaughlin and Rorty 1988: 559). In one of the chapters, Rorty, as "Tov-Ruach," commends the work published under her real name both in the body of the text and in the notes, and refers to herself in the third person, adding in the last note "I am grateful to Amélie Rorty for the hospitality that made the writing of this paper possible." The note continues, "She was kind enough to make available to me a number of the papers published in this volume: stimulated by them, and by her skeptical questions, I was able to work through some problems [...]," (Tov-Ruach 1980: 488; see also 476). Commentary of this sort would understandably impede most readers from suspecting the real identity of the chapter's author. Rorty confirmed in print the use of the florid pseudonym with a note appended to a revised version of one of her essays, stating: "Another version appeared under the pseudonym of Leila Tov-Ruach" (Rorty 2000: 209).

In this situation, there are two authors of record, but only one genuine author, for four chapters appearing in the two volumes. In late September 2017, I wrote to the press to request published clarifications of authorship. With remarkable speed, University of California Press issued two errata to correct the scholarly record. Less than a week after receiving my request, two dated notices appeared on the publisher's website that confirmed "UC Press would like to clarify that 'Leila Tov-Ruach' is a pseudonym used by the editor of the volume, Amélie Oksenberg Rorty" (University of California Press 2017a, b [with slight variation]). These two errata, issued for volumes published 37 and 29 years earlier, provide a clear correction of the scholarly record (Stern 2017).

The pseudonym had caused confusion for additional parties. Psychologist Anne Thompson was stricken to find similarities in the 1988 Tov-Ruach chapter to an unpublished paper of her own, referenced merely with a compare citation ("cf. Anne Thompson") in the penultimate footnote to the chapter (Tov-Ruach 1988: 262). A few years earlier, Thompson had asked Rorty to comment on the unpublished paper. In 2017, Thompson reflected on the incident, stating:

> I did not know how an Israeli psychiatrist could have got hold of my paper, but obviously she had it. I had no suspicion of any identity between Rorty and Tov-Ruach […]. I was shocked to find out that Tov-Ruach was actually Rorty. (Thompson 2017a, b)

Prior to learning about the use of the pseudonym, Thompson mistakenly assumed that her manuscript had been inappropriately shared by an unknown party who had attended a conference presentation or by colleagues who had a copy of the manuscript.

In another volume she edited, Rorty authored a chapter on Plato's views of education that appeared under the name "Zhang LoShan (pseudonym)." LoShan is described on the contributors' page as "a professor of ancient philosophy in the People's Republic of China" (Rorty 1998: x). In an editorial note at the end of the chapter, however, Rorty explains that she herself wrote the chapter, "assuming the persona of a Chinese scholar who had studied ancient philosophy in Canada and the United States and who had returned to the PRC to teach," and she added further that references to secondary literature in the chapter are limited "to those that would be available to Zhang LoShan, and to whatever articles and books colleagues in the West might have sent him" (LoShan 1998: 44). Given this detailed account, the pseudonym is fully—though bizarrely—disclosed, and hence the authorship condition is not violated. Nevertheless, as Rorty is respected for her work on the history of philosophy, it is unsurprising that sometimes the article is cited by other researchers with attribution to her under her real name rather than to "Zhang LoShan" and therefore there are inconsistencies in citation in the downstream literature.

The use of pseudonyms admits of degrees of disclosure. Some have suggested the logician Alfred Tarski used a family name to publish under the thinly veiled pseudonym "Al. Tajtelbaum" in the journal *Analysis* in 1957 in responding to a logical puzzle popularized by Elizabeth Anscombe (Gaskin and Hill 2013: 202; Sanford 2005: 765). The use of the pseudonym may have been a humorous reflection of the content of the article, which was titled, "'It is Impossible To Be Told Anyone's Name'." (Tajtelbaum 1957).

That philosophy often falls short of the practices of the sciences regarding authorship integrity is curious. Scientific disciplines as well as philosophy have enjoyed long histories in which the use of pseudonyms was largely accepted, but the sciences have moved away from such a practice in the present-day version of the scholarly record. Philosophy has a distinguished tradition of pseudonymous writing, with Søren Kierkegaard perhaps providing the best-known example. In line with current expectations of authorship, as noted above, Kierkegaard's pseudonymous works are invariably published today under the name *Søren Kierkegaard*. Similarly, some of the works produced by ancient and medieval scholars who appropriated the famous names of established authorities are presently published in editions with a correct name given as the author of record for each instance. For those cases in which name of the genuine author of a forgery is unknown, a prefix is attached to the author of record, as in the case of the present-day publication of the works of Pseudo-Dionysius the Areopagite. During the early modern period, not just philosophers but even publishers of philosophical texts have used pseudonyms to avoid censors and other authorities (Singer 1937).

Nevertheless, the ongoing acceptance of pseudonymous publication in some sectors of contemporary philosophy is particularly surprising. Since the history of philosophy is a major area of philosophical activity, the present-day allowance of pseudonymity seems antithetical to the future writing of a genuine history of the discipline. Furthermore, editors who allow pseudonymous submissions lose, in the words of one theorist, the "deterrence of authorial accountability," which guards against fraud, falsification, and error (Fulda 2007: 86). The appearance of undisclosed pseudonyms in philosophy occurs not only in articles and collections of essays in philosophy for advanced students and researchers, but can also be found in introductory student textbooks, where editors at times secretly write chapters under assumed names to increase the range of positions offered (e. g., Gay-Williams 2012).

The allowance of pseudonyms in contemporary philosophy generates problems that last for decades, including confusion about authorship and infelicities in the downstream literature. To take a distinctive example: using the pseudonym "Diodorus Cronus" (the name of an obscure ancient Greek philosopher), Steven Cahn and Richard Taylor co-authored in *Analysis* what would become an influential article titled "Time, Truth and Ability" (Diodorus Cronus 1965). Early responses in print to this article referenced the author of record obliquely as "a latter-day reincarnation residing in New York City" (Bar-Hillel 1965: 54) and as "someone purporting to be 'Diodorus Cronus'" (Bertolet and Rowe 1979: 137). Cahn himself defended Diodorus Cronus in a separate article in *Analysis* the following year under his own name (Cahn 1966), without revealing that he was really defending himself, and throughout his discussion he referenced Diodorus Cronus in the third person. The pseudonymous 1965 article was authoritatively credited to both Cahn and Taylor in the bibliography of Taylor's philosophical works that appeared in a *Festschrift* for Taylor a decade and a half later (van Inwagen 1980: 304, 305). In subsequent works by others, authorship of the 1965 article is referenced in three inconsistent ways: (1) sometimes to Diodorus Cronus alone; (2) often to both Cahn and Taylor; (3) occasionally to Taylor but without a reference to Cahn. Notably, Cahn 1966 is discussed in the later philosophical literature without acknowledgment that Cahn is providing a defense of his position, under two names, in the same journal, in consecutive years.

A second article was published under the name "Diodorus Cronus" in 1971. Appearing in *The Southern Journal of Philosophy*, this pseudonymous article was written by Richard Taylor alone (Diodorus Cronus 1971). The article provides some subtle hints at its true authorship, as it was published with a short biographical blurb for "Diodorus Cronus" that mentions several professional positions held by Taylor himself (1972: 113). The article is now usually referenced in subsequent literature as authored by Taylor. The use of the pseudonym "Diodorus Cronus" in both the 1965 and 1971 articles is in one sense a disclosed pseudonym, since the repurposed ancient name is not likely to be taken as the genuine name of a twentieth-century author by present-day readers; the authors literarily assume the identity of an ancient philosopher through use of the pseudonym, but the act is one of shielding the identities of the true authors rather than attempting deception. Nevertheless, the confusion in some sectors about the true authorship of the articles is reflected in the inconsistent various attributions of authorship in the downstream literature.[3] In late October 2017,

[3] Both the 1965 and 1971 articles have been republished in anthologies with "Diodorus Cronus" as the author of record. The republication of the 1971 article four years later states that Richard Taylor wrote the work "under a pseudonym" (Diodorus Cronus 1975: 96). The republication of the 1965

the editor of *The Southern Journal of Philosophy* presented to the editorial board a request for a corrigendum that I had sent to the journal the previous month. A correction titled "Editorial Corrigendum" for the article was published a few months later, stating that the "author's true name is Richard Taylor" (Anonymous 2018: 156).

In contemporary philosophy, there are divergent views about maintaining strict authorship requirements as well as divergent views about how to respond when works are revealed to have been published anonymously. The results of 11 requests sent in 2017 for published clarifications of authorship for the abovementioned pseudonymously written articles and book chapters is shown in Table 2.1.

Table 2.1 Results of 11 requests for published clarifications of authorship at philosophy journals for pseudonymously published articles

	Journal or publisher	Author of record	Genuine author(s)	Status of request	Type of clarification
1	*Analysis*	"Diodorus Cronus"	Steven Cahn; Richard Taylor	Unresolved	
2	*Analysis*	"Al. Tajtelbaum"	Alfred Tarski	Unresolved	
3	*Australasian Journal of Philosophy*	"Bruce Le Catt"	David Lewis	Granted	Erratum (Anonymous 2017d)
4	*Journal of Philosophy*	"Carmen de Macedo"	Neven Sesardić	Granted	Corrigendum (Anonymous 2017e)
5	*Mind*	"R. E. Hobart"	Dickinson S. Miller	Denied	–
6	*Mind*	"R. E. Hobart"	Dickinson S. Miller	Denied	–
7	*Mind*	"R. E. Hobart"	Dickinson S. Miller	Denied	–
8	*Philosophical Studies*	"M. Lisagor"	Joseph Margolis	Granted	Correction (Margolis 2018)
9	*Southern Journal of Philosophy*	"Diodorus Cronus"	Richard Taylor	Granted	Corrigendum (Anonymous 2018)
10	*University of California Press*	"Leila Tov-Ruach"	Amélie Oksenberg Rorty	Granted	Erratum (University of California Press 2017a)
11	*University of California Press*	"Leila Tov-Ruach"	Amélie Oksenberg Rorty	Granted	Erratum (University of California Press 2017b)

article appeared with a new title "The Necessity of Everything That One Does," with "Diodorus Cronus" as the author of record, but with an introductory statement hinting at the true authorship by referencing Taylor's apiculture interests: "Diodorus Cronus is a metaphysician and beekeeper who lives in Trumansburg, N. Y." (Diodorus Cronus 1978: 148). It was republished twice more, but without reference to the original author of record "Diodorus Cronus" (Cahn 2007: 19–24; Cahn et al. 2011: 137–144).

At the time of this writing, 6 of the 11 articles have been subject to correction by the journals and publishers. Three requests were denied, and the remaining two cases are presently unresolved. In my view, all articles appearing in academic journals under undisclosed pseudonyms should be corrected with errata or corrigenda that disclose the true authorship. The need to issue a correction in a given case increases to the extent that: (1) readers of the pseudonymous articles are deceived; and (2) inconsistent authorial attribution occurs in the downstream literature. On this twofold heuristic, several of the cases examined above would warrant published corrections. In those cases in which an author writes on a topic both under a pseudonym and under a genuine name, thereby creating the appearance of multi-authored discussion or controversy, readers may be led to overestimate the significance of the debate in the contemporary philosophical scene. When those with insider information later cite the pseudonymous works with authorial credit to the genuine authors, problems are created in the downstream literature when the same works are credited to more than one author of record. The articles published with authorial attribution to Le Catt and Tov-Ruach score highly on both elements of this twofold heuristic. In contrast, the articles published with authorial attribution to Diodorus Cronus would not score highly on the first element of this heuristic, since readers likely will not be deceived into thinking that the repurposed ancient name is the real name of the genuine authors of the articles. Nevertheless, the attribution inconsistencies in the downstream literature for the Diodorus Cronus articles cause those articles to score highly on the second element of this heuristic.

In an interview in the summer of 2017, Peter Singer discussed the possibility of a new philosophy journal in which no contributors would publish under their real names. The point of such a journal would be to promote the free discussion of controversial topics. Singer explains:

> I've recently been sounded out about a proposal for a journal that would allow people to publish anonymously. The journal would keep a record of authorship that could, on request by the author, be sent to committees considering appointments and promotions. (Sosis 2017)

If articles in this journal were scored in light of these two elements, they would not likely deceive readers about authorship—given the unusual authorial policy of such a journal—but eventually there could be problems of inconsistent authorial attribution, should those with insider information later cite the works with attribution to the genuine authors. Furthermore, if the editors of such a journal were to promise the protection of anonymity to all its contributors, this policy could create some difficulties for maintaining author accountability, especially if allegations of suspected plagiarism, falsification, or fabrication were ever brought forth for one of the published articles. Additionally, a straightforward presentation of conflict-of-interest statements could pose a challenge.

The use of undisclosed pseudonyms is not the only kind of violation of the *Authorship Condition* that can lead a publisher to withdraw a work from the scholarly record. In recent years, the exposure of acts of ghostwriting, guest or gift authorship, and various kinds of name appropriation have led to corrections of the scholarly record in a variety of disciplines (Weber-Wulff 2014: 14–15; Johal et al. 2017; Öchsner 2013: 85). Unscrupulous researchers will sometimes add the name of a prolific researcher in a field to a submitted manuscript, with the hope that it will thereby have a greater chance of avoiding desk-rejection. The phenomenon of established researchers having to disavow articles appearing under their names is becoming more common (Crama et al. 2016; Weber-Wulff 2014: 16). Additionally, many hoax articles, in addition to violating the *Knowledge Condition*, violate the *Authorship Condition* insofar as they are published under pseudonyms.

A relatively recent phenomenon that challenges the *Authorship Condition* has been termed "hyperauthorship." This occurs when a single published item appears with multiple authors of record reaching numbers well beyond the traditional notions of co-authorship (Cronin 2001). New forms of collaboration have prompted examinations of the kind of contribution an individual must make to justify inclusion as one of several authors of record for a published work. According to some researchers, the "term 'author' has gained an expansive, ambiguous, and in many ways, diluted meaning" (Borenstein and Shamoo 2015: 279). In an editorial titled, "It is Time to Restore Rules for Authorship of Scientific Publications," three medical researchers argued that "One reason for the decrease of confidence in science is the decay of rules for authorship of scientific publications" (Ludvigsson et al. 2018: 586). In extreme cases, single articles have been published with hundreds and even thousands of authors. In one case, 24.5 pages of a 33-page research article consisted in the identification of the more than 5000 purported authors of record and their institutional affiliations (Aad et al. 2015). Even though cases where the authors of record for a published item exceed 1000 are rare, papers with 50 or more authors are becoming increasingly more common (King 2012).

Hyperauthorship causes problems before and after publication, not only weakening accountability for the final product but also generating problems in the earlier publication process. In describing the prevalence of hyperauthorship in the field of nuclear and particle physics, one commentator asks, "who is left to peer review a paper when just about all the experts in a given field are among its authors?" (Mallapaty 2018). The *Authorship Condition* can be strained in two directions: the use of an undisclosed pseudonym and the practice of hyperauthorship each brings about a loss of authorial accountability for published works. In 2017, Elsevier withdrew a scientific volume that consisted mostly of several years' worth of discussion board posts by over 300 contributors on *LinkedIn*. Some of the discussion board participants were surprised to see their online comments put forward as authoritative recommendations in the volume, and some of the surrounding controversy involved the problem of tenuous authorship accountability (Hall 2017).

In recent years, there have been calls to strengthen author recognition through unique digital author-identifier systems. Two researchers sharing the name "A. M. Harrison" (Andrew Marc Harrison and Anthony Mark Harrison) have explained that such systems would disambiguate authors with identical or similar names, organize an author's entire publication history, and allow the documentation of an author's work through any name changes (Harrison and Harrison 2016). Another example of research in the area of authorship disambiguation is an article co-authored by four economists who share the same surname: "A Few Goodmen: Surname-Sharing Economist Coauthors" (Goodman et al. 2015). One of the more successful author identification systems has been the *Open Researcher & Contributor ID* (ORCID) registry (Haak et al. 2012).

The implementation of author identification systems might also help to prevent other abuses of the *Authorship Condition*, as when articles have appeared with attribution to fictitious co-authors to give the appearance of a more diverse or better-staffed research group (Ghorayshi 2017). In an article titled, "Why Fake Data When You Can Fake a Scientist?" Adam Marcus and Ivan Oransky warn that "there is a tiny but growing horde of scientists and collaborators who are figments of someone's imagination" (Marcus and Oransky 2016).

The *Authorship Condition* is subject to a wide variety of violations beyond the ones mentioned here. In a hoax meant to demonstrate the proliferation of dubious publishing outlets and the ease of manipulating citation metrics, Cyril Labbé created a fictitious scientist named Ike Antkare, whose "works" promoted Antkare to such a stupendously high h-index on *Google Scholar* (n.d.) that Antkare achieved the 21st position of most highly cited researchers, above Einstein but below Freud (Labbé 2010). Notable examples of scientific papers appearing with fictitious coauthors whose names are not-so-subtle expletives again violate the *Authorship Condition*, but in obvious cases perhaps few are fooled (see Tartamella 2014 for examples).

2.1.3 The Publication Condition

The *Publication Condition* requires that scholarly works are edited, endorsed, and widely disseminated by an academic publisher in electronic or print form after some degree of vetting. Most often the vetting involves some form of a peer-review process, which admits of a variety of levels of review (Tennant et al. 2017). With the advent of various technologies and the proliferation of certain open-access initiatives, the role of traditional publishers as gatekeepers of the scholarly record has in recent years been subject to increasing levels of criticism (Eve 2014). One critique of traditional publishing appeared in a retrospective titled, "Becoming Professor with Almost no Publications" (Flensburg 2017). Recent years have seen various criticisms of the dependency of researchers on traditional publishers. A reform advocate has characterized the matter by stating that the research goal of academia "is in direct competition with the publisher's goals of making profits" (Logan 2017: 2).

In response to such views, defenders of the traditional publishing industry have attempted to articulate more fully the value-added elements that academic publishers bring to research (Anderson 2018). On one account titled, "Why Do We Still Have Journals?" an editor proposes that a publisher accomplishes four indispensable tasks: (1) certifying that an article has been subject to a vetting process, (2) convening engaged scholars who possess specialized interests and expertise, (3) selecting worthwhile manuscripts for distribution to the relevant subsection of a discipline, and (4) providing high-quality editing services to manuscript authors (Davis 2014). But even publishers have acknowledged that their relationship to researchers varies according to discipline, with some fields (e.g., high energy physics) enjoying greater degrees of independence from traditional publishers in the area of managing peer review due to extensive pre-publication collaborations among researchers (Reller 2012).

The *Publication Condition* indicates that a publisher has established a permanent or fixed instance of an item, which is usually designated as the "version of record." The version of record is the stable, citable version, in contrast to pre-publication iterations such as a submitted manuscript, a manuscript under review, an accepted manuscript, or proof version (Acreman et al. 2008). The published version of record is also the version that is subject to later corrections by a publisher, such as a retraction, erratum, corrigendum, or expression of concern.

The distinction between the version of record issued by a publisher and the various pre-publication iterations is crucial. Theorist Roberto Casati has noted that "publication has a huge significance in the life" of a manuscript, since the act of publication generates a set of entitlements:

> A published paper can be mentioned in the CV of its authors. The results of the paper can be quoted in other published papers, for the purpose of discussing them, criticizing them, replicating them. The paper's impact can be measured according to various metrics (citations, links, impact of the journal it is published in, impact of its authors). The paper contributes to the reputation of its author but also of the institution(s) she belongs to, eventually of that of her country or even continent. (Casati 2010: 194)

These various entitlements do not necessarily hold for the other iterations of a manuscript that exist prior to the publication of the version of record.

The existence of the *Knowledge*, *Authorship*, and *Publication* conditions can be highlighted indirectly through works that violate more than one of them at the same time. In 1976, an "ironic science" paper was published under a pseudonym and purported to discuss empirical research about what causes one person to pass the salt (or other condiments) to others at a meal. The ironic humor of the paper would seem to be obvious from the paper's thesis, to the effect that "at present social science has not found firm evidence to support the validity of the folk belief that the utterance, 'Please pass the salt,' is causally linked to the movement of salt from one end of a table to another" (Pencil 1976: 36). The "salt-passage" literature in psychology has increased significantly in recent years, and at least six articles have been published on the topic that (1) are apparent instances of ironic science, (2) use undisclosed pseudonyms and fake institutional affiliations, and (3) are published in vanity or pay-to-publish journals. In an overview of the so-called "salt passage" literature, one researcher has remarked that the "use of parody and satire invite us to question how the scientific method is applied and how its results are disseminated" and that the apparent lack of peer-review processes at such journals is "undermining the credibility of scientific publishing" (McKelvie 2017: 5). The problems generated from "salt-passage" literature come from failures to observe the *Knowledge*, *Authorship*, and *Publication* conditions.

2.1.4 The Library and Database Conditions

Access to items in the scholarly record is furthered by both the *Library Condition* and the *Database Condition*. The former indicates that items of the scholarly record are collected, housed, and catalogued by academic or research libraries. The latter indicates that such items are listed within particularized, searchable academic databases created by indexing providers that serve a discipline or set of closely related disciplines. Both the *Library* and *Database* conditions ensure that practitioners have efficient pathways to items within the scholarly record. Items that fail these two conditions may be difficult for practitioners to discover, and therefore such items are unlikely to contribute on a large scale to the work of students and researchers.

The appearance of free, non-subscription electronic catalogues of articles has affected the way some scholars learn about new publications in their fields. Both *Google Scholar* (n.d.) and *Microsoft Academic* (n.d.) are non-specialized databases that list academic articles. As these databases are uncurated, they include articles that fall outside of traditional conceptions of the scholarly record. As one critic frames the issue, "Google Scholar indexes much pseudoscience. It is perhaps the world's largest index of junk science" (Beall 2018: 293). Some disciplines have seen the launch of free, web-based specialized databases. In philosophy, the free online user-edited database *PhilPapers* (n.d.) is now a respected alternative to the traditional subscription-based curated database, *The Philosopher's Index* (n.d.).

2.1.5 The Discipline Condition

Lastly, the *Discipline Condition* requires that a work belong to a recognized academic field. Writings entirely unrelated to any existing academic discipline—such as those purporting to present current views of unicorn migration patterns, or details of angel metabolism—would be precluded from entering the scholarly record. As some works belong securely to more than one discipline, it is possible to reference a relevant portion or definable section of the scholarly record as "the scholarly record [*in discipline x*]". On this view, articles in a given academic journal could be said to belong to the scholarly record in discipline *y* and also to the scholarly record in discipline *z*. Articles appearing in multidisciplinary journals thereby belong to the respective scholarly records of the disciplines served by the journal. Since articles appearing in multidisciplinary venues will appear in the particularized research databases of each relevant discipline, the way in which a work satisfies the *Discipline Condition* can affect how it satisfies the *Database Condition.*

2.2 Works at the Margin of the Scholarly Record

Many research items that are different in kind belong securely to the scholarly record by fulfilling the six aforementioned conditions. The following generally satisfy them all:

- A single-authored monograph with a university press
- An edited collection with a commercial academic publisher

- A guest editor's introductory essay in a special issue of an established journal
- An annotated translation of a classic work with a commercial academic publisher
- A published annotated bibliography (e.g., an entry in *Oxford Bibliographies Online*)
- A book review in an annual with a university press
- A regularly updated electronic encyclopedia entry with a commercial academic publisher

Each of these items, despite representing different genres within the scholarly record, can maximally satisfy all six conditions. They belong to the scholarly record because each sets forth a putative knowledge claim in at least one recognized academic discipline under identifiable authorship; each appears with the imprint of an academic press; and each enjoys catalogue entries in library holdings and in specialized research databases.

There are items, however, that only imperfectly fulfill these six conditions or even fail to fulfill certain of these conditions altogether. How are they related to the scholarly record? One might consider an established scholar's unpublished essay that circulates informally and becomes extensively cited and discussed. Although uncommon in some disciplines, such situations do happen, and in one case like this in philosophy the author himself later noted, "It was only when refutations of the paper I have not yet written began to appear in print that I became aware of the extent to which copies of the draft were multiplying" (Burnyeat 1995: 15). Similarly, translations of texts or the detailed scholarly notes of researchers have circulated widely for years in mimeographed, photocopied, or electronic form without finding an academic publisher, all the while being discussed in published articles and used extensively by students and scholars. One could imagine attempts to argue that individual items that maximally satisfy the *Authorship*, *Discipline*, and *Knowledge* conditions but fail the *Publication*, *Library*, and *Database* conditions should still be said to belong safely within the scholarly record.

In philosophy, the subfield of Aristotelian studies provides distinctive examples. Many "pro manuscripto" English translations of the Aristotelian commentaries by Thomas Aquinas appeared in typescript form under the aegis of "Alum Creek Press" in the 1960s and have circulated widely. The "Alum Creek Press" was really a mimeograph machine on the third floor of Erskine Hall at what is now Ohio Dominican University in Columbus, Ohio, USA. The production of the mimeographed copies was organized by Dominican friar Pierre Conway, who translated the commentaries. Another Dominican, James Weisheipl, authored "Aristotelian Methodology: A Commentary on the 'Posterior Analytics' of Aristotle," and this typescript has enjoyed a clandestine life among Thomistically inclined researchers since its completion in 1958 (Weisheipl 1958; see Osborne 2012). Similarly, the two collections of comments appearing under the respective titles "Notes on Book Zeta of Aristotle's 'Metaphysics', Being the Record by Myles Burnyeat and Others of a Seminar held in London, 1975–1979" (Burnyeat 1979) and "Notes on Books Eta and Theta of Aristotle's 'Metaphysics', Being the Record by Myles Burnyeat and Others of a Seminar held in London,

1979–1982" (Burnyeat 1982) have circulated extensively, and have even been subject to book reviews in *The Philosophical Review* and *The Classical Review*. In cases such as these, some of the content of the marginal items is indirectly brought within the scholarly record when cited and discussed by other items that are firmly established in the scholarly record. Indeed, indirect participation through citation also allows relatively newer media (e.g., relevant social media postings, blog entries, discussion lists, online editorials) as well as older media (e.g., newspaper articles from the journalistic record) to be brought into the scholarly record in a qualified way.

Questions over the status of items that fail to satisfy the publication requirement are likely to continue as present-day researchers make greater use of online preprint servers and institutional repositories to disseminate their work prior to (or even, in lieu of) submission to traditional publishers. Doctoral dissertations have also endured a kind of liminal existence with respect to the scholarly record, as the requirement that they be published was virtually abandoned in the United States in the first half of the twentieth century, even though some European universities still preserve a publication requirement as a condition for receiving the doctoral degree. The gradual substitution of the publication requirement with the request that dissertations be submitted instead to an indexed microfilm depository was controversial in some circles during the last century, not only because of a fear for a decline in quality but also because of the tenuous relationship unpublished dissertations would have toward the scholarly record. One longtime university administrator of last century worried that dissertations "relegated to the limbo of microfilm" rather than being published will bring about "the complete loss of the scholarly product itself" (Deferrari 1962: 432). Unpublished, well-known dissertations are rare, but occasionally some dissertations achieve some status by being cited by other works that are firmly established in the scholarly record. Again, these cases involve marginal items that are indirectly brought into to the scholarly record through citation and discussion in items indisputably within the scholarly record.

Other items at the boundary of the scholarly record today include those in the relatively recent genre of digital humanities. Online hypertexted critical editions of classic texts or content-rich interactive scholarly websites have led some academics to call for new peer-review mechanisms that would provide a kind of academic imprimatur to those digital works that demonstrate scholarly competence and enduring value. Such validating initiatives are still in their incipient stages and have yet to receive significant scholarly support (Bates et al. 2006).

Non-traditional venues for post-publication peer review contain other items at the margin of the scholarly record. The website PubPeer, for example, is a platform for anonymous comments on published scholarly book chapters and articles. Originally, only those with established scholarly credentials (e.g., a publication record and an affiliation with a research institution) could register on the site and leave anonymous critiques of published work, but now unregistered anonymous comments are allowed. Authors themselves can reply and defend their work, with their postings designated with an "Author Response" logo. Such post-publication reviews have exercised a tremendous influence in exposing research fraud and have led to many retractions of deficient articles. Some opponents have contended that unmoderated post-publication review venues are tantamount to abuse (Blatt 2015; Fiske 2016; Teixeira da Silva 2017b). PubPeer postings clearly fail the *Authorship*, *Publication*, *Library*, and *Database* conditions. Nevertheless, the commercial academic publisher Taylor & Francis includes links to PubPeer postings on webpages of their journal articles that have received treatment on the PubPeer website (Townsend 2013). While this relationship between PubPeer and Taylor & Francis fails to satisfy the *Publication Condition*, it prompts a consideration of whether new, emerging venues of post-publication review will eventually come to be recognized as part of the scholarly record. For now, the safer position is to consider these post-publication venues to be instrumentally useful for maintaining the integrity of the scholarly record, without considering the postings themselves to fall within the scholarly record. On this view, the venues draw attention to items of the scholarly record but remain extrinsic to it.

Recent years have seen controversies regarding which commercial publishers should be counted as academic. Separating a low-quality but genuine academic commercial publisher from a vanity or predatory publisher may be difficult to professional outsiders as well as to some insiders. Such a separation is crucial, however, as the distinction between the two is one of kind rather than degree. The proliferation of vanity publishing venues, which generate profits primarily by charging fees to authors after little or most often no peer review, has led to attempts to identify legitimate scholarly commercial publishers. Dubious publishers exploit the open-access model of publishing by purporting to conduct peer review, and they generate profits directly from authors. On some accounts, the abundance of pseudo-academic, predatory, or vanity publishers entails the potential corruption of science as well as the erosion of scholarly communication (Al-Khatib 2016; Beall 2015). In the words of one editor,

> Publishing in such journals is not a matter of academic freedom. Predatory publishers are frauds and criminals. To knowingly use them is to engage in fraudulent and criminal activity. (Watson 2017)

Various lists have appeared from several sources to demarcate publishing venues that provide an acceptable entryway for academic work into the scholarly record. The *Directory of Open Access Journals* (DOAJ) (2003) is a free vetted list of open-access journals, and *Cabells Scholarly Analytics* (2017) offers subscription-based lists that attempt to separate genuine academic journals from questionable ones. An influential list covering online open-access publishers and journals was edited by librarian Jeffrey Beall and ran from 2012 to 2017, but it was shuttered due to pressures from publishers, librarians, and the editor's home institution (Beall 2017; Silver 2017). Recent years have seen offerings of journal rankings by the European Science Foundation, the Australian Research Council, and the National Agency for the Evaluation of Universities and Research Institutes, among others, and these lists have generated much controversy (Howard 2011). Anonymous lists of vanity or predatory venues have appeared online, and in the view of some they provide an invaluable service to researchers (Chawla 2018).

As a rule, works appearing with vanity publishers also generally fail the *Library* and *Database* conditions. Well-informed librarians will not allocate funds for their purchase, and open-access vanity publications generally are absent from discipline-specific research databases, appearing perhaps only in maximally inclusive search mechanisms such as Google Scholar.[4] Since articles published in pseudo-academic, predatory, or vanity publishers are not generally discoverable through traditional curated academic databases, legitimate scientific discoveries reported in those outlets may remain unrecognized by researchers. This phenomenon has led some to propose the notion of a "lost science," since the work of genuine researchers, when taken by a pseudo-academic, predatory, or vanity venue, will remain unregistered within the scholarly community (Clark and Smith 2015). Major grant agencies, such as the United States National Institutes of Health, counsel that "authors are encouraged to publish papers arising from NIH-funded research in reputable journals," not only to "protect the credibility of published research" but to ensure "effective communication of scientific results" (National Institutes of Health 2017).

Questionable outlets capture the work of two kinds of authors, however. As one commentator explains:

> Of course, a journal is only predatory to the extent that an author is unaware of the quality of a journal and the publishing services it provides. When an author is fully aware of a predatory journal and the nature of the publishing services it provides, the journal and its publications then become fraudulent and unethical. (Ray 2016: 311)

Both those who are swindled and those who submit to such venues with knowledge of their services increase the visibility of such venues. Authors who publish with institutional affiliations further raise the appearance of credibility to them. The institutional rewards of publishing in such outlets can bring significant financial benefits to authors, particularly if the authors are employed by an institution that fails to distinguish between legitimate publication venues and non-legitimate ones. In a study titled, "The Rewards of Predatory Publications at a Small Business School," one researcher concluded that "there are few incentives not to publish in predatory journals," because predatory publications "produce greater rewards than many non-predatory journal publications" and positively correlate with internal research awards at some institutions (Pyne 2017: 156, 138).

Whether a given commercial publisher is an academic one is often a matter of dispute, and extreme cases can involve legal action in addition to decades of debate. For example, over the years the publisher Edwin Mellen Press has filed separate lawsuits against an academic trade newspaper, a librarian, and a university over claims made about the quality of its books, its pricing, and its editorial processes. The first lawsuit concerned the appearance of a critical article titled, "Vanity's Fare: How One Tiny Press made $2.5 Million Selling Opuscules to your University

---

[4]A 2017 editorial in *The Lancet*, titled, "PubMed should Raise the Bar for Journal Inclusion," reported a study showing that predatory journals have infiltrated databases in some medical subdisciplines, including neuroscience, neurology, and rehabilitation (Manca et al. 2017a: 734, see b). For a qualified defense of some of the benefits that predatory publishers provide to the research community, see Eve and Priego 2017.

Library" (St. John 1993, reprinted in Reid 2006). An overview of this publisher's more recent legal actions and threats against individuals and institutions notes that they are "not isolated incidents" (Anderson 2016). *The New York Times* described the founder of the press in the following way: "The idiosyncratic Mr. Richardson, 81, has long been a figure of controversy within some scholarly precincts for the quality of the books he published and for suing critics of his press for libel" (Cohen 2013). Over the last decade, the large open-access Swiss publisher Frontiers has also been subject to criticism, not only for its inclusion of questionable scientific content in some journals but also for spamming potential authors and for its unusual editorial policies, yet the publisher still has a significant following in some sectors of the research community (Marcus and Oransky 2018: 302–303; Schneider 2015a, b, 2017). Occasionally a publisher of international standing will suffer reputational damage for significant failures in certain areas of its publications. The publishing house Elsevier issued a statement of regret after it was reported that the company had presented works sponsored by pharmaceutical companies under the guise of six peer-reviewed medical journals in the early- to mid-2000s, without disclosing the industry sponsorship (Grant 2009).

Often a nonprofit learned society will function as a publisher for its flagship journal, the society's proceedings, annuals, or other volumes, without the assistance of an independent academic publisher. In such cases, the institutional weight of the society carries the authoritative imprint brought by a traditional publisher to fulfill the *Publication Condition*, as the society takes on the role of reviewing, editing, endorsing, and disseminating the publication under its authority. Works appearing in this way generally satisfy the other remaining five conditions as well.

An unusual case occurs when the editors-in-chief and editorial board of a specialized journal leave a major publisher to start an independent, free-standing open-access journal with the same coverage as the first. The new journal might have a similar or near-identical title. In a well-publicized case, the entire editorial staff—six editors and 31 editorial board members—of the linguistics journal *Lingua* left the publisher Elsevier in 2015 to start the journal *Glossa* as an open-access initiative (Wexler 2015). In 2017, the members of the editorial staff of *Journal of Algebraic Combinatorics* announced they were departing from the publisher Springer to found *Algebraic Combinatorics* as an open-access journal (Anonymous 2017b). The conversion (or "flipping") of journals from traditional online and print subscription format through a publisher to electronic open access provides a modern-day equivalent of the Ship of Theseus paradox. The ancient biographer Plutarch records that the Athenians had preserved the ship on which Theseus had returned to Athens by replacing the old timbers to keep the ship intact, leading to a debate among philosophers over whether the ship venerated by Athens was really the same as the one in which Theseus had sailed (*Theseus*, XXIII). Writing a millennium and a half later, Thomas Hobbes popularized the example by asking whether the discarded timbers, if taken up and reconstituted in the same order to produce a second ship, could be said to be the real ship of Theseus (*De corpore*, 11.7). Such a paradox of unity and identity can apply to journals. New journals that incorporate scope of coverage, editorial outlook, title, and entire staff of an existing subscription print journal to replicate another in electronic form may be viewed by some as the continuation of the same journal in a new venue, particularly when the continuing print version of the journal is believed by some to have gone into rapid decline in quality as a result of the editorial departure (for one account, see Eve 2017). Such journal metamorphoses can be followed by legal challenges. Harvard Library's study of such transitions, the *Journal-Flipping Project*, warns, "It is difficult to know if the initial enthusiasm can sustain the new journal long enough for the editors to reclaim the prestige of the former journal" (Solomon et al. 2016: 59). The new versions of the journals can face both economic and reputational challenges.

The role of a traditional publisher as the gatekeeper of articles admitted to the scholarly record is lessened with the transfer of journals to open-access initiatives. Such moves are not always successful. To take one example: *Medieval Philosophy and Theology*, which began as an annual print volume published by University of Notre Dame Press in 1991 and continued as a biannual print journal published by Cambridge University Press in 1996, never successfully converted to life without a traditional publisher. In 2003, after struggling to get its subscription base above 500, dealing with an "irregular manuscript flow" and "publication delays," and holding a 20% acceptance rate, the journal separated from its publisher and transformed to an online open-access journal (MacDonald 2006). No new issues ever appeared in the open-access, publisher-free format, however, but all past issues are archived on the journal's homepage maintained by Cornell University Library's "Initiatives in Publishing" program (*Medieval Philosophy and Theology* 1991–2003).

In light of these considerations, it can be seen that the *Publication Condition* is a source for a great number of the controversies that surround works at the margin of the scholarly record. Works that fail the *Publication Condition*, or fulfill it only imperfectly, will likely also have difficulties fulfilling other conditions, such as the *Library* and *Database* conditions. Even though what constitutes publication in a digital age is a matter of continuing discussion, the traditional publisher has not relinquished its role in the dissemination of scholarly work. Items that are deficient with regard to the *Publication Condition* but strong in the other conditions are at the boundary of the scholarly record.

2.3 The Scholarly Record in Transition

The challenging cases suggest that unanimity regarding the precise boundary of the scholarly record is difficult to achieve. Ongoing attempts to expand the boundary in light of new genres in digital scholarship and the emerging varieties of post-publication review will likely continue. Perhaps in the not-too-distant future, conditions 1–6 may be viewed no longer as the individually necessary and jointly sufficient conditions that must be satisfied for an item to be admitted indisputably to the scholarly record. Instead, the scholarly record might be said to consist of items that possess shared characteristics according to Ludwig Wittgenstein's notion of family resemblance. On such a view, such items could be said to share many characteristics (including, perhaps, 1–6 noted above) without the requirement that a whole set or even a defined subset of the characteristics be possessed by each item in the scholarly record.

Active debate about the boundary of the scholarly record can often be found within the tenure and promotion committees at those universities with substantive research expectations for their faculty members. Whether a given item "counts" for tenure or for promotion is oftentimes a discussion, by proxy, of what constitutes the limits of the scholarly record. Marginal items, whose inclusion as part of the scholarly record is open to debate, do not always advance an academic's career. Such items will fail to count as positive contributions at a university that is conservative in following traditional models of scholarly achievement. Analogous debates can be found in graduate schools that face proposals for alternative research projects to replace the traditional doctoral dissertation. The post-graduation form of such projects is often difficult to predict, and it is challenging to gauge how novel projects might gain eventual inclusion in the scholarly record in some revised, publishable form.

Even though conditions 1–6 represent the hallmarks that are possessed by present-day undisputed items of the scholarly record, the requirements of earlier ages have differed significantly. In the medieval period, many works were compilations of earlier material from a variety of sources, and the methods of the time exhibited conceptions of authorship far removed from present-day standards. Influential scientific and philosophical works of the early modern period appeared in the form of self-published, anonymous, or pseudonymous works. Furthermore, discipline boundaries were somewhat blurry or more generous in times past. There have been significant modifications regarding what constitutes scientific or scholarly achievement in more recent centuries, and these changes largely result from evolving institutions and more refined conceptions of academic disciplines. If works were produced today, modeled after the style and production methods of yesteryear, they would likely remain outside the scholarly record.

As many of the items that were admitted to the scholarly record in the distant past fall short of the six conditions met by items indisputably admitted today, it is necessary to distinguish an analysis of the scholarly record at a given time from an analysis of the scholarly record as a temporally evolving system. The concepts of synchronic and diachronic analysis that are used in linguistics can be helpful in considering the scholarly record. A language system can be considered as it exists statically at a given time (synchronically) and also as it develops or evolves over time (diachronically). The scholarly record admits of a similar twofold analysis. Conditions 1–6 are relevant markers for understanding the scholarly record synchronically at the present time, since items that are indisputably added to the scholarly record today meet these conditions. A diachronic analysis of the scholarly record, however, would need to consider how these conditions have evolved and how earlier conceptions of the scholarly record have accounted for works that were added to the scholarly record according to different conceptions of how successful research activity is expressed. A diachronic approach to the scholarly record would identify genetic relationships between later and earlier additions to the scholarly record, highlighting the dependencies of later works on their predecessors.

Due to the differing requirements of the scholarly record of times past, one might be tempted to think that works of distant ages that belonged to the scholarly record of yesteryear should be considered as grandfathered into today's scholarly record. To be sure, calls to expurgate from the scholarly record those past works that fail present-day requirements are rare, and most often are offered merely in jest. In response to a humorous call for retraction of a theologically inspired epistle dedicatory that inaugurated the seventeenth-century scientific journal *Philosophical Transactions of the Royal Society*, the journalists at *Retraction Watch* tweeted, "1665 paper finally subjected to post-publication peer review, 350 years later" (*Retraction Watch* 2016). Two considerations, however, count against a need to grandfather works of the past into a synchronic conception of the present-day scholarly record. First, there is evidence that the significant works of the past that are relevant to the scientific and scholarly activities of researchers and students today still successfully come to fulfill today's six conditions as they are reissued in new editions according to the expectations of present-day scholarship. In many disciplines, classic works of the past are reborn in critical editions issued by academic publishers, often supplemented with prefatory essays, annotations, and other scholarly accoutrements. In some disciplines, such as those in humanities, researchers dedicate a significant part of their scholarly efforts to updating or editing earlier works. In this process of re-editing, works of the past come to satisfy the present-day conditions for admittance into the scholarly record. These new editions of past works certainly meet the *Knowledge*, *Authorship*, *Publication*, *Library*, *Database*, and *Discipline* conditions. A present-day scholarly edition of Euclid's *Elements* with a critical apparatus, or the new magisterial multi-volume editions such as *The Cambridge Ben Johnson* or *The Oxford Francis Bacon*, all preserve past works in light of present-day requirements of items in the scholarly record. Furthermore, careful researchers often spend time locating and referencing the most reliable editions when engaging the works of past ages. Book reviewers often criticize monograph authors who have cited unreliable or outdated editions when more authoritative ones exist, especially when an author has failed to cite a more recent volume that more clearly meets conditions 1–6 noted above. Various initiatives and book series by academic publishers are explicit projects of textual recovery, as they bring forgotten, lost, or little-known works into the scholarly record under today's conditions.

A second reason why grandfathering of past works into the scholarly record is unnecessary is that such works often gain indirect representation within the present-day scholarly record insofar as they are discussed, referenced, or excerpted in recent works. This indirect admittance to the scholarly record allows these past contributions to endure without violating the present-day conditions that govern the scholarly record. Such indirect and partial admittance into the scholarly record may eventually lead the item to be brought into the scholarly record through eventual re-editing and publication.

2.4 The Identity of Items of the Scholarly Record

There has been a longstanding temptation to privilege the printed object, such as a printed monograph or journal article, when conceptualizing the scholarly record and endeavoring to identify its boundary. This temptation may be furthered by the experience of walking amidst the stacks of a library at a major university, where the seemingly endless rows of print books and journals present a compelling visual testament to human knowledge. This outlook must be strongly resisted. The scholarly record is not reducible to tangible objects such as hardcover books or articles in printed journals. A growing proportion of the items of the scholarly record are never printed, and even for those that are, the paper version may be the least frequently used.

The relatively recent appearance of so-called "bookless libraries" in academic institutions testifies to the need for care in consideration of what constitutes a discrete item of the scholarly record. The manifold ways in which a single item of the scholarly record is accessed by users demonstrates that an item is irreducible to any of its print or even electronic formats; in short, an item of the scholarly record maintains its identity across both classes of formats. The print version may not even be temporally prior, since an article appearing in the most recent printed issue of an academic journal might have already been available electronically for several years in an online-first repository on the publisher's website. It may have also have already been cited extensively in the literature in several venues prior to finding final expression in a paper format.[5] A similar temporal priority of an electronic version to a print version is now often found with the publication of many academic books. Some publishers issue monographs or edited collections electronically first, and then offer print versions only as print-on-demand individual purchases. Issuance in paper form is no longer essential to the publication of an item, and some traditional academic publishers have minimized or discontinued altogether print publication of some new book series and journals.

To warn that the print version of an item should not be privileged in attempts to conceptualize the items of the scholarly record may not go far enough; one may err in the opposite direction and misidentify the item with a particular electronic version. The identity of a single item within the scholarly record can be found across many platforms and in many formats. One might consider the following scenario for one item of the scholarly record, which is typical for a relatively successful contribution by a researcher in some disciplines. The item first appears as an article in a high-profile journal, and it is issued both in print and electronically. Several years later, the journal's content is made available as downloadable PDFs in the JSTOR repository. The item is then anthologized in a themed multi-authored collection with a major press, but now with entirely new formatting and pagination, both in print and electronically. Later, the author incorporates the same item as a chapter—with

[5] The delay between prior online accessibility and print publication may be intentional by some editors who seek to implement an 'online queue' stratagem for boosting impact factors. The delay creates a longer window for increasing citations within an impact factor calculation timeframe (Martin 2016a: 40–41, b: 4–5).

minimal or no revisions—in a monograph with a major university press. The monograph is issued simultaneously in both print and e-book formats, and the pagination and formatting is altered again. It is dispersed across various electronic platforms, and the print versions include hardcover and paperbound, with various binding and paper quality changes to reflect pricing strategies for the years following initial publication. The press then makes this latest version of the item for sale by individual download, shorn from the other chapters of the book. (Some academic publishers now request that authors submit manuscripts with separate bibliographies for each chapter, rather than at the end of the book, presumably in light of the potential to apportion each chapter separately as a distinct marketable and saleable object.) Finally, as many of the largest academic publishers have proprietary subscription-based platforms for distributing their works to academic institutions (e.g., Oxford Scholarship Online, SpringerLink, Brill Online, Cambridge Core) the book is made accessible through several electronic library subscription platforms, where the formatting and pagination are changed once again.[6] In short, the same item can exist in many electronic and print versions. Such a proliferation of a single item of the scholarly record—in so many venues and formats—is typical today for work appearing through major academic publishers.

The above example shows that an adequate conceptualization of the scholarly record requires that one recognize the identity of an item across a variety of print and electronic formats. It is the *same* item that appears or is expressed in the various formats. The identity of the item of the scholarly record is sustained through all its presentations (e. g., reformatting, repagination, file type changes). Focusing on the unity or identity of the item across the various formats assists one in resisting the temptation to identify an item with any of its print versions or any one of its electronic versions, which are simply distinct ways in which one accesses the same item. The same item appears *as* printed, *as* anthologized in a collection, *as* downloadable, *as* viewable in a variety of electronic file formats (and perhaps even *as* cited, *as* remembered, *as* discussed, etc.). The identity of the item in the scholarly record is sustained through all these domains, and the item is only knowable as encountered through its domains. In short, a discrete item of the scholarly record is irreducible to any one of its print or electronic formats.

2.5 A Definable Scholarly Record

The present chapter has attempted to define the scholarly record using four conceptual tools. First, the *Knowledge*, *Authorship*, *Publication*, *Library*, *Database*, and *Discipline* conditions were set forth as the criteria for identifying what is indisputably admitted to the scholarly record today. Items that are widely and uncontroversially recognized now as belonging to the scholarly record maximally fulfill these

[6] The ongoing changes in formatting from the print version of a work has been subject to critique. For one approach, see "The Evils of Oxford Scholarship Online" (Pasnau 2014).

conditions. Works that fall short of satisfying these six conditions to various degrees are questionably related to the scholarly record. Items that fulfill most of them, or fulfill all of them only imperfectly, can be said to be marginal insofar as they exist at the boundary of the scholarly record. To say that an item is marginal is not a denial that it possesses much utility to the practitioner of a discipline. Research instruments, for example, can be useful without belonging to the scholarly record.

Secondly, the concept of indirect participation was used to explain how marginal works that are in some way deficient in fulfilling conditions 1–6 may be brought within the scholarly record indirectly when they are cited, referenced, discussed, or explained in other items that are firmly within the scholarly record. This concept of indirect participation in the scholarly record can clarify debates over works at the margin of the scholarly record.

Thirdly, the distinction between synchronic and diachronic approaches to the scholarly record solves some of the perceived impediments to attaining precision about the scholarly record. Separating these two typologies allows one to isolate the scholarly record as a present system not only from various past versions that have operated under different rules but also from future versions that are only anticipated.

Finally, the insight that a discrete item of the scholarly record is irreducible to any particular one of its print or electronic forms inoculates one from the misstep of privileging one medium in attempting to define the scholarly record.

These four conceptual tools allow for a clear approach to the scholarly record. Clarity is a prerequisite for understanding the methods of correcting the scholarly record in the aftermath of serious instances of research misconduct. A failure to recognize clearly the boundaries of the scholarly record will result in the misapplication of these methods of correction in response to violations of research integrity. At stake is nothing less than maintaining the reliability of the scholarly record. There is evidence that students and researchers face an impending crisis, insofar as there is currently confusion about the scholarly record and a resultant pattern of failure with regard to the judicious use of the standard methods of correction. The presence of uncorrected plagiarized books and articles in the scholarly record threatens the integrity of the research enterprise.

Occasionally one finds the claim that plagiarized articles or book chapters that are published in obscure or in non-elite venues can be ignored and need not be retracted, on the grounds that since most researchers focus their attention elsewhere, few researchers will likely read the deficient works. In my view, this approach is wanting, as the scholarly record should not be conflated with the set of publications that have reputations as being elite or top venues. The reputations of venues can change, and not all researchers wish to judge journal or book publishers at a given time according to the slippery criterion of prestige. Furthermore, valuable scholarly contributions can appear in newly established venues, and scholars may support up-and-coming venues by placing their research in them. The safer course is to pursue corrections of the scholarly record for all substantially deficient works falling within it. That is, all substantially deficient works that fulfill the six hallmarks of items belonging to the scholarly record—including plagiarized books and articles—

should be corrected to ensure a reliable body of published literature for students and researchers. On other occasions, one finds the claim that older articles as well as articles in journals that are no longer producing new issues need not be corrected, since those articles are claimed to be of little interest to present-day researchers. Again, this view reduces the items in scholarly record to a subset of such items. Articles in many humanities disciplines—including philosophy—sometimes have a long shelf-life. A clear conception of the scholarly record is necessary for the successful deployment of the various instruments for correcting it in the aftermath of research misconduct.

References

Aad, G., et al. 2015. Combined measurement of the Higgs boson mass in *pp* collisions at $\sqrt{s}$ = 7 and 8 TeV with the ATLAS and CMS experiments. *Physical Review Letters* 114 (191803): 1–33.

Abbott, Alison. 2007. Academic accused of living on borrowed lines. *Nature* 448 (7154): 632–633. https://doi.org/10.1038/448632b. Accessed 6 July 2018.

Acreman, Beverley, et al. 2008. *Journal article versions (JAV): Recommendations of the NISO/ ALPSP JAV Technical Working Group*. Baltimore: National Information Standards Organization. http://www.niso.org/publications/rp/RP-8-2008.pdf. Accessed 6 July 2018.

Al-Khatib, Aceil. 2016. Protecting authors from predatory journals and publishers. *Public Research Quarterly* 32 (4): 281–285.

Anderson, Rick. 2016. Edwin Mellen Press ends its lawsuit against librarian. *The Scholarly Kitchen*, February 6. https://scholarlykitchen.sspnet.org/2015/02/06/edwin-mellen-press-ends-its-lawsuit-against-librarian. Accessed 6 July 2018.

Anderson, Kent. 2018. Focusing on value – 102 things journal publishers do (2018 update). *The Scholarly Kitchen*, February 6. https://scholarlykitchen.sspnet.org/2018/02/06/focusing-value-102-things-journal-publishers-2018-update. Accessed 6 July 2018.

Anonymous. 2002. Important notice. *The Dickensian* 458 (98.3): 232A–235A.

———. 2009. Retraction. *The British Medical Journal* 338.379. https://doi.org/10.1136/bmj.b379. Accessed 6 July 2018.

———. 2012. Retraction notice. *Computers and Mathematics with Applications* 63 (9): 1424.

———. 2014. Withdrawal: 'A new type of misconduct in the field of the physical sciences: The case of the pseudonyms used by [—] to anonymously criticize other people's works on arXiv'. *Journal of the Association for Information Science and Technology* 65 (11): 2375.

———. 2017a. Retraction. *Cogent Social Sciences* 3: 1. https://doi.org/10.1080/23311886.2017.1336861. Accessed 6 July 2018.

———. 2017b. *Journal of Algebraic Combinatorics* flips to fair open access. *MathOA: Mathematics in Open Access*, July 27. https://svpow.files.wordpress.com/2017/07/jaco-flips-to-fair-open-access.pdf. Accessed 6 July 2018.

———. 2017c. Philosophy journal corrects 35-year-old article 'written' by a cat. *Science*, August 1. https://www.sciencemag.org/news/sifter/philosophy-journal-corrects-35-year-old-article-written-cat. Accessed 6 July 2018.

———. 2017d. Erratum. *Australasian Journal of Philosophy*. http://www.tandfonline.com/doi/full/10.1080/00048402.2017.1358248. Accessed 6 July 2018.

———. 2017e. Corrigendum. *The Journal of Philosophy* 114 (7): 392.

———. 2018. Editorial corrigendum. *The Southern Journal of Philosophy* 56 (1): 156.

Axelrod, Jeremy. 2006. Caustic acrostic. *The New Criterion*, September 27. http://www.newcriterion.com/posts.cfm/caustic-acrostic-4303. Accessed 6 July 2018.

Badiou Alain, 2012. *The adventure of French philosophy*. Trans: Bruno Bosteels. London: Verso.

Bar-Hillel, Yehoshua. 1965. Et tu, Diodorus Cronus? *Analysis* 26 (2): 54–56.

Bates, David et al. 2006. Peer review and evaluation of digital resources for the arts and humanities. Arts and Humanities Research Council. http://www.history.ac.uk/sites/history.ac.uk/files/Peer_review_report2006.pdf. Accessed 6 July 2018.

Beall, Jeffrey. 2015. Predatory journals and the breakdown of research cultures. *Information Development* 31 (5): 473–476.

———. 2017. What I learned from predatory publishers. *Biochemia Medica* 27 (2): 273–279.

———. 2018. Scientific soundness and the problem of predatory journals. In *Pseudoscience: The conspiracy against science*, ed. Allison B. Kaufman and James C. Kaufman, 283–299. Cambridge, MA: The MIT Press.

Bertolet, Rod, and William L. Rowe. 1979. The fatalism of 'Diodorus Cronus'. *Analysis* 39 (3): 137–138.

Blatt, Michael R. 2015. Vigilante science. *Plant Physiology* 169: 907–909. https://doi.org/10.1104/pp.15.01443. Accessed 6 July 2018.

Bohannon, John. 2016. Fight over author pseudonyms could flare again. *Science* 351 (6276): 902.

Borenstein, Jason, and Adil E. Shamoo. 2015. Rethinking authorship in the era of collaborative research. *Accountability in Research* 22: 267–283.

Burnyeat, Miles. 1979. Notes on Book Zeta of Aristotle's 'metaphysics', being the record by Myles Burnyeat and others of a seminar held in London, 1975–1979. Unpublished manuscript.

———. 1982. Notes on Books Eta and Theta of Aristotle's 'metaphysics', being the record by Myles Burnyeat and others of a seminar held in London, 1979–1982. Unpublished manuscript.

Burnyeat, M.F. 1995. Is an Aristotelian philosophy of mind still credible (a draft). In *Essays on Aristotle's 'De anima'*, ed. Martha C. Nussbaum and Amélie Oksenberg Rorty, 15–26. Oxford: Clarendon Press.

Cabells Scholarly Analytics. 2017–. http://www.cabells.com. Accessed 6 July 2018.

Cahn, Steven. 1966. An unanswered paradox. *Analysis* 26 (6): 203–206.

Cahn, Steven M. 2007. *Puzzles and perplexities: Collected essays*. 2nd ed. Lanham: Lexington Books.

Cahn, Steven M., Robert B. Talisse, and Scott F. Aikin. 2011. *Thinking about logic: Classic essays*. Boulder: Westview Press.

Casati, Roberto. 2010. On publishing. *Social Epistemology* 24 (3): 191–200.

Chawla, Dalmeet Singh. 2016. U.S. gov't researchers withdraw climate paper after using pseudonyms. *Retraction Watch*, September 13. http://retractionwatch.com/2016/09/13/u-s-govt-researchers-withdraw-climate-paper-after-using-pseudonyms. Accessed 6 July 2018.

———. 2018. The undercover academic keeping tabs on 'predatory' publishing. *Nature*. https://doi.org/10.1038/d41586-018-02921-2. Accessed 6 July 2018.

Chopwhittle, Darcy. 2001. Review of Philinda Blank, when the cows come home: barn architecture and changes in bovine public space. *Social Science History* 25 (4): 609–613.

Clark, Jocalyn, and Richard Smith. 2015. Firm action needed on predatory journals. *The British Medical Journal* 350: 210. https://doi.org/10.1136/bmj.h210. Accessed 6 July 2018.

Cohen, Patricia. 2013. Diary of a Hitler aide resurfaces after a hunt that lasted years. *The New York Times*, June 13. http://www.nytimes.com/2013/06/14/world/europe/diary-of-a-hitler-aide-resurfaces-after-a-hunt-that-lasted-years.html. Accessed 6 July 2018.

Crama, Yves, Michael Grabisch, and Silvano Martello. 2016. A brand new cheating attempt: A case of usurped identity. *4OR: A Quarterly Journal of Operations Research* 14 (4): 333–336.

Cronin, Blaise. 2001. Hyperauthorship: A postmodern perversion or evidence of a structural shift in scholarly communication practices? *Journal of the American Society for Information Science and Technology* 52 (7): 558–569.

Czepel, Robert. 2017. Korrektur für Fachartikel – von einer Katze. *Österreichischer Rundfunk*, August 8. http://science.orf.at/stories/2859196. Accessed 6 July 2018.

Davis, Gerald F. 2014. Why do we still have journals? *Administrative Science Quarterly* 59 (2): 193201.

de Macedo, Carmen. 2008. Guilt by statistical association: revisiting the prosecutor's fallacy and the interrogator's fallacy. *The Journal of Philosophy* 105 (6): 320–332. [Corrigendum issued in Anonymous. 2017e.].
Deferrari, Roy J. 1962. Reproducing the doctoral dissertation through microfilm. In *Memoirs of The Catholic University of America 1918–1960*, 431–434. Boston: Daughters of St. Paul.
Diodorus Cronus. 1965. Time, truth and ability. *Analysis* 25 (4): 137–141.
———. 1971. The governance of the kingdom of darkness: A philosophical fable. *The Southern Journal of Philosophy* 9 (2): 113–118. [Corrigendum forthcoming as Anon. 2018].
———. 1975. The governance of the kingdom of darkness: A philosophical fable. In *Philosophical Explorations*, ed. Peter A. French, 96–104. Morristown: General Learning Press.
———. 1978. The necessity of everything that one does. In *Introductory readings in metaphysics*, ed. Richard Taylor, 147–153. Englewood Cliffs: Prentice-Hall.
Directory of Open Access Journals (DOAJ). 2003. https://doaj.org. Accessed 6 July 2018.
Dynes, Wayne R., et al. 2016. *Encyclopedia of homosexuality*., 2 vols. New York: Routledge.
Easton, Loyd D., ed. 1975. *Miller S. Dickinson: Philosophical analysis and human welfare. Selected essays and chapters from six decades*. Dordrecht: D. Reidel Publishing Company.
Eve, Martin Paul. 2014. *Open access and the humanities: contexts, controversies and the future*. Cambridge, UK: Cambridge University Press. https://doi.org/10.1017/CBO9781316161012. Accessed 6 July 2018.
———. 2017. If I still subscribed to Elsevier's Lingua, I'd demand a refund. Paul Martin Eve, May 29. https://www.martineve.com/2017/05/29/lingua-refund. Accessed 6 July 2018.
Eve, Martin Paul, and Ernesto Priego. 2017. Who is actually harmed by predatory publishers?, *TripleC: Communication, Capitalism & Critique* 15 (2): 755–770. https://doi.org/10.31269/triplec.v15i2.867. Accessed 6 July 2018.
Faulkes, Zen, ed. 2018. *Stinging the predators: A collection of papers that should never have been published*, Version 4.1. https://doi.org/10.6084/m9.figshare.5248264. Accessed 6 July 2018.
Fiske, Susan T. 2016. Mob rule or wisdom of crowds? *APS Observer*, [in Press] 1–2. http://datacolada.org/wp-content/uploads/2016/09/Fiskepresidential-guest-column_APS-Observer_copy-edited.pdf. Accessed 6 July 2018.
Flensburg, Per. 2017. Becoming professor with almost no publications. *Scandinavian Journal of Information Systems* 29 (1): 1–26.
Florence, Maurice. 1994. Foucault, Michel, 1926–. Trans. Catherine Porter. In *The Cambridge companion to Foucault*, ed. Gary Gutting, 314–320. Cambridge, UK: Cambridge University Press.
Fulda, Joseph S. 2007. The ethics of pseudonymous publication. *Journal of Information Ethics* 16 (2): 75–89.
Gamut, L.T.F. 1991. *Logic, language, and meaning. Volume 1: Introduction to logic*. Chicago: University of Chicago Press.
Gardin, Harrett. 2005. Game preserve ethics: The case for hunting the poor. *Southwest Philosophy Review* 21 (1): 103–110.
Gaskin, Richard, and Daniel J. Hill. 2013. Reach's puzzle and mention. *Dialectica* 67 (2): 201–222.
Gay-Williams, J. 2012. The wrongfulness of euthanasia. In *Intervention and reflection: Basic issues in bioethics*, ed. Ronald Munson, 9th ed., 593–596. Boston: Wadsworth Cengage Learning.
Gentleman, A. 2010. A letter from a gentleman in Dunedin to a lady in the country. In *Hume on is and ought*, ed. Charles R. Pigden, 76–91. New York: Palgrave Macmillan.
Geradt, Max, Sara Jahnke, Julia Heinz, and Jürgen Hoyer. 2018. Is contact with children related to legitimizing beliefs toward sex with children among men with pedophilia? *Archives of Sexual Behavior* 47 (2): 375–387.
Ghorayshi, Azeen. 2017. Caltech professor who harassed women was also investigated for creating an imaginary female researcher. *BuzzFeed News*, May 25. https://www.buzzfeed.com/azeenghorayshi/harasser-faked-female-scientist. Accessed 6 July 2018.
Goldingay, Sarah. 2009. Plagiarising theory: Performance and religion? *Studies in Theatre and Performance* 29 (1): 5–14.

Goodman, Allen C., Joshua Goodman, Lucas Goodman, and Sarena Goodman. 2015. A few Goodmen: Surname-sharing economist coauthors. *Economic Inquiry* 53 (2): 1392–1395.

Google Scholar. n.d. https://scholar.google.com. Accessed 6 July 2018.

Grant, Bob. 2009. Elsevier published 6 fake journals. *The Scientist*, May 7. http://www.the-scientist.com/?articles.view/articleNo/27383/title/Elsevier-published-6-fake-journals. Accessed 6 July 2018.

Grove, Jack. 2017. Cash-for-publications is 'common practice'. *Times Higher Education*, January 19. https://www.timeshighereducation.com/news/cash-publications-common-practice. Accessed 6 July 2018.

Guarino, Ben. 2016. Scientists published climate research under fake names. Then they were caught. *The Washington Post*, September 19. https://www.washingtonpost.com/news/morning-mix/wp/2016/09/19/scientists-published-climate-research-under-fake-names-then-they-were-caught. Accessed 6 July 2018.

Haak, Laurel L., Martin Fenner, Laura Paglione, Ed Pentz, and Howard Ratner. 2012. ORCID: A system to uniquely identify researchers. *Learned Publishing* 25 (4): 259–264.

Hall, Matt. 2017. Attribution is not permission. *Agile Scientific*, August 31. https://agilescientific.com/blog/2017/8/31/attribution-is-not-permission. Accessed 6 July 2018.

Harrison, Andrew Marc, and Anthony Mark Harrison. 2016. Necessary but not sufficient: Unique author identifiers. *BMJ Innovations* 2 (1): 141–433. https://doi.org/10.1136/bmjinnov-2016-000135. Accessed 6 July 2018.

Harvey, Stephanie. 2002. Dicken's villains: A confession and a suggestion. *The Dickensian* 458 (98.3): 233–235. [Corrigendum issued in Anonymous. 2002.].

Hobart, R.E. 1930a. Hume without scepticism (I). *Mind* 39 (155): 273–301.

———. 1930b. Hume without scepticism (II). *Mind* 39 (156): 409–425.

———. 1934. Free will as involving determination and inconceivable without it. *Mind* 43 (169): 1–27.

Howard, Jennifer. 2011. Journal-ranking system gets dumped after scholars complain. *The Chronicle of Higher Education*, June 1. http://www.chronicle.com/article/Journal-Ranking-System-Gets/127737. Accessed 6 July 2018.

Jahnke, Sara, Alexander F. Schmidt, Max Geradt, and Jürgen Hoyer. 2015. Stigma-related stress and its correlates among men with pedophilic sexual interests. *Archives of Sexual Behavior* 44 (8): 2173–2187.

Johal, Jaspreet, Marios Loukas, Rod J. Oskouian, and R. Shane Tubbs. 2017. 'Political co-authorships' in medical science journals. *Clinical Anatomy* 30 (6): 831–834.

Kierkegaard, Søren. 2009. *Repetition* and *philosophical crumbs*. Trans. M. G. Piety. Oxford: Oxford University Press.

King, Christopher. 2012. Multi-author papers: Onward and upward. *ScienceWatch Newsletter*. http://archive.sciencewatch.com/newsletter/2012/201207/multiauthor_papers. Accessed 6 July 2018.

Labbé, Cyril. 2010. Ike Antkare: One of the great stars in the scientific firmament. *International Society for Scientometrics and Informetrics Newsletter* 6 (2): 48–52.

Lavoie, Brian, and Constance Malpas. 2015. *Stewardship of the evolving scholarly record: From the invisible hand to conscious coordination*. Dublin: OCLC Research. http://www.conference-center.oclc.org/content/dam/research/publications/2015/oclcresearch-esr-stewardship-2015.pdf. Accessed 6 July 2018.

Lavoie, Brian, Eric Childress, Ricky Erway, Ixchel Faniel, Constance Malpas, Jennifer Schaffner, and Titia van der Werf. 2014. *The evolving scholarly record*. Dublin: OCLC Research http://www.oclc.org/content/dam/research/publications/library/2014/oclcresearch-evolving-scholarlyrecord-2014.pdf. Accessed 6 July 2018.

Lawrence, Amanda, John Houghton, Julian Thomas, and Paul R. Weldon. 2014. *Where is the evidence? Realising the value of grey literature for public policy & practice: A discussion paper*. Melbourne: Swinburne Institute for Social Research. https://doi.org/10.4225/50/5580B1E02DAF9. Accessed 6 July 2018.

Lawrence, Amanda, Julian Thomas, John Houghton, and Paul Weldon. 2015. Collecting the evidence: Improving access to grey literature and data for public policy and practice. *Australian Academic & Research Libraries* 46 (4): 229–248.
Le Catt, Bruce. 1982. Censored vision. *Australasian Journal of Philosophy* 60 (2): 158–162. [Erratum issued in Anonymous. 2017d.].
Lewis, David. 1986. *Philosophical papers*. Vol. II. New York: Oxford University Press.
Lewis, Stephanie R. 2015. Bibliography of the work of David Lewis. In *A companion to David Lewis*, ed. Barry Loewer and Jonathan Schaffer, 562–571. Malden: Wiley Blackwell.
Logan, Corina J. 2017. We can shift academic culture through publishing choices. *F1000Research* 6 (518): 1–22. https://doi.org/10.12688/f1000research.11415.2. Accessed 6 July 2018.
LoShan, Zhang. 1998. Plato's counsel on education. In *Philosophers on education: New historical perspectives*, ed. Amélie Oksenberg Rorty, 30–48. London: Routledge.
Ludvigsson, Johnny, Pierre Lefebvre, and Jørn Nerup. 2018. It is time to restore rules for authorship of scientific publications. *Pediatric Diabetes* 19 (3): 586.
MacDonald, Scott. 2006. Medieval philosophy and theology. *Proceedings of the 149th ARL membership meeting*. http://old.arl.org/arldocs/resources/pubs/mmproceedings/149/macdonald.htm. Accessed 6 July 2018.
Maisonneuve, Hervé. 2017. Après 35 ans, une revue corrige un article écrit sous le pseudonyme d'un chat. *Rédaction Médicale et Scientifique*, August 3. http://www.h2mw.eu/redactionmedicale/2017/08/après-35-ans-une-revue-corrige-un-article-écrit-sous-le-pseudonyme-dun-chat.html. Accessed 6 July 2018.
Mallapaty, Smriti. 2018. Paper authorship goes hyper: A single field is behind the rise of thousand-author papers. *Nature Index*, January 30. https://www.natureindex.com/news-blog/paper-authorship-goes-hyper. Accessed 6 July 2018.
Manca, Andrea, et al. 2017a. PubMed should raise the bar for journal inclusion. *The Lancet* 390: 734. https://doi.org/10.1016/S0140-6736(17)31943-8. Accessed 6 July 2018.
———. 2017b. The surge of predatory open-access in neurosciences and neurology. *Neuroscience* 353 (14): 166–173.
Marcus, Adam. 2017. After 35 years, philosophy journal corrects article…by a cat. *Retraction Watch*, August 1. http://retractionwatch.com/2017/08/01/35-years-philosophy-journal-corrects-article-cat. Accessed 6 July 2018.
Marcus, Adam, Ivan Oransky. 2016. Why fake data when you can fake a scientist? *Nautilus*, November 24. http://nautil.us/issue/42/fakes/whyfake-data-when-you-can-fake-a-scientist. Accessed 6 July 2018.
———. 2018. Pseudoscience: Coming to a peer-reviewed journal near you. In *Pseudoscience: The conspiracy against science*, ed. Allison B. Kaufman and James C. Kaufman, 301–305. Cambridge, MA: The MIT Press.
Margolis, Joseph. 2018. Correction to: On Harman's theory of knowledge. *Philosophical Studies* 175 (7): 1827.
Martin, Ben R. 2007. Keeping plagiarism at bay—a salutary tale. *Research Policy* 36 (7): 905–911.
———. 2016a. Ethics and integrity in publishing. In *How to get published in the best management journals*, ed. Timothy Clark, Mike Wright, and David J. Ketchen Jr., 29–48. Cheltenham: Edward Elgar Publishing.
———. 2016b. Editors' JIF-boosting stratagems – which are appropriate and which not? *Research Policy* 45: 1–7.
McCook, Alison. 2016. Philosophy journal spoofed, retracts hoax article. *Retraction Watch*, April 7. http://retractionwatch.com/2016/04/07/philosophy-journal-spoofed-retracts-hoax-article. Accessed 6 July 2018.
McKelvie, Stuart J. 2017. 'A case for treatment': What do research reports on salt and pepper passage reveal about research and publication practices? *Current Psychology*: 1–6 https://doi.org/10.1007/s12144-017-9620-x. Accessed 6 July 2018.
McLaughlin, Brian P., and Amélie Oksenberg Rorty, eds. 1988. *Perspectives on self-deception*. Berkeley: University of California Press.

McMillen, Liz. 1995. Pseudonym or hoax? *The Chronicle of Higher Education*, May 26. http://www.chronicle.com/article/Pseudonym-or-Hoax-/85818. Accessed 6 July 2018.

Medieval Philosophy and Theology. 1991–2003. https://cip.cornell.edu/mpat. Accessed 6 July 2018.

Microsoft Academic. n.d. http://academic.research.microsoft.com. Accessed 6 July 2018.

Miller, Dickinson S. 1936. James's philosophical development: Professor Perry's biography. *The Journal of Philosophy* 33 (12): 309–318.

Moss, Stephen. 2013. The man behind the great Dickens and Dostoevsky hoax. *The Guardian*, July 10. https://www.theguardian.com/books/2013/jul/10/man-behind-dickens-dostoevsky-hoax. Accessed 6 July 2018.

Naiman, Eric. 2013. When Dickens met Dostoevsky. *The Times Literary Supplement*, April 10. https://www.the-tls.co.uk/articles/public/whendickens-met-dostoevsky. Accessed 6 July 2018.

National Institutes of Health. 2017. Statement on article publication resulting from NIH funded research. https://grants.nih.gov/grants/guide/notice-files/NOT-OD-18-011.html. Accessed 6 July 2018.

Neuroskeptic. 2012. The nine circles of scientific hell. *Perspectives on Psychological Science* 7 (6): 643–644.

———. 2013. Anonymity in science. *Trends in Cognitive Sciences* 17 (5): 195–196.

Öchsner, Andreas. 2013. *Introduction to scientific publishing: Backgrounds, concepts, strategies.* Heidelberg: Springer.

Oltermann, Philip. 2016. Human-animal studies academics dogged by German hoaxers. *The Guardian*, March 1. https://www.theguardian.com/world/2016/mar/01/human-animal-studies-academics-dogged-by-german-hoaxers. Accessed 6 July 2018.

Oransky, Ivan. 2015. So-kalled research: French sociology journal retracts hoax article. *Retraction Watch*, March 12. http://retractionwatch.com/2015/03/12/so-kalled-research-french-sociology-journal-retracts-hoax-article. Accessed 6 July 2018.

———. 2018. Author who lied to journals about his identity slated to have four articles on vaccines retracted. *Retraction Watch*, May 27. https://retractionwatch.com/2018/05/27/author-who-lied-to-journals-about-his-identity-slated-to-have-four-articles-on-vaccines-retracted. Accessed 6 July 2018.

Osborne, Thomas. 2012. Weisheipl's commentary on the posterior analytics online. *Thomistica: A site for the academic study of St. Thomas Aquinas*, October 25. https://thomistica.net/news/2012/10/25/weisheipls-commentary-on-the-posterior-analytics-online.html. Accessed 6 July 2018.

Pasnau, Robert. 2014. The evils of Oxford Scholarship Online. *In Medias PHIL: News and notes for the medieval philosophy community*, August 22. https://inmediasphil.wordpress.com/2014/08/22/the-evils-of-oso. Accessed 6 July 2018.

Pencil, Murdock. 1976. Salt passage research: The state of the art. *Journal of Communication* 26 (4): 31–36.

PhilPapers. n.d. https://philpapers.org. Accessed 6 July 2018.

Pyne, Derek. 2017. The rewards of predatory publications at a small business school. *Journal of Scholarly Publishing* 48 (3): 137–160.

Ray, Margaret. 2016. An expanded approach to evaluating open access journals. *The Journal of Scholarly Publishing* 47 (4): 307–327.

Die Redaktionsleitung. 2016. An die Leser und Abonnenten. *Totalitarismus und Demokratie*. https://web.archive.org/web/20160306132833/http:/www.hait.tu-dresden.de/TD/artikel.asp?art=636&nr=24&m=1. Accessed 6 July 2018.

Reid, Paul H., Jr. 2006. *The Edwin Mellen Press versus 'Lingua Franca': A case study in the law of libel*. Lewiston: The Edwin Mellen Press.

Reller, Tom. 2012. An interview with Tom Reller, Vice President, Global Corporate Relations, Elsevier. https://adametkin.wordpress.com/2012/09/04/an-interview-with-tom-reller-vice-president-global-corporate-relations-elsevier. Accessed 6 July 2018.

Retraction Watch (@RetractionWatch). 2016. 1665 paper finally subjected to post-publication peer review. August 8, 5:46 PM. https://twitter.com/RetractionWatch/status/762812280312659968. Tweet. Accessed 6 July 2018.
Ring, Richard R., and Richard Kay. 2003. *Proceedings of the Pseudo Society. First series: 1986–93*. Kalamazoo: Medieval Institute Publications.
Ronagh, Maryam, and Lawrence Souder. 2015. The ethics of ironic science in its search for spoof. *Science and Engineering Ethics* 21 (6): 1537–1549.
Rorty, Amélie Oksenberg, ed. 1980. *Explaining emotions*. Berkeley: University of California Press.
———, ed. 1998. *Philosophers on education: Historical perspectives*. London: Routledge.
———. 2000. Freud on unconscious affects, mourning and the erotic mind. In *The analytic Freud: Philosophy and psychoanalysis*, ed. Michael P. Levine, 195–210. London: Routledge.
Sanford, David H. 2005. Pseudonyms, philosophical. In *The Oxford companion to philosophy*, ed. Ted Honderich, 765. Oxford: Oxford University Press.
Schechner, Richard, Talia Rodgers, Claire L'Enfant, Judith Butler, Marvin Carlson, Tracy C. Davis, David Savran, Shannon Jackson, Branislav Jakovljevic, Jill Dolan, Zarrilli Phillip, W.B. Worthen, Joseph Roach, and Peggy Phelan. 2009. Concerning theory for performance studies. *TDR: The Drama Review* 53 (1): 7–46.
Schneider, Leonid. 2015a. Publisher problems at frontiers (part 1). *Lab Times Online*, August 9. http://www.labtimes.org/editorial/e_631.lasso. Accessed 6 July 2018.
———. 2015b. Publisher problems at frontiers (part 2). *Lab Times Online*, August 25. http://www.labtimes.org/editorial/e_635.lasso. Accessed 6 July 2018.
———. 2017. Frontiers: Vanquishers of Beall, publishers of bunk. *For Better Science*, September 18. https://forbetterscience.com/2017/09/18/frontiers-vanquishers-of-beall-publishers-of-bunk. Accessed 6 July 2018.
Sesardić, Neven. 2011. Where in the world is Carmen de Macedo? *Inside Higher Ed*, January 31. https://www.insidehighered.com/views/2011/01/31/where-world-carmen-de-macedo. Accessed 6 July 2018.
Silver, Andrew. 2017. Controversial website that lists 'predatory' publishers shuts down. *Nature*. https://doi.org/10.1038/nature.2017.21328. Accessed 6 July 2018.
Singer, Charles. 1937. The pseudonym of Spinoza's publisher. *Journal of the Warburg Institute* 1 (1): 77b–78b.
Smart, J.J.C. 2001. David Kellogg Lewis (1941–2001). *Australian Academy of the Humanities: Proceedings* 26: 1–3. https://www.humanities.org.au/wp-content/uploads/2017/04/AAH-Obit-Lewis-2001.pdf. Accessed 6 July 2018.
Solomon, David J., Mikael Laakso, Bo-Christer Björk. 2016. Converting scholarly journals to open access: A review of approaches and experiences. https://dash.harvard.edu/handle/1/27803834. Accessed 6 July 2018.
Sorensen, Roy. 2008. *Seeing dark things: The philosophy of shadows*. Oxford: Oxford University Press.
Sosis, Cliff. 2017. Peter Singer. *What is it like to be a philosopher*. http://www.whatisitliketobeaphilosopher.com/#/peter-singer. Accessed 6 July 2018.
St. John, Warren. 1993. Vanity's fare: How one tiny press made $2.5 million selling opuscules to your university library. *Lingua Franca*, September/October: 22–25, 62.
Stern, Victoria. 2017. An accomplished philosopher invented a pseudonym. Why? *Retraction Watch*, October 13. http://retractionwatch.com/2017/10/13/accomplished-philosopher-invented-pseudonym. Accessed 6 July 2018.
Tajtelbaum, Al. 1957. It is impossible to be told anyone's name. *Analysis* 17 (3): 52–53.
Tartamella, Vito. 2014. The true story of [—] (and other scientific jokes). *Parolacce*, October 5. www.goo.gl/84Szn6. Accessed 6 July 2018.
Teixeira da Silva, Jaime A. 2017a. Are pseudonyms ethical in (science) publishing? Neuroskeptic as a case study. *Science and Engineering Ethics* 23 (6): 1807–1810.

———. 2017b. Ethical exceptionalism: Can publishing rules be manipulated to give the impression of ethical publishing? *Bangladesh Journal of Medical Science* 16 (4): 610–614. https://doi.org/10.3329/bjms.v16i4.33623. Accessed 6 July 2018.

Tennant, Jonathan P., et al. 2017. A multi-disciplinary perspective on emergent and future innovations in peer review. *F1000Research* 6 (1151): 1–40. https://doi.org/10.12688/f1000research.12037.3. Accessed 6 July 2018.

The Philosopher's Index. n.d. http://philindex.org. Accessed 6 July 2018.

Thompson, Anne. 2017a. Comment. *Retraction Watch*, November 12. http://retractionwatch.com/2017/10/13/accomplished-philosopherinvented-pseudonym/#comment-1496302. Accessed 6 July 2018.

———. 2017b. Comment. *Daily Nous: News for and about the Philosophy Profession*, November 12. http://dailynous.com/2017/10/13/amelierortys-use-pseudonym/#comment-126902. Accessed 6 July 2018.

Tov-Ruach, Leila. 1980. Jealousy, attention, and loss. In *Explaining emotions*, ed. Amélie Oksenberg Rorty, 465–488. Berkeley: University of California Press. [Erratum issued in University of California Press 2017a.].

———. 1988. Freud on unconscious affects, mourning, and the erotic mind. In *Perspectives on self-deception*, ed. Brian P. McLaughlin and Amélie Oksenberg Rorty, 246–263. Berkeley: University of California Press. [Erratum issued in University of California Press 2017b.].

Townsend, Fiona. 2013. Post-publication peer review: PubPeer. *Editor's Bulletin* 9 (3): 45–46.

Tripodi, Benedetta. 2016. Ontology, neutrality and the strive for (non-)being-queer. *Badiou Studies* 4 (1): 72–102.

University of California Press. 2017a. Erratum. https://www.ucpress.edu/book.php?isbn=9780520039216. Accessed 6 July 2018.

———. 2017b. Erratum, https://www.ucpress.edu/book.php?isbn=9780520061231. Accessed 6 July 2018.

van Inwagen, Peter. 1980. A bibliography of the philosophical writings of Richard Taylor. In *Time and cause: Essays presented to Richard Taylor*, ed. Peter van Inwagen, 301–306. Dordrecht: Springer.

Van Noorden, Richard. 2014. Publishers withdraw more than 120 gibberish papers. *Nature*. https://doi.org/10.1038/nature.2014.14763. Accessed 6 July 2018.

Vision, Gerald. 1997. *Problems of vision: Rethinking the causal theory of perception*. New York: Oxford University Press.

Watson, Roger. 2017. Publishing in fraudulent journals is criminal. *Times Higher Education*, August 10. https://www.timeshighereducation.com/opinion/publishing-fraudulent-journals-criminal. Accessed 6 July 2018.

Weber-Wulff, Debora. 2014. *False feathers: A perspective on academic plagiarism*. Heidelberg: Springer.

Weisheipl, James. 1958. Aristotelian methodology: A commentary on the *Posterior Analytics* of Aristotle. https://archive.org/details/PosteriorAnalyticsCommentary. Accessed 6 July 2018.

Wexler, Ellen. 2015. What a mass exodus at a linguistics journal means for scholarly publishing. *The Chronicle of Higher Education*, November 5. http://www.chronicle.com/article/What-a-Mass-Exodus-at-a/234066. Accessed 6 July 2018.

Willoughby, Myrtle. 2012. Stoppism: Retrospects and prospects. *The Journal of Philosophy* 109 (4): 281–294.

Chapter 3
What Is Academic Plagiarism?

Abstract This chapter defends a fourfold heuristic for determining when academic plagiarism has occurred. Drawing from contemporary literature on research integrity, I propose that academic plagiarism has been committed when there is: (1) a non-trivial appropriation of words, images, or formulas, (2) with inadequate credit, (3) that generates an appearance of original authorship, (4) in a discrete item belonging to the scholarly record. This approach is sufficiently general to include a wide range of text manipulations, and yet it is sufficiently narrow to express to the particularities of plagiarism in the context of published research findings. In defending this heuristic, I argue that intent is not required for academic plagiarism, and I propose that academic plagiarism should be treated as a strict-liability offense. The presence or absence of a guilty mind or *mens rea* is irrelevant to the need to correct the scholarly record when publications themselves are deficient. Intent may be an important element to be considered by institutions that have the role of investigating and punishing wrongdoers for scientific misconduct, but intent should be considered immaterial by members of the research community who have the privilege and responsibility of maintaining the reliability of publications for the world of learning. Too often the role of correcting the scholarly record is conflated with the role of investigating and issuing punishments for research misconduct, but the two are quite different. The chapter also considers the topic of duplicate or redundant publication (often called "self-plagiarism"), and it distinguishes academic plagiarism from copyright violation.

Keywords Academic plagiarism · Self-plagiarism · Copyright · Research misconduct · Fraud

Successful plagiarism is one activity—among many—that parallels genuine scholarly achievement. A wide range of academic impostures exist, as the following extreme scenario shows:

> A man purchases a fraudulent doctoral degree from a diploma mill. Under this credential, he presents a plagiarized manuscript at a vanity conference held in a popular vacation venue. The plagiarized manuscript is then published for a fee in a vanity press's journal.

M. V. Dougherty, *Correcting the Scholarly Record for Research Integrity*, Research Ethics Forum 6, https://doi.org/10.1007/978-3-319-99435-2_3

This plagiarized article quickly appears on the journal webpage of the vanity press's website, accompanied with a false metric generated by a bibliometric company selling fake impact factors. Afterwards, the plagiarized article is indexed in a popular all-inclusive search engine. Learning of the plagiarism, the genuine author files a complaint with the vanity press, only to drop it after being told that investigations for plagiarism require complainants to pay a significant fee. The plagiarist, however, has no qualms in paying the journal more money for the privilege of being identified as one of its new editorial board members, and soon his name appears on the masthead of the journal that first published his plagiarized article. With reference to all these apparent credentials, the man is listed as a faculty member on the website of the diploma mill from which he originally obtained his fraudulent doctoral degree. He then agrees to serve as the editor-in-chief of another vanity journal by the same vanity press, one that has hijacked the title of a legitimate established journal. In this new role, he accepts for publication many other plagiarized articles submitted by people like him.

The man in this scenario is fully immersed in what has rightly been called "the false academy" (Eriksson and Helgesson 2017), and the publication of the plagiarized article is simply one of several missteps.[1] All his activities fall entirely outside of the genuine scholarly enterprise. An astute student or researcher, who knows the limits of the scholarly record, will see the vanity publication for what it is. Even those new to academic culture will be able identify the vanity publication as such, using established criteria for distinguishing genuine from non-genuine academic journals (Shamseer et al. 2017; Richtig et al. 2018; Umlauf and Mochizuki 2018). The imposture of false metrics, the questionable editorial board of the journal, and the like will be recognized as a parody of academic achievement, entirely removed from the genuine scholarly work of students and researchers. As long as the limits of the scholarly record and accepted research norms are clearly recognized, each of the academic *Doppelgänger*'s activities, patterned after genuine ones, will offer a minimal threat to the work of students and researchers.

Consider how greatly changed the situation would be, however, if the academic *Doppelgänger* were to succeed in publishing a plagiarized article in an established, reputable journal. There would be a serious corruption of the scholarly record. The fraudulent article would masquerade as a genuine item, and it would likely fool even the astute student and researcher. After all, the plagiarized article would appear to satisfy the *Knowledge*, *Authorship*, *Publication*, *Library*, *Database*, and *Discipline* conditions that demarcate items that belong safely to the scholarly record; the plagiarized article would likely be trusted because of these hallmarks. Students and researchers would likely cite the plagiarized article in their research and publications, which would result in the unwitting contamination of their scholarship and in increased visibility for the plagiarized article. The genuine author, whose work was

[1] Critics have provided a diagnosis of the false academy's basic elements, which include: diploma mills (Contreras and Gollin 2009), vanity conferences (Grove 2017; Carey 2016), predatory and vanity publishers (Kurt 2018; Beall 2016a, 2018; Pyne 2017; Darbyshire et al. 2017; Beninger et al. 2016; Djuric 2015; Stevenson 2004), fraudulent impact factors (Gutierrez et al. 2015), non-discriminating article databases (Beall 2018: 293), retraction charges by dubious publishers (Cobey 2017; McCook 2016), the purchase of fraudulent editorial board memberships (Beall, 2016c), and the phenomenon of journal highjacking (Dadkhah and Borchardt 2016).

misappropriated, would not only be denied scientific credit for any original contributions to knowledge, but the true genealogy of ideas and the authentic history of discovery would be concealed. If the scholarly record were to remain uncorrected, there would be an ongoing progressive corruption for succeeding generations of students and researchers. If the academic *Doppelgänger* were to be employed in academia, and his undiscovered academic impostures were to be successfully repeated over time, he would likely take up honors, promotions, and grants that should have been awarded to upright researchers, thereby corrupting the larger academic system in addition to the scholarly record.

3.1 Identifying Academic Plagiarism

The United States federal government defines plagiarism as "the appropriation of another person's ideas, processes, results, or words without giving appropriate credit" (Code of Federal Regulations 2005: §93.103, 589). Similarly, the European Science Foundation speaks of plagiarism as "the appropriation of other people's material without giving proper credit" (European Science Foundation 2011: 6). While certain misdeeds in a variety of contexts can accurately be described as plagiarism, my concern here is plagiarism as it is found within the published research literature. I propose the following four-part heuristic for identifying academic plagiarism, where academic plagiarism is understood stipulatively or restrictively as plagiarism within the published research literature.

Academic Plagiarism:

1. A non-trivial appropriation of words, images, or formulas,
2. With inadequate credit,
3. That generates an appearance of original authorship,
4. In a discrete item belonging to the scholarly record.

As it is understood here, a finding of academic plagiarism requires that each of the four parts of this heuristic be met. This fourfold requirement expresses the individually necessary and jointly sufficient conditions that must be fulfilled to warrant a finding of academic plagiarism in the published literature. The heuristic is meant to guide discussions of post-publication responses to academic plagiarism by indicating when a published correction of the scholarly record is justified.

3.1.1 Appropriation

That academic plagiarism involves the taking of something from another is reflected in the etymology of the word *plagiarism*, as *plagiarius* is the Latin term for kidnapper (Glare 1982: 1386). This ancient expression was originally used in reference to one who abducts a slave or a child, and then by extension, to one who engages in literary appropriation. The Roman poet Martial is the earliest recorded author to use the term *plagiarius* in the extended sense of one who abducts the words of others (Seo 2009). The Greek adjective πλάγιος means "askew" or "crooked," and is at times identified as a precedent to the Latin *plagiarius* (Liddell et al. 1953: 1410). The etymology of these terms underscores the meaning of plagiarism as a disordered appropriation of something that belongs to another. In medieval Latin, *plagiator* designated one who instigates slaves to run away from their masters (Niermeyer et al. 2002: 1049). By the Renaissance, *plagiarius* had the present-day sense of "plagiarist" (Hoven 1994: 269), as the fifteenth-century humanist and philosopher Lorenzo Valla used the term to designate another who had taken his work (McGill 2012: 9).

In the earlier Greek philosophical tradition, members of rival philosophical schools at times leveled charges of plagiarism against the venerable figures of each school. In his detailed study of plagiarism in ancient philosophy, Anton-Hermann Chroust observes that "beginning with the fifth and especially the fourth century before Christ, the charge of 'philosophical plagiarism' was made with increasing frequency, and, it may be added, with mounting recklessness" (Chroust 1961: 220). He notes that major early Greek thinkers endured such accusations: Heraclitus was said to have plagiarized Orpheus, Empedocles to have stolen from Parmenides, and Democritus to have plagiarized Pythagoras. Socrates, Plato, and Aristotle would similarly be denounced for a variety of appropriations by various detractors. In his account of the lives of the ancient philosophers, Diogenes Laertius uses the term λογοκλοπία or "word-plunder" in reporting an accusation of plagiarism leveled against a philosopher, and the traditional term for plagiarism was generally κλοπή, which carried the sense of fraud, pilfering, or theft (*Vitae philosophorum*, VIII, 54; Liddell et al. 1953: 962). Despite the ultimate baselessness of many such charges among rival philosophical schools, the existence of these accusations offers evidence that the unacknowledged appropriation of the words of others was viewed with opprobrium among the learned communities of the ancient period.

According to Chroust, only in late antiquity did "some general but still rather vague standards" come into existence "concerning an author's obligation to give proper recognition to his authorities or sources of information" (221). In the medieval period, however, some influential works in philosophy, theology, and canon law incorporated the works of others verbatim and without attribution, and this practice reflected conceptions of authorship quite different from later periods (Hathaway 1989; Schabel 2005; Zahnd 2015). In the words of one commentator, the word *plagiarism* "should indeed probably be dropped in reference to the Middle Ages, since it expresses a concept of literary individualism and property that is distinctively modern" (Constable 1983: 39). By early modernity, notions of authorship were sufficiently developed to allow for the practice of plagiarism allegations to take hold once again among learned communities. In the seventeenth century, treatises on plagiarism appeared that characterized the topic as a moral failing, and they provided speculations about the disordered motivations of plagiarists (Kivistö 2014: 118–134). A growing body of literature now questions the originality of major figures of early modern philosophy, and the notion of a plagiarism of ideas (if not words) is invoked in such re-assessments of the history of philosophy (Mercer 2017; Ares et al. 2018). Even some of the major nineteenth-century text editing and publishing ventures were embroiled in controversies about authorship (Bloch 1994: 61–66). In short, the notion of the unacknowledged appropriation of the works of others was firmly reestablished as an act worthy of opprobrium in the early modern period.

3.1.2 Non-triviality

The requirement that academic plagiarism involve a *non-trivial* appropriation occasions the question of what counts as non-trivial. One can distinguish between two general approaches. A permissive approach to plagiarism maintains that academic plagiarism occurs only in cases where what is appropriated is substantial in terms of quantity or quality. A restrictive approach maintains that academic plagiarism can occur even when what is appropriated is not substantial in terms of quantity or quality.

Defenders of a permissive approach are not difficult to find. In the words of one such theorist, academic plagiarism requires that what is appropriated "must have significant value" and must represent "the crux, the core, an entity, a unified or artistic whole" (St. Onge 1988: 60). On this view, plagiarism charges should be limited to those cases involving a portion of a text that concerns the fundamental, original findings of a study. The undocumented appropriation of the other parts of a published work is not held to the same standard and is not considered to be serious. On a permissive approach, it would be impossible to plagiarize the purportedly less important or supplementary parts of an academic work, such as a summary of a research method, a literature review, introductory material, footnotes, or a bibliography. To appropriate what is not of significant value, or what is not the crux or the core of an article, does not warrant a designation of plagiarism according to this permissive approach. Another defender of the permissive approach contends that the word *plagiarism* should not be used to refer to the copying of texts "that contain no original idea," such as those texts that constitute an "introduction" or "background section" of a published scholarly work (Bouville 2008: 311, 312). On this view, "the core of science are facts and theories, not words," so it is possible to be "copying words without copying ideas" and thereby avoid "real plagiarism" (314).

Since English is considered by many to be the *lingua franca* of scientific and academic writing, the mastery of English is judged by some as a 'language tax' unfairly assessed on those who are non-native speakers. One researcher accused of plagiarism defended himself in an editorial in *Nature* published under the title, "Plagiarism? No, We're just Borrowing Better English" (Yilmaz 2007: 658). The editors of the journal in which one of the researcher's articles appeared later published an editorial acknowledging that "there has been cutting and pasting of introductory material from papers by other authors" (Ellis and Nicolai 2007: 1969–1970). Acknowledging that such an approach "constitutes plagiarism of that material, even if there is no plagiarism of research results," the editors still declined to issue a retraction or correction, presumably because of confidence in the remaining portion of the paper. Similarly, some researchers argue that "scientists are not writers," so those papers that present the results of original experiments amidst appropriated subsidiary portions can be corrected but should not be retracted (Chaddah 2014: 127). Others have argued that in recognition of the collaborative character of science and in the interests of efficiency, the scientific community should adopt a positive concept of 'remixing' whereby it becomes acceptable to use "text and parts of earlier publications" from various sources "to build a novel publication" (Heller et al. 2014: 197). The adoption of such a view would require major overhauls in copyright law and in traditional understandings of authorship.[2]

Certain institutional research integrity offices use such a permissive approach in determining which received claims of suspected plagiarism should trigger a full investigation. The U.S. Office of Research Integrity of the Department of Health and Human Services (ORI) declares that

> ORI generally does not pursue the limited use of identical or nearly-identical phrases which describe a commonly-used methodology or previous research because ORI does not consider such use as substantially misleading to the reader or of great significance. (The Office of Research Integrity 1994)

The ORI restricts the consideration of plagiarism to the appropriation of core findings or the presentation of data in order to limit the types of cases that are subject to investigation. On this view, the appropriation of a method statement in an article or wholesale borrowing a literature review would not trigger a formal investigation for plagiarism, as what is appropriated is not considered to be the core of the article. In his handbook on ethical scientific writing, Miguel Roig contextualizes ORI's position on the matter by noting that "highly technical descriptions of a methodology, phenomena, etc., can be extremely difficult to properly paraphrase" and that not all writers possess the "thorough conceptual understanding of the concepts and processes being described" (Roig 2015:14). Much like the ORI, the Swiss National Science Foundation (SNSF) states that a "trivial case" of plagiarism can be deemed to have occurred when "only a few citations are missing" or "only a small amount of text is uncited compared to the entire text," or "the content of the uncited text is of a general nature or concerns the state of research" (Swiss National Science Foundation 2016: 6). On this account also, plagiarism sufficient to trigger a formal investigation (with resultant sanctions) is restricted to plagiarism in the 'core' part of an article that presents new findings. In short, this permissive view holds that one can commit the undocumented appropriation of the work of others in the professedly non-essential parts of a research article without fearing that an investigating body will make a judgement that academic plagiarism has been committed.

Even if one is in principle committed to the division of an article into its core parts (e.g. research findings) and non-core parts (e.g., literature review, methodology

[2]The concept of 'remixing' has antecedents in the medieval notions of authorship discussed earlier in this chapter.

statements, footnotes, or bibliography) such a division does not transfer seamlessly to all genres of published academic research. Some review articles offer a précis of a subfield, or present the state of a question, consisting only of a literature review without presenting original empirical findings. Furthermore, the research of many disciplines, such as those in humanities, generally does not consist in the reporting of empirical findings. In the humanities, one can find cases where determinations of plagiarism have been made, for example, with regard to a translation (Arenson 2002), a bibliography (Vincent 2007), or a book review (Dodaro 2014). Even dictionaries have been determined to have been plagiarized, either in whole (Considine 2008: 95) or in part (Zgusta 1988: 157). The ORI and the SNSF do not generally promote work in humanities, so perhaps the division of articles into core and non-core parts is more intelligible for some fields than others.

The *Committee on Publication Ethics* recognizes that plagiarism can occur in both major and minor forms. Its flowchart for editors who receive evidence of suspected misconduct recommends that editors should consider publishing a retraction in cases where there is "unattributed use of large portions of text and/or data, presented as if they were by the plagiarist" (Wager 2013). The flowchart also recommends that editors should consider publishing a correction even for "minor copying of short phrases" and should consider publishing a "correction giving reference to original paper(s) if this has been omitted" (ibid.). The twofold division between major and minor levels of gravity for plagiarism is not to excuse the latter one, but to specify the kind of correction. Retractions are reserved for major acts of plagiarism, but lesser corrections such as corrigenda are reserved for minor ones.

In some discussions of academic plagiarism, there have been efforts to demarcate what counts as plagiarism by focusing on quantitative rather that qualitative factors. Some approaches attempt to determine a maximum number of unreferenced verbatim or near-verbatim words that can appear in a work before a designation of academic plagiarism is warranted. In a handbook on scientific ethics, David Koepsell advises, "Technically, whenever one uses six or more words in a row written by another person, one has a scientific and moral duty to attribute the source" (2017: 32). A survey of academic journal editors across many disciplines outside humanities concluded that "no more than 7 words should be copied without citation and quotation marks" (Zhang 2016: 28). Nevertheless, a more common view is that there exist "no set rules about how much reused text constitutes plagiarism" (Morris et al. 2013: 360). According to some theorists, "plagiarism (in principle) can consist in as little as one word," provided that it is a "very special, novel word or expression creatively used," such as a new concept or name (Helgesson and Eriksson 2015: 97, 96).

Other quantitative approaches involve setting percentage thresholds for textual overlap using text-matching software programs as the primary determinant for whether academic plagiarism has been committed. Plagiarism detection software programs generate originality reports and highlight text matched with other sources to yield a percentage of overlap with other texts in a given manuscript. The reputation of journals that advertise an allowable percentage of identity with other previously published works is not high, however (Beall 2016b, 2013; cf. Prasad 2017).

The originality reports generated by plagiarism detection programs are not guarantees regarding whether academic plagiarism has or has not occurred; such reports always need an expert human judgment to render a decision (Weber-Wulff 2014: 14, 78, 86). The appropriation of a text in a different language or even a close paraphrase can go undetected by text-matching software systems. Such systems are particularly vulnerable to documents that have been subtly modified using electronic paraphrase tools (Rogerson and McCarthy 2017). What is appropriated from another text need not involve words at all, as one can appropriate an image, table, graph, or numerical data. Data integrity programs exist that consider image files for manipulation, duplication, and various irregularities, and again reports generated by such programs always require a human element of analysis (Koppers et al. 2017; Butler 2018).

In my view, attempts to establish minimum word lengths for acceptable unquoted sentence appropriation, or minimum allowable percentages of unsourced material, or degrees of acceptable image duplication, are untenable. Furthermore, attempts to distinguish more important from less important parts of an article, for the sake of restricting academic plagiarism findings to the more "important" parts, are also entirely misguided. The reason is clear: only a restrictive approach to academic plagiarism will preserve an academic record of the highest quality for students and researchers. A restrictive approach by editors who have stewardship over academic publishing venues and by authors who submit manuscripts will improve the reliability of the published literature. One can still acknowledge that "the extent of plagiarism can be very wide, from copying a few phrases to wholesale duplication of text and experiments" (Hames 2007: 177), thereby granting that acts of academic plagiarism can admit of degrees of severity, and yet insist that academic plagiarism in all its variations damages the reliability of the scholarly record and harms those researchers who have unknowingly taken plagiarized works as trustworthy.

There is another reason for the irrelevance of a distinction between the purported core and non-core parts of an article in making a determination of academic plagiarism. Established standard typologies of plagiarism include some forms that consist of appropriating small portions of text, such as a few sentences, or a few paragraphs, from many sources to produce a single work. In such instances, the hidden original sources can be many, with no one individual source forming the "crux" or "core" of either the source works or the plagiarizing article. In cases as these, the resultant work is a tissue of unacknowledged borrowings from a variety of sources. Plagiarism theorists have designated various kinds of multi-sourced academic plagiarism as "mosaic," "clause-quilt," or "patchwork" plagiarism (Weber-Wulff 2014: 9).

A permissive approach that would restrict all determinations of academic plagiarism to cases where the "core" or original findings of an article have been misappropriated is entirely misguided. The undocumented verbatim or near-verbatim appropriation of an introductory literature review in an article does not cease to be academic plagiarism just because the original research findings later in the same article are not also plagiarized. It is better to recognize that acts of plagiarism admit of degrees of seriousness. In the words of some commentators, "plagiarism of research results, and also discussion, is seen by many as considerably worse than plagiarism from the introduction or methods sections" (Helgesson and Stefan

Eriksson 2015: 99). That one could have potentially committed a more egregious act of plagiarism in a 'more significant' part of an article does not render an actual act of plagiarism in a 'lesser' part of an article to be blameless. That an article is not deficient in one part is not a successful defense against deficiencies in another; plagiarism of a literature review is still plagiarism even if plagiarism of original research findings would have been a more serious offense. Again, a better approach is to acknowledge that academic plagiarism admits of degrees of gravity, rather than to insist that academic plagiarism should be restricted to the "core" parts of a scholarly work.

One difficulty with some permissive approaches to academic plagiarism that divide an article into core and non-core parts is that they often call for a continued presumption of genuine authorship for all portions of the article not covered in a particular plagiarism complaint. One often sees the response by authors of record to an accusation of suspected plagiarism in a method statement or literature review to be along the lines of: "yes, there is sloppiness in the introductory parts, but the data and research findings are good." For some editors and research integrity officers, such a reply is sufficient to bring plagiarism investigations to a close. Such continued presumption of integrity by editors and research integrity officers is unwarranted once any portion of an article is exhibited to be misappropriated without credit to the original, genuine authors.

When an author of record has been significantly deficient in one part of the article, he or she should not continue to enjoy the presumption of trust for the remaining parts of the article on behalf of editors, readers, and research integrity officers. This more prudent view has been expressed by one journal editor as:

> When plagiarized text is discovered, a high level of suspicion is in order to look for other forms of inappropriate scientific behaviour such as falsification of the accompanying data (i.e., alterations of data) or fabrication (i.e., making up data). (Lose 2011: 903)

This view can be understood as a present-day application of the medieval scholastic maxim *bonum ex integra causa, malum ex quocumque defectu* (a thing is good if what goes into it is entirely good, but bad if it is defective in any way).[3] In practical terms, the presumption of originality for the entire article should not longer be maintained when part of the article is demonstrably defective. An investigation of the article by the editor and the relevant integrity offices should be started. A high burden of proof for originality should be placed on the author of record regarding portions of the article not directly covered by the plagiarism complaint, as the defects in one part of the article should render the credibility of the author of record into question. Editors and research integrity officers who assume *ab initio* that the submitted evidence of suspected defects of an article offered by a particular whistleblower is necessarily exhaustive of the potential problems with that article—and who restrict their investigations accordingly—offer a presumption of originality to an author of record that is unwarranted. A given whistleblower may not have found all of the defects with an article, so further scrutiny should be given to the other portions of the article in question.

---

[3] The maxim is commonly attributed to Pseudo-Dionysius, as expressed in *De divinis nominbus*, IV.30. For an account of this maxim, see Bretzke (2013: 26–27).

3.1.3 Inadequate Credit

When a published item belonging to the scholarly record appropriates from earlier work in a non-trivial manner, and no reference of any kind is offered to the original source material, then it is quite uncontroversial that inadequate credit has been rendered. Since "no credit" unquestionably fulfills the aforementioned condition of "inadequate credit" for academic plagiarism, there should be no controversy in cases such as these. The challenging cases are those where some reference has been offered to the original source material, but the credit does not fully reflect the level of dependency upon it.

The mere presence of a bibliographical reference to the original source material should not inoculate an item of the scholarly record from a judgment that it is plagiarized. Recent typologies of plagiarism have categorized the ways that items of the scholarly record can offer some reference to the original source material but still warrant a designation of plagiarism. One currently recognized type of plagiarism has come to be called "Pawn Sacrifice Plagiarism," described by Debora Weber-Wulff as a situation where "the source citation is either given in a footnote or only listed in the bibliography" but the author has "not made clear, however, exactly how much has been taken (2014: 10)." A variant occurs when there is

> a proper attribution of a sentence, but then the text copy continues on, copying the source for additional sentences or even paragraphs without making clear that this is the author of the source speaking and not the purported author. (ibid.)

There has been an increase in scholarly interest in types of plagiarism where source material is explicitly but inadequately identified. This increase may be a response to the view occasionally defended in some quarters that plagiarism—in principle—cannot occur if the original source is mentioned somewhere, in some way, in the article itself. While such a view remains a minority position, it is vigorously championed by authors of record trying to fend off charges of academic plagiarism. The weakness of such a position is that it fails to distinguish inadequate credit from no credit; it implausibly treats attribution as a strict either/or, denying that attribution admits of degrees. The theorist K. R. St. Onge holds a minority position among present-day commentators on plagiarism by arguing that "any evidence pointing to the original author or source mitigates the intent to deceive even if source identification is textually distant from the materials copied or less than adequate on other grounds" (St. Onge 1988: 68). The principle is easily open to abuse.

Determining whether or not an item of the scholarly record adequately offers credit to its original source material is not an intractable problem. Such a scenario need not be open to endless debate or be resolvable only through an arbitrary judgment call that could go either way. A reasonable standard can be expressed as: would the average practitioner of a discipline be able to determine what is original, and what is not, on the basis of indicators offered in the text? I would like to propose the following litmus test, hereafter designated as the "Reverse Engineering Standard for Inadequate Credit."

Reverse Engineering Standard for Inadequate Credit:

Credit to sources is inadequate when a practitioner of a discipline is unable to distinguish, on the basis of indicators offered in the text, the part of the item that is original from the part that is not.

Such a test is straightforward, as scholarly conventions include a bevy of indicators available to authors to signal when portions of an item originate elsewhere. These signals include quotation marks, footnotes, in-text citations, extract formatting, text-boxes, and italicization, among others. In the case of a published article, for example, the *Reverse Engineering Standard for Inadequate Credit* can be applied by asking: are the normal ways of signaling to the reader the various dependencies on prior works sufficient to allow a researcher, on the basis of markers in the article, to separate those words, images, or formulas that are original to the article of record from those that are not? If the reader has no way of determining that the author of record is not the originator of some of the material in the article because of the absence of the conventional signals or markers, then the condition of inadequate credit has been met.

Such a litmus test shows that the matter of determining adequate credit need not be reduced simply to inquiring whether the original source material is somewhere indicated in the text. Adequate credit must manifest to the reader the parts that originate elsewhere. The presence or absence of the conventional indicators of dependence is crucial for determining whether academic plagiarism has occurred. In the words of two experts on research integrity, "The difference between plagiarism and proper citation may come down to the placement of quotation marks" (Shamoo and Resnik 2015: 22). The great variety of conventions available to authors for indicating dependencies on source materials makes their absence in acts of academic plagiarism quite striking.

3.1.4 An Appearance of Original Authorship

When the *Reverse Engineering Standard for Inadequate Credit* is met, readers have no way of knowing, at least from the text itself, which parts consist of unreferenced portions of the work of genuine authors. In this case, readers cannot work back or reverse engineer the text to the original sources on which it relies. In the absence of adequate signals of dependence, a trusting reader will falsely be led to assume that the author of record has produced the words, images, or formulas when they really are re-appearing from undisclosed or deficiently-disclosed sources. The threshold of inadequate credit will have been met, and the reader will have been left with the false impression that the stated author of record is the genuine one. The item of the scholarly record thereby generates an appearance of original authorship when none is warranted. The genuine author will be deprived of the credit for the work, and the academic plagiarist may benefit from the illusion of research productivity, receiving—undeservedly—the promotions, grants, and esteem that attend academic success.

The world of learning is a context where there is an expectation of genuine, original scholarship. Plagiarism theorists have emphasized that this contextual expectation is a crucial feature; if original authorship is not expected in a given context, then the invocation of the word *plagiarism* in such a context is a category mistake. Teddi Fishman argues persuasively that plagiarism occurs "in a situation in which there is a legitimate expectation of original authorship" (Fishman 2009: 5). Research communities value originality, and the expectations and standards of published research provide a different context from many other forms of writing. Private personal copying in a nonpublic venue (e.g., a personal diary), removed from an act of publication, is generally not judged to constitute plagiarism; one should "only speak of plagiarism when there is an expectation of original work" (Weber-Wulff 2014: 5). Furthermore, a host of documents that are published do not generally carry with them the expectation that the author of record is the sole or genuine author; law clerks often write the opinions of the judges for whom they work and politicians typically employ speechwriters (Posner 2007, 20–21, 36). Ecclesiastical documents often are the culmination of the contributions by many individuals, even though they appear under the name of a single church figure. In law, politics, and ecclesiastical contexts, accusations of plagiarism are usually restricted to cases when ghostwriters are found to have turned in material to their employers that have already appeared elsewhere in print. When several passages of two journal articles from 1995 to 2001 by theologian Archbishop Victor Fernandez appeared without citation in Pope Francis's 2016 apostolic exhortation *Amoris lætitia*, some observers were led to believe that Fernandez must have been the papal ghostwriter for the document. The charges of plagiarism that followed were not put forward because of an apparent use of a ghostwriter, but because the apparent ghostwriter was incorporating his previously published articles without citation into the papal document (Pakaluk 2017).

3.1.5 A Discrete Item of the Scholarly Record

Even though the issuance of a work with a vanity publisher, a predatory publisher, or a pseudo-scholarly outlet may give to an outsider the appearance of contributing to the scholarly record, such a publication, as noted above, is merely a *Doppelgänger* activity that parodies genuine scholarly activity. The normal mechanisms for correcting the scholarly record after violations of scientific integrity, such as a published statement of retraction, are not generally employed for vanity publications. Vanity projects also fail a number of the conditions that mark a genuine scholarly publication, including the *Publication*, *Library*, and *Database* conditions. Plagiarism can occur in a variety of contexts (e.g., journalism, music, art, student essays), but my concern here is plagiarism in the context of published research; that is, plagiarism in items belonging to the scholarly record.

3.2 The Intent to Plagiarize

The absence of any reference to the intent of a plagiarist is a distinctive feature of the fourfold heuristic for academic plagiarism: (1) the non-trivial appropriation of words, images, or formulas, (2) with inadequate credit, (3) that generates an appearance of original authorship, (4) in a discrete item belonging to the scholarly record. Many standard definitions of plagiarism in use today include some explicit requirement of a *mens rea* or guilty mind, on the view that to plagiarize successfully one must intend to plagiarize. One often hears of plagiarism described as an act of theft or stealing, and the use of terms borrowed from criminal law or moral philosophy has led some to identify malicious intent as essential to plagiarism. A recent account is representative of that view:

> Pretending that someone else's work is one's own, i.e., using it without attribution, is plagiarism [...]. Plagiarism is thus a form of theft (stealing someone else's work) and dishonesty (passing it off as one's own). (Morris 2016: v)

On such a view, a bad intent (in this case, specified as "pretending" or "dishonesty") is set forth as a necessary element of plagiarism. At times the *mens rea* in acts of plagiarism is further specified as the attempt "to gain credit that was not legitimately earned" (Fishman 2009: 5). Other commentators promote "imposture" or "false pretense" as an essential element found in all acts of plagiarism (Scanlon 2007: 61). Nevertheless, support for the *mens rea* condition has been diminishing in plagiarism scholarship in recent years. There have been recent efforts to minimize or altogether exclude intent from the definition of plagiarism (Weber-Wulff 2014: 5, 14; Gipp 2014: 10; Helgesson and Eriksson 2015: 94). In surveying the recent literature on plagiarism, some commentators state that theorists currently disagree whether intention is important in determining if an act of plagiarism has been committed (Carroll 2016: 201), but others go further and say that now "in the literature, there seems to be a clear majority for understanding plagiarism in a way that does not presuppose intention" (Helgesson 2016: 2241). The relevance of intent can now be approached as a disputed, and no longer favored, position among current theorists working on the topic of plagiarism.

The approach to academic plagiarism proposed here is entirely "intent-free." In my view, intent should not be considered in determining whether academic plagiarism has occurred. Academic plagiarism should be treated like a strict liability offense such as acts of speeding or selling alcohol to minors, which can generally be adjudicated without any reference to intent. On the view that academic plagiarism is like a strict liability offense, whether the plagiarism occurs due to malicious intent, reckless disregard, carelessness, sloppiness, or ignorance is entirely irrelevant for its finding. Intentional plagiarism may nevertheless be more blameworthy than unintentional plagiarism, as the former involves fraudulent behavior. The position here agrees with Gert Helgesson and Stefan Eriksson, who write, "plagiarism does not necessarily involve an intention to deceive" (2015: 94).

The primary reason for treating academic plagiarism as strict liability offense is that the need to correct the scholarly record remains, whether or not the plagiarism was intentional. The reliability of the scholarly record is damaged irrespectively of whether the putative author acted with or without a *mens rea*. If the scholarly record is left uncorrected, it will corrupt the research of students and scholars who later unwittingly incorporate the defective work in their scholarly endeavors. A secondary reason for leaving intent outside of considerations of academic plagiarism is the difficulty in assessing it in many cases. As Debora Weber-Wulff puts it:

> It is usually impossible to tell from the text itself if the text parallel is the result of unintended, honest mistakes, sloppy referencing skills, or intent to cut corners and to deceive. (2014:14)

Similarly, Elizabeth Wager notes:

> Intention to deceive is often considered a factor distinguishing misconduct from careless work or honest error. However, it is usually impossible to prove intent and therefore may be less useful in practice than in theory. (Wager 2014: 37)

With an intent-free approach to plagiarism, one can determine whether academic plagiarism has occurred and mend the scholarly record accordingly, without ever entering into the oftentimes intractable difficulty of discerning the private intent of the author of record for a deficient article.

My experience with investigations into academic plagiarism conducted by journal editors and publishers is that individuals accused of suspected plagiarism on the basis of significant and overwhelming evidence invariably focus their defenses on intent. In such contexts, one often finds the accused responding to the charges of suspected plagiarism along the lines of "I did not intentionally plagiarize" or "I did not seek, deliberately or with malicious forethought, to 'steal'." Along the above-proposed fourfold heuristic of academic plagiarism, such common defenses by plagiarists based on autobiographical descriptions of intent become entirely irrelevant to the determination of whether plagiarism has occurred and whether the scholarly record requires correction.

Excluding intent to treat plagiarism as a strict liability offense allows one to separate the issue of commission from the issue of culpability. These two should not be conflated. The need to correct the scholarly record in a given case can be demonstrated by establishing the first issue without ever entering into questions about the second. Ideally the two issues will be adjudicated by different bodies. A journal editor, focused on the integrity of articles appearing under her or his stewardship of a journal, should be less interested in culpability and more interested in the journal's reliability for readers. On the other hand, the institutional home of the author of record of a plagiarized article may be quite interested in the issue of culpability. However, any increased insight into the precise mental state that may have led a would-be author to plagiarize will not add or detract from the harm inflicted on the world of learning when the scholarly record is damaged through academic plagiarism.

The irrelevance of intent for defining academic plagiarism can further be shown by noting that the correction of the scholarly record need not be viewed as essentially punitive. A journal editor who issues a statement of retraction, for instance, can be motivated solely to warn students and researchers that a certain previously published article is deficient in crediting its sources, independently of any consideration of whether such a declaration could have any extrinsic, secondary effects on the professional life of the plagiarist. Those secondary effects might include the diminishment of the plagiarist's professional reputation or even the dismissal of the plagiarist from a university or research post, but concerns about such matters are incidental to the editor's essential role of insuring the integrity of a journal for its readers. Any considerations or predictions about the professional (or even personal) ramifications that a published correction may have for the plagiarist will likely provide a distraction to an editor with regard to the obligation of stewardship over a journal and the quality of its contributions to the scholarly record. As one commentator puts it, "The main purpose of retractions is to correct the literature and ensure its integrity rather than to punish authors who misbehave" (Wager et al. 2009: 2). The deficient published work of a longtime serial plagiarist and the deficient published work of a budding graduate student can each be plagiarized and each requires a correction of the scholarly record, but the issue of punishment may be different for each case and should fall outside an editor's ambit of responsibility.

The feelings of the immediate victims of plagiarism—those whose works have been misappropriated by the plagiarist—are also likely to be a distraction and should be considered irrelevant by editors and publishers in determining the best way to correct the scholarly record when academic plagiarism has been discovered. There have been times when a true victim of plagiarism, due to friendship with the plagiarist, or a desire to be uninvolved with scandal, or even personal animosity toward a whistleblower, has opposed the correction of the scholarly record in a case of demonstrated academic plagiarism. By analogy to strict liability offences, the feelings of a victim are irrelevant to whether the offense has been committed. Again, an editor should be guided by the obligation to maintain a reliable scholarly record rather than by extrinsic considerations such as the feelings of the victim of plagiarism or the future professional ramifications that a published correction might have for a plagiarist. The point of the retraction is neither to assuage any feelings the victim of plagiarism may have nor to chastise a plagiarist, but to repair the damage inflicted on the scholarly record in order to insure a reliable body of published research for present and future students and researchers.

The irrelevance of intent can be shown also from still another perspective. There are times when circumstances necessitate a correction of the scholarly record for plagiarism and there is no possibility of investigating the mind of the putative author or of contacting the victims of plagiarism. If significant time has passed between publication and discovery, the author of record and the primary victims whose works were misappropriated may be outside of the research community without contact information, non-responsive to inquiries, or deceased. Such scenarios are not insurmountable impediments to correcting the scholarly record. Best practices call for editors to contact those accused of suspected plagiarism to provide an opportunity to respond to the charges when possible, and to notify victims of plagiarism when possible, but a lack of a reply by such parties does not rule out a published correction.

Even though the presumed right for the accused to respond is discussed in the scholarship on plagiarism (e. g., Wager 2013), there has been little discussion of the presumed right of a victim of plagiarism to be informed of the misappropriation of her or his work. Even in cases where retractions are issued for acts of academic plagiarism, the genuine authors whose works have been misappropriated are at times not informed by the editors and publishers about the misappropriation and subsequent published correction.

Of course, the scientific integrity offices at the institutional homes of suspected plagiarists may be quite interested in questions concerning intent, culpability, and the possession of a *mens rea*. Additionally, a journal editor might subsequently be very interested in these very issues when considering whether to accept any future manuscript submissions from suspected plagiarists. Nevertheless, corrections of the record for academic plagiarism are essentially reparative and only contingently or incidentally punitive.

Given the high burden of proof for establishing bad intent in a given situation, one may wonder why so many of the traditional definitions of plagiarism have required a *mens rea*. Perhaps such an emphasis on intent originates in the desire to distinguish situations of fraud from situations of error. Even though this distinction is surely important in many contexts, it is not relevant with regard to the maintenance of a reliable scholarly record. If a purported author of a publication has appropriated the work of others without adequate credit, the presence or absence of bad intent is not relevant insofar as either way the need to correct the scholarly record remains. Whether the academic plagiarism is the result of extreme negligence in failing to provide quotation marks and references to original source material, or the result of a fraudulent intent to pass off another's discoveries as one's own, the scholarly record remains unreliable for students and researchers as long as the record remains uncorrected.

When an item of the scholarly record is found to appropriate from sources without offering adequate credit, there has been a longstanding assumption that the two possibilities of "intentional misconduct" and "honest error" exhaust the range of intents on the part of the authors of record (Resnik and Stewart 2012). This assumption is often supplemented with the provision that a finding of plagiarism can be made only after "intentional misconduct" has been established. On this view, it has been contended that editors must "always be very careful to distinguish between genuine errors and the intention to deceive; the latter constitutes misconduct, the former does not" (Hames 2007: 173). This bifurcation of intent into "intentional misconduct" or "honest error" may indeed be useful for institutional investigations into research misconduct, and for any subsequent consideration of appropriate penalties to be applied to guilty employees, yet it would be a mistake to think the same set of assumptions transfers seamlessly to investigations by editors and publishers. Non-trivial inadequate citation need not be the result either "intentional misconduct" or "honest error" but could result from just negligent omission. There has been some improvement in recognizing that this dichotomy of "intentional misconduct" or "honest error" is insufficient. The United States Federal Policy on Research Misconduct states that "research misconduct is defined as fabrication, falsification, or plagiarism" and that "the misconduct be committed intentionally, or knowingly, or recklessly" (Federal Register 2000: 65.235, 76262). This acknowledgment that recklessness can suffice for a finding of plagiarism expands the range of approaches to scientific misconduct (see Resnik et al. 2017).

Growing dissatisfaction with intent-based approaches have led some theorists to redefine the terms of research integrity discussions. Magne Nylenna and Sigmund Simonsen have argued that "scientific misconduct—or rather, conduct inconsistent with accepted scientific standards—is a continuum ranging from honest errors to outright fraud" (Nylenna and Simonsen 2006: 1883). On their view, "there is a long grey zone from white to black, with laxity, negligence, and recklessness all being used to characterise non-intentional, but still egregious, deviations from accepted standards" (ibid). Similarly, in his compendium on authorial ethics, Robert Hauptman argues that "inadvertent error repeated *ad nauseum* is tantamount to misconduct" (Hauptman 2008: 328). Somewhat profanely, but arrestingly, statistician and political scientist Andrew Gelman has argued that "any sufficiently crappy research is indistinguishable from fraud" (Gelman 2016). He argues that cheating cannot be distinguished from incompetence in many high-profile cases, and that it is not always possible to discern "what was going through their minds when they were doing what they were doing." The traditional divide between the three major research offences of Fabrication, Falsification, and Plagiarism (FFP) and a host of other offences oftentimes designated as Questionable Research Practices (QRPs) is no longer taken as obvious in some quarters.[4]

3.3 Self-Plagiarism

The term *self-plagiarism* has become popular in recent years to refer to the undocumented re-use of one's published texts in subsequent publications. Depending upon the quantity of reused material, varieties of so-called self-plagiarism have been recognized in the scholarship on research ethics. A situation involving the republication of substantially the same article in a new venue without reference to the original publication is commonly designated as *redundant* or *duplicate publication*. The re-use of smaller amounts of text is at times described as *text recycling* (Smolčić 2013; Norman and Griffiths 2008; BioMed Central n.d.). Some scholars find the concept of self-plagiarism to be an oxymoron or to involve a contradiction, whereas others find the concept to be useful (Broome 2004). Given the variety of kinds of textual re-use that are included under the term, one theorist has proposed abandoning the expression *self-plagiarism* in all discussions of research integrity:

> My suggestion is to eliminate the notion altogether from the jargon of scholarly ethics, and either employ a blander generic term in its stead (e.g., "inappropriate reuse of ideas, text or data"), or simply focus the effort on the relevant lower-level categories (duplicate publication, "salami slicing", textual recycling, cheating in evaluations etc.). (Andreescu 2013: 796)

[4] An updated catalogue of *QRPs* is offered in Bouter et al. (2016).

The heuristic of academic plagiarism proposed in this chapter is neutral on the issue of self-plagiarism, and there are distinct disciplinary tendencies regarding the issue. A recent study concludes that "the extent of text recycling varies substantially between research fields" (Horbach and Halffman 2017: 1) In the natural sciences, the republication of the same material under a new title and in a new venue without acknowledgment of its original publication can create the false impression that a particular finding had been replicated or reconfirmed. For this reason, many if not most scientific fields commonly issue retractions on the basis of "redundant publication" or "duplicate publication." In extreme cases, duplicate publications can create the illusion that an experimental intervention has become well-established as an evidence-based practice, or that the efficacy of a certain drug regimen has been revalidated.

Academic plagiarism violates the trust of the reader, who is duped into thinking that the author of record and the author of origin are the same. In self-plagiarism, however, the violation of trust is of a different order: there is a strict identity of the author of record and the author of origin, but the default assumption of readers that new work is being presented is violated. The reduplicated text is original to the author of record, but not original in the sense that it is the second iteration of the same work.

The republication of one's work without acknowledging prior publication may create a host of secondary problems, such as the illusion of productivity, the distortion of future meta-analyses, the violation of copyright, and the wasting of the valuable time and resources of readers, reviewers, and publishers. Other scholars are also harmed, as duplicate publications take up valuable space in journals, displacing original work that otherwise would have been published. Furthermore, the presentation of the same material in more than one venue creates research inefficiencies within the scholarly record, as the same findings in several articles will be indexed individually in disciplinary research databases, thereby unnecessarily increasing the work of researchers who are forced to wade through the same material in tracking down the redundant publications in various venues (see Norman and Griffiths 2008; Roig 2015: 20–21; Teixeira da Silva 2017). Other commentators emphasize that the undocumented republication of one's work is a way of "gaming the current reward system of science" (Horbach and Halffman 2017: 1).

In a well-publicized case of redundant publication in the field of economics by an established researcher, the same article was published in at least four venues under different titles and without cross-referencing. An editor of one of the affected journals published an open letter to the offender. In it, he set forth the evidence carefully and expressed dismay over the author's behavior, writing "we find your conduct in this matter ethically dubious and disrespectful" but also stating that the journal would not be pursuing legal action for the deception (Autor 2011: 239).

Despite such concerns regarding the practice of duplicate or redundant publication, it occasionally finds its defenders. Under the subtitle, "Publish the Same Results Several Times," the mathematician and founding editor of several academic journals Gian-Carlo Rota observed that academic sub-disciplines often work in extreme isolation. He then remarked,

> It may soon be indispensable to present the same result in several versions, each one accessible to a specific group; the price one might have to pay otherwise is to have our work rediscovered by someone who uses a different language and notation, and who will rightly claim it as his own. (Rota 1997a: 23–24, b: 199)

In the humanities disciplines, there appears to be a greater tolerance of duplicate publication with minor changes, and at times such a practice is defended in terms of the need to communicate findings to wider audiences or to receive criticism of work that can then later be refined. One ethicist, after entertaining the question "May Scholars Republish the Same Material?" concluded that "there can be good reasons for various sorts of duplicate publishing" and "it often is obligatory in order to fulfill one's responsibility as a scholar and teacher," as long as some reference to the original publication is offered (Grisez 1997: 687). In some fields, the written corpus of some scholars has been caricatured as the constant unacknowledged republishing of the same article or book over and over again, under different titles and in different venues. There are extremes, of course, as reflected in a quip made to one seemingly productive but narrowly focused scholar: "I see you have another book out. What are you calling it this time?" (Kaczor 2011: 3).

Despite differences across disciplines, recent years have seen a near-universal tightening of rules on the unacknowledged republication of one's work. In the view of one commentator, authors who re-use their work without attribution to the original published version "may soon find themselves facing retractions, as publishers are starting to crack down on duplicate publication" (Weber-Wulff 2014: 13). As a standard practice, academic publishers now require authors to sign a disclosure warranting that either manuscripts have not been published previously or that clear attribution has been given for any selections that have already appeared in print. At times one sees references to an older unofficial rule that an author may include 30% of one's previously published material in an otherwise new work without acknowledgment, but such a view rarely has present-day defenders and has more recently come under criticism (Bretag and Mahmud 2009). Recently, two philosophers have proposed that a professional code of conduct for philosophy should include a prohibition of self-plagiarism as well as the endorsement of the general principle that one should "not artificially divide ideas and discussion across papers that would better serve the discipline in one paper" (Davies and Felappi 2017: 759).

In the discipline of philosophy, retractions on the basis of duplicate or redundant publication have been unheard of until recently. One reason for this shift may be the expansion of some traditional scientific publishers into the area of philosophy. *The Journal of Value Inquiry* issued two retractions of articles (Anonymous 2014, 2017) after receiving reports that the respective articles had been published previously in other venues. In covering the 2014 case, a journalist at *Retraction Watch* described the situation as, "The author of a 2003 study on 'the ethics of being first' is retracting it because it turns out he had already published it elsewhere—making it, well, not first" (Oransky 2013a).

For the 2017 retraction, I had contacted the journal, noting that the article, which happened to be on the topic of "double-effect," had itself already been published verbatim two times previously in other venues. The published retraction notice that later appeared explained that "After concerns were raised that the article contained redundant material, the editor investigated and found that the paper constituted the third verbatim publication of the same text" (Anonymous 2017: 361). The publisher of *The Journal of Value Inquiry* is Springer—a historically scientific publisher—that inherited significant coverage in the discipline of philosophy with its acquisition of Kluwer Academic Publishers in 2004. It seems likely that this merger has assisted the moving of publication norms in philosophy more closely in alignment with those of the natural sciences, at least regarding duplicate publication, and retractions have also appeared in other Springer philosophy journals for duplicate publication (e.g., Anonymous 2013a).

I had noticed that the author of the 2017 retracted paper in *Journal of Value of Inquiry* also had twice published two other articles, and I so notified the journals involved. The first of these two cases was resolved with the publication in 2018 of a retraction notice that states, "The author has retracted this article because it shows significant overlap with a previous publication by the same author. This article is therefore redundant" (Anonymous 2018). For the second case, however, the journal editor believed that the issuance of a retraction for the duplication would be too harsh, and the editor opted instead to send a warning to the author (See McCook 2018a). In my view, such a response was inadequate, as it did not provide any measure of correction to the scholarly record.

In 2018, the editor of *Business Ethics Quarterly* published a detailed three-page expression of concern acknowledging that in 2002 the journal had published an article that had already been published twice before—with the same title—in other journals in 2001 and 2002. In its three journal versions, the influential article had amassed thousands of citations. The expression of concern explains that "it appears that editors of all three journals were more or less simultaneously vetting versions of the [...] article" (Barry 2018: 238). In both the expression of concern and in a subsequent interview with a journalist, the editor explained that undisclosed redundant publication had occurred, but in the passage of time a clear account of how it had happened at the various journals was difficult to construct (McCook 2018b).

Allegations of self-plagiarism are generally restricted to the undisclosed re-publication of the same work. With a disclosure of prior publication, some philosophy journals will republish articles that first appeared elsewhere, even when such journals advertise their content as "original research" or instruct readers that only unpublished original manuscripts are considered for publication. For example, an article first published in *Philosophical Perspectives* reappeared in *Pacific Philosophical Quarterly* a year later, but with a disclosure in the first footnote that the re-publication is "with the consent of the editors" of both journals (Bealer 2000: 24, n. 1, see 1999). In another case, one article appeared first in *The Review of Metaphysics* and then under a different title the same year in *Synthese*, with a footnote disclosure in the latter stating that "portions of the present text have appeared previously" in the other journal and are "reprinted with permission" (Rota 1991: 177, see 1990). The *Synthese* version of the article was later published two more times in anthologies, with disclosures of previous publication (Rota 1997c, 2006).

Particular notoriety attends articles in ethics that are retracted for self-plagiarism when no disclosure of previous publication appears with the republication. A distinctive case involves the recent retraction of an article on the topic of publishing ethics that was originally titled "Ethics and Integrity of the Publishing Process: Myths, Facts, and a Roadmap." In a lengthy statement appearing in the journal about that now-retracted article, the editors emphasized the emotional toll such cases can take on both editors and authors, stating that self-plagiarism creates "an unpleasant if not painful experience" that "brings embarrassment to authors and disappointment to the editors" (Tsui et al. 2014: 157, 158). The letter expressed that the duplication was reported by the instructor of a graduate course on research ethics, who had discovered that the article in question had been published in a forum on research and publishing ethics in another journal 2 years earlier. The editors noted that "doing nothing would be the worst outcome" and stated that a lengthy description of the affair would assist in bringing greater attention the problem of self-plagiarism in academic writing (159).[5]

Even investigations into duplicate or redundant publication do not require that one first establish intent on the part of authors or editors. R. B. C. Huygens recounts an unusual situation where a journal editor unwittingly allowed the publication of the same edited fragments of a medieval manuscript twice in distinct articles. When Dom Alban Dold sought to publish an edited transcription of the fragments in *Revue Bénédictine* prior to World War II, the editor, Dom Germain Morin, responded positively, noting that "I am satisfied I can assure you that nowhere have I found the slightest trace of these texts," and "they really deserve to be published" (Huygens 2000: 24). However, the editor's memory was not accurate in this one instance, as Huygens relates:

> Unfortunately, both Dold *and* Morin overlooked the fact that seventeen years earlier Dom Morin himself had already published the very same texts in the same *Revue Bénédictine*, from the same Vienna manuscript in which his fellow-Benedictine Alban Dold had just rediscovered them. (ibid.)

This episode involved an omission both on the part of the manuscript editor and the journal editor, occasioning the republication of a text without reference to the original publication.

In some humanities fields, such as philosophy, there is evidence that acquisitions editors generally allow monographs to consist of 20–40% of previously published material, provided that prior publication is clearly disclosed (Weinberg 2015). Important articles often are republished in anthologies, and the complete or selected works of major scholars are frequently republished in uniform multi-volume edi-

[5] Retractions of publications on ethical topics on the basis of authorship violations are not as uncommon as one might expect. Another distinctive case involves the retraction of a book chapter titled "On the Role and Function of Ethics Committees" that was issued by the publisher "due to proven plagiarism by the author" (Anonymous 2015: E1). See also the retraction of the article titled "Truth, Deception, and Lies" that was issued due to "serious plagiarism" (Visker 2010: 5–6). For other examples, see Anonymous (2013b, 2010b, c).

tions. Accusations of self-plagiarism, however, only come about when already-published material is republished without acknowledgment of the previous venue.

A recent article has argued that covert text recycling should be allowable in biomedical disciplines, as long as disclosure is made to an editor even though such disclosure would not be made to readers (Moskovitz 2017). On this view, journal editors are encouraged to consider the acceptability of uncited text re-use, perhaps starting with up to "10% of Introduction and Methods sections and 5% elsewhere" (ibid. 7). Miguel Roig has counseled that editors exercise a certain flexibility and avoid pre-determined thresholds, with particular latitude in cases for authors who are writing in their non-native languages (2017a, b).

3.4 Euphemisms for Academic Plagiarism

While arguably there are no exact synonyms in English for the term *plagiarism*, there are plenty of euphemisms available, and many of them appear in published corrections of the scholarly record. These euphemisms reduce the overtly negative connotations of the unadorned term. To be called a *plagiarist* is surely not a compliment, and the name is a particularly strong term of opprobrium in academic and research settings. For a putative researcher to be branded as a plagiarist does not enhance her or his reputation. Since there are considerable professional and reputational risks to being described as a plagiarist, and furthermore, since there are analogous risks in charging someone else with an offense of suspected plagiarism, it is not surprising that a host of terms and expressions have been used to substitute for the term *plagiarism*. To some observers, these substitutions are euphemisms at best and deceptions at worst.

Certain expressions appear in published corrections of the scholarly record where one might have expected to see the term *plagiarism* used instead, and often they are characterized as euphemisms by those critical of questionable editorial practices (Marcus and Oransky 2015; Oransky 2013b). These euphemisms for academic plagiarism include:

- Unacknowledged collaboration
- A failure in originality
- Unexplained textual overlap with previously published material
- Erroneous citations
- Inadvertent copying
- Use of unattributed material
- Close resemblance to the work of others
- Lack of appropriate citations
- Similarity with previously published articles
- Unreferenced text duplication
- Deficiencies in originality
- A violation of an author agreement with the publisher

Even though the use of euphemisms for plagiarism may be viewed as beneficial in light of the demands of etiquette or for the sake of *bella figura*, their use can be harmful when appearing in published corrections of the scholarly record, such as in a statement of retraction. Many readers may be left unclear regarding the precise deficiencies of the article in question when the relevant word *plagiarism* is omitted from a published statement of retraction.

Optimal corrections of the scholarly record are explanatory, and the use of vague substitutes for the term *plagiarism* limits the efficacy of published corrections. Some placeholders for the term 'plagiarism' are more synonymous than euphemistic. When publishers use the expressions "unattributed copying" or "unattributed dependence," many readers are likely to discern that plagiarism has been committed, but some readers may be puzzled over the meaning of expressions such as "unacknowledged collaboration" or "unprofessional quoting." Retraction statements should not require a sophisticated hermeneutic on the part of readers for determining when articles are being retracted for plagiarism; the use of euphemisms should be avoided to ensure a clear and unambiguous correction of the scholarly record.

Euphemisms not only obscure the precise reasons for a retraction, but can also be quite aggravating to the victims of plagiarism who are seeking a full accounting of the wrongs committed against them through the misappropriation of their works. A *Retraction Watch* article containing an interview with a victim of academic plagiarism reported that the victim "is 'very dissatisfied' with the retraction notice for its failure to use the word plagiarism" (Marcus 2010). Furthermore, the victim took exception to the retraction statement's confident declaration that there was "no intention to use pre-existing work without appropriate attribution" (Anonymous 2010a). The victim explained:

> To say that it wasn't intentional is mind-boggling. You cannot systematically lift someone else's text without intending to do it. It seems not possible. A sentence or two, maybe, but not paragraphs. (Marcus 2010)

The use of euphemisms for plagiarism victimizes a second time the genuine authors whose works have been misappropriated in acts of plagiarism, as euphemisms publicly downplay or mischaracterize the acts that have harmed them.

3.5 Plagiarism vs. Copyright Infringement

Some might think that the term *copyright infringement* should be counted among the euphemisms for plagiarism. On straightforward use, the two are not interchangeable. A writer can plagiarize without violating copyright (as when one publishes under one's name a work of another that has happened to fall out of copyright). Likewise, a writer can violate copyright without plagiarizing (as when one publishes extensive extracts of another's work, with attribution, but in violation of fair use restrictions). Of course, the same act can be both an act of plagiarizing and an act of violating copyright (as when one publishes a recent copyrighted work of someone else under one's name). Even though the terms *copyright infringement* and *plagiarism* are not co-extensive, they are not always clearly delineated. In short, one can consider plagiarism "a failure to acknowledge" and copyright infringement "a failure to obtain authorization" (Saunders 2010: 280).

Whether a given misappropriation constitutes a violation of copyright must be investigated according to the norms of a particular region. Copyright laws can vary according to country, despite advances made in international copyright law (Bannerman 2016). Given the plurality of copyright laws, the differing countries of origin for scholarly publishers and authors, the lengthy time commitment of legal proceedings, the expense of lawyers, and the lack of profitability for authors in most sectors of academic publishing, most claims of copyright violation for academic works are often quite difficult to adjudicate (Sonfield 2014). In contrast to this variability, basic academic and research norms are transnational rather than regional, and therefore their apparent violation through acts of suspected plagiarism—at least in principle—are more easily adjudicated.

One strategy for someone accused of suspected academic plagiarism is to defend herself or himself against the charge by stating that the matter is only one of minor copyright violation. By shifting the domain of inquiry, someone who is accused might forestall a judgment that she or he has acted improperly. Theorist Dan Harms has observed that there are two different value systems or two different "languages" that impede corrections of the scholarly record for plagiarism. The first, embraced by some publishers, views matters "from a strictly legal standpoint" and limits correction of the scholarly literature to cases where laws appear to have been broken, potential financial damages may be assessed, or there is a perceived risk of liability if no action is taken. The second, embraced by ethical researchers, concerns "a duty to the truth; to the quoted authors, whose work had been distributed without their permission; and to the readers, who deserved accurate information from a reputable source" (Harms 2006: 9). The first weakens traditional goals of the academic life, while the second is often proposed to be contrary to the financial interests of publishers. In the judgment of Harms, the first is growing in influence to the detriment of traditional research values.

3.6 Statutes of Limitation for Plagiarism

The disclosure in recent years of plagiarized dissertations by some high-profile government officials in Europe has led some to call for a statute of limitations for investigations into plagiarized dissertations. Debora Weber-Wulff has carefully discussed and critiqued such an approach, noting that it requires the preposterous assumption that "plagiarism could be 'healed' by the passage of time alone" (2014: 57; see 171–172). Defenders of statutes of limitation for plagiarism argue for an analogy with the statutes of limitation that restrict the prosecution of certain criminal offences. Those limits exist because if too much time has passed, the accused may have difficulty in providing an adequate defense and the prosecution may have difficulties in identifying reliable evidence or witnesses to the original crime. Nevertheless, in some countries there are no statutes of limitation for the prosecution of certain serious crimes (e.g. kidnapping, murder).

The Office of Research Integrity of the Department of Health and Human Services in the United States endorses a 6-year time limit for reporting academic misconduct, but the limit is subject to wide exceptions (Code of Federal Regulations 2005: §93.105, 589). These exceptions govern cases where the misconduct involves the health and safety of the public. Furthermore, the 6-year statute of limitations is renewable if an author of record for potentially fraudulent work ever cites it in a later publication or receives some benefit because of it.

In my view, there should be no statute of limitations for the correction of published articles that are demonstrated to have been plagiarized. In some humanities disciplines, years and even decades may elapse before new work is undertaken on certain specialized research topics, so acts of plagiarism may not be discovered until much time has passed and new researchers take up the topic anew. Again, the purpose of the correction of the scholarly record is not to punish the author of record, but to insure a reliable scholarly record for students and researchers. Perhaps a statute of limitations would be beneficial if a finding of academic plagiarism were to require the establishment of a *mens rea*, but as argued above, bad intent by an author of record is irrelevant to the need to correct the scholarly record. The determination of a *mens rea* may be relevant for both legal proceedings and an academic institution's inquiry into research fraud or scientific misconduct, but editors or publishers need not worry about the intent of the author of record when using the proposed heuristic for determining whether plagiarism has been committed. To impose a statute of limitations for issuing published corrections of the scholarly record for plagiarism is a mistake, as any uncorrected plagiarized work will continue to damage the scholarly record and harm students and researchers who may use and cite the defective publications.

3.7 In Sum

The wholesale undocumented appropriation of sentences and paragraphs—and also images or formulas—in a scholarly work cannot be justified as a practice of "poetic allusion" of the type one might find in a creative or literary work. In scholarly, academic, and scientific writing, there are strict authorship expectations and a variety of conventions for indicating where text is taken from the work of others. This chapter has defended the utility of a heuristic for academic plagiarism as: (1) the non-trivial appropriation of words, images, or formulas, (2) with inadequate credit, (3) that generates an appearance of original authorship, (4) in a discrete item belonging to the scholarly record. Articles and books that satisfy the four conditions of this heuristic warrant a published correction of the scholarly record.

References

Andreescu, Liviu. 2013. Self-plagiarism in academic publishing: The anatomy of a misnomer. *Science and Engineering Ethics* 19 (3): 775–797.

Anonymous. 2010a. Retraction. *BMC Medical Ethics* 11 (20): 1. https://doi.org/10.1186/1472-6939-11-20. Accessed 6 July 2018.

———. 2010b. Retraction note: End-of-life discontinuation of destination therapy with cardiac and ventilatory support medical devices: Physician-assisted death or allowing the patient to die? *BMC Medical Ethics* 11(20). https://doi.org/10.1186/1472-6939-11-20. Accessed 6 July 2018.

———. 2010c. Erratum to: Modern business ethics research: Concept, theory and relationships. *Journal of Business Ethics* 93(1): 495.

———. 2013a. Retraction note: Corporate social responsibility in purchasing and supply chain. *Journal of Business Ethics* 113(1): 183.

———. 2013b. Retracted article: The "strong" versus "weak" premise of stakeholder legitimacy and the rhetorical perspective of diffusion. *Journal of Business Ethics* 113(3): 561.

———. 2014. Retraction note to: 'Surgical research and the ethics of being first'. *The Journal of Value Inquiry* 48(1): 171.

———. 2015. *Retraction note*, E1. Springer. https://doi.org/10.1007/978-3-531-94257-5_5. Accessed 6 July 2018.

———. 2017. Retraction note to: Strategic bombing, causal beliefs, and double effect. *The Journal of Value Inquiry* 51(2): 361.

———. 2018. Retracted article: Habits, priming and the explanation of mindless action. *Minds and Machines* 28. https://doi.org/10.1007/s11023-016-9410-5. Accessed 29 Aug 2018.

Arenson, Karen W. 2002. SUNY classics professor is accused of plagiarism. *The New York Times*, February 22. http://www.nytimes.com/2002/02/22/nyregion/suny-classics-professor-is-accused-of-plagiarism.html. Accessed 6 July 2018.

Ares, J., J. Lara, D. Lizcano, and M.A. Martínez. 2018. Who discovered the binary system and arithmetic? Did Leibniz plagiarize Caramuel? *Science and Engineering Ethics* 24 (1): 173–188.

Autor, David H. 2011. Letter. *Journal of Economic Perspectives* 25 (3): 239–240. https://doi.org/10.1257/jep.25.3.239. Accessed 6 July 2018.

Bannerman, Sara. 2016. *International copyright and access to knowledge*. Cambridge, UK: Cambridge University Press.

Barry, Bruce. 2018. Expression of Concern. *Business Ethics Quarterly* 28 (2): 237–239.

Bealer, George. 1999. A theory of the a priori. *Philosophical Perspectives* 13: 29–55.

———. 2000. A theory of the a priori. *Pacific Philosophical Quarterly* 81 (1): 1–30.

Beall, Jeffrey. 2013. Publisher requires only 20% original content in article submissions. In *Scholarly open access: Critical analysis of scholarly open-access publishing*. October 10. http://web.archive.org/web/20160203064300/https://scholarlyoa.com/2013/10/10/original-content-in-articlesubmissions. Accessed 6 July 2018.

———. 2016a. Dangerous predatory publishers threaten medical research. *Journal of Korean Medical Science* 31 (10): 1511–1513. https://doi.org/10.3346/jkms.2016.31.10.1511. Accessed 6 July 2018.

———. 2016b. Bogus Iran-based journal allows up to 40% plagiarism. In *Scholarly open access: Critical analysis of scholarly open-access publishing*. January 29. http://web.archive.org/web/20161113193046/https://scholarlyoa.com/2016/01/28/bogus-iran-based-journal-allows-up-to-40-plagiarism. Accessed 6 July 2018.

———. 2016c. Open-access journal charges people to serve on its editorial board. In *Scholarly open access: Critical analysis of scholarly open-access publishing*. September 8, https://web.archive.org/web/20160910085339/https://scholarlyoa.com/2016/09/08/open-access-journal-charges-people-to-serve-on-its-editorial-board. Accessed 6 July 2018.

———. 2018. Scientific soundness and the problem of predatory journals. In *Pseudoscience: The conspiracy against science*, ed. Allison B. Kaufman and James C. Kaufman, 283–299. Cambridge, MA: The MIT Press.

Beninger, Peter G., Jeffrey Beall, and Sandra E. Shumway. 2016. Debasing the currency of science: The growing menace of predatory open access journals. *Journal of Shellfish Research* 35 (1): 1–5. https://doi.org/10.2983/035.035.0101. Accessed 6 July 2018.

BioMed Central. n.d. Text recycling guidelines. *BioMed Central*. https://publicationethics.org/files/Web_A29298_COPE_Text_Recycling.pdf. Accessed 6 July 2018.

Bloch, R. Howard. 1994. *God's plagiarist: Being an account of the fabulous industry and irregular commerce of the Abbé Migne*. Chicago: University of Chicago Press.

Bouter, Lex, M. Joeri Tijdink, Nils Axelsen, Brian C. Martinson, and Gerben ter Riet. 2016. Ranking major and minor research misbehaviors: Results from a survey among participants of four World Conferences on Research Integrity. *Research Integrity and Peer Review* 1 (17): 1–8. https://doi.org/10.1186/s41073-016-0024-5. Accessed 6 July 2018.

Bouville, Mathieu. 2008. Plagiarism: Words and ideas. *Science and Engineering Ethics* 14 (3): 311–322.

Bretag, Tracey, and Saadia Mahmud. 2009. Self-plagiarism or appropriate textual re-use? *Journal of Academic Ethics* 7 (3): 193–205.

Bretzke, James T. 2013. *Consecrated phrases: A Latin theological dictionary. Latin expressions commonly found in theological writings*. 3rd ed. Collegeville: The Liturgical Press.

Broome, Marion E. 2004. Self-plagiarism: Oxymoron, fair use, or scientific misconduct? *Nursing Outlook* 52 (6): 273–274.

Butler, Declan. 2018. Researchers have finally created a tool to spot duplicated images across thousands of papers. *Nature*. February 23. https://www.nature.com/articles/d41586-018-02421-3. Accessed 6 July 2018.

Carey, Kevin. 2016. A peek inside the strange world of fake academia. *The New York Times*. December 29. https://www.nytimes.com/2016/12/29/upshot/fake-academe-looking-much-like-the-real-thing.html. Accessed 6 July 2018.

Carroll, Jude. 2016. Making decisions on management of plagiarism cases where there is a deliberate attempt to cheat. In *Handbook of academic integrity*, ed. Tracey Bretag, 199–210. Singapore: Springer.

Chaddah, Praveen. 2014. Not all plagiarism requires a retraction. *Nature* 511 (7508): 127. https://doi.org/10.1038/511127a. Accessed 6 July 2018.

Chroust, Anton-Hermann. 1961. Charges of philosophical plagiarism in Greek antiquity. *The Modern Schoolman* 38 (3): 219–237.

Cobey, Kelly. 2017. Illegitimate journals scam even senior scientists. *Nature* 549: 7. https://doi.org/10.1038/549007a Accessed 6 July 2018.

Code of Federal Regulations. 2005. *42 parts 1 to 399: Public health*. Washington, DC: U. S. Government Printing Office. https://ori.hhs.gov/sites/default/files/42_cfr_parts_50_and_93_2005.pdf. Accessed 6 July 2018.

Considine, John. 2008. *Dictionaries in early modern Europe: Lexicography and the making of heritage*. Cambridge, UK: Cambridge University Press.

Constable, Giles. 1983. Forgery and plagiarism in the middle ages. *Archiv für Diplomatik: Schriftgeschichte Siegel- und Wappenkunde* 29: 1–41.

Contreras, Alan, and George Gollin. 2009. The real and the fake degree and diploma mills. *Change: The Magazine of Higher Learning* 41 (2): 36–43.

Dadkhah, Mehdi, and Glenn Borchardt. 2016. Hijacked journals: An emerging challenge for scholarly publishing. *Aesthetic Surgery Journal* 36 (6): 739–741.

Darbyshire, Philip, Lisa McKenna, Susan Fiona Lee, and Christine E. East. 2017. Taking a stand against predatory publishers. *Journal of Advanced Nursing* 73 (7): 1535–1537.

Davies, Benjamin, and Giulia Felappi. 2017. Publish or perish. *Metaphilosophy* 48 (5): 745–761.

Djuric, Dragan. 2015. Penetrating the omerta of predatory publishing. *Science and Engineering Ethics* 21 (1): 183–202.

Dodaro, Robert. 2014. Notification. *Augustinianum* 54 (1): 237.

Ellis, George F.R., and Hermann Nicolai. 2007. Editorial note: The issue of plagiarism. *General Relativity and Gravitation* 39 (12): 1969–1970.

Eriksson, Stefan, and Gert Helgesson. 2017. The false academy: Predatory publishing in science and bioethics. *Medicine, Health Care and Philosophy* 20 (2): 163–170.

European Science Foundation. 2011. *The European code of conduct for research integrity*. Berlin: ALLEA. http://ec.europa.eu/research/participants/data/ref/h2020/other/hi/h2020-ethics_code-of-conduct_en.pdf. Accessed 6 July 2018.

Federal Register. 2000. Federal policy on research misconduct. *Federal Register* 65 (235): 76262. https://www.federalregister.gov/documents/2000/12/06/00-30852/executive-office-of-the-president-federal-policy-on-research-misconductpreamble-for-research#print. Accessed 6 July 2018.

Fishman, Teddi. 2009. 'We know it when we see it' is not good enough: Toward a standard definition of plagiarism that transcends theft, fraud, and copyright. 4th Asia Pacific Conference on Educational Integrity (4APCEI) 28–30 September, University of Wollongong, NSW Australia. http://ro.uow.edu.au/cgi/viewcontent.cgi?article=1037&context=apcei. Accessed 6 July 2018.

Gelman, Andrew. 2016. Clarke's law: Any sufficiently crappy research is indistinguishable from fraud. In *Statistical modeling, causal inference, and social science*. June 20. http://andrewgelman.com/2016/06/20/clarkes-law-of-research. Accessed 6 July 2018.

Gipp, Bela. 2014. *Citation-based plagiarism detection: Detecting disguised and cross-language plagiarism using citation pattern analysis*. Wiesbaden: Springer Vieweg.

Glare, P.G.W., ed. 1982. *Oxford Latin dictionary*. Oxford: Clarendon Press.

Grisez, Germain. 1997. *Difficult moral questions*. Quincy: Franciscan Press.

Grove, Jack. 2017. Predatory conferences 'now outnumber official scholarly events'. *Times Higher Education*. October 26. https://www.timeshighereducation.com/news/predatory-conferences-now-outnumber-official-scholarly-events. Accessed 6 July 2018.

Gutierrez, Fredy R.S., Jeffrey Beall, and Diego A. Forero. 2015. Spurious alternative impact factors: The scale of the problem from an academic perspective. *BioEssays* 37 (5): 474–476.

Hames, Irene. 2007. *Peer review and manuscript management in scientific journals: Guidelines for good practice*. Malden: Blackwell Publishing.

Harms, Dan. 2006. Plagiarism, publishing, and the academy. *Journal of Scholarly Publishing* 38 (1): 1–13.

Hathaway, Neil. 1989. Compilatio: From plagiarism to compiling. *Viator: Medieval and Renaissance Studies* 20: 19–44.

Hauptman, Robert. 2008. Authorial ethics: How writers abuse their calling. *Journal of Scholarly Publishing* 39 (4): 323–353.

Helgesson, Gert. 2016. Plagiarism. In *Encyclopedia of global bioethics*, ed. Henk ten Have, 2239–2249. Cham: Springer.

Helgesson, Gert, and Stefan Eriksson. 2015. Plagiarism in research. *Medicine, Health Care and Philosophy* 18 (1): 91–101.

Heller, Lambert, Ronald The, and Sönke Bartling. 2014. Dynamic publication formats and collaborative authoring. In *Opening science: The evolving guide on how the internet is changing research, collaboration and scholarly publishing*, ed. Sönke Bartling and Sascha Friesike, 197–211. Cham: Springer.

Horbach, S.P.J.M., and W. Halffman. 2017. The extent and causes of academic text recycling or 'self-plagiarism'. *Research Policy*. https://doi.org/10.1016/j.respol.2017.09.004. Accessed 6 July 2018.

Hoven, René, ed. 1994. *Lexique de la prose Latin de la Renaissance*. Leiden: E. J. Brill.

Huygens, R.B.C. 2000. *Ars Edendi: A practical guide to editing medieval Latin texts*. Turnhout: Brepols.

Kaczor, Christopher. 2011. Introduction. In *O rare Ralph McInerny: Stories and reflections about a legendary Notre Dame professor*, ed. Christopher Kaczor, 1–5. South Bend: St. Augustine's Press.

Kivistö, Sari. 2014. Plagiarism and academic thieves. In *The vices of learning: Morality and knowledge at early modern universities*, 118–134. Leiden: Brill.

Koepsell, David. 2017. *Scientific integrity and research ethics: An approach from the ethos of science*. Cham: Springer.

Koppers, Lars, Holger Wormer, and Katja Ickstadt. 2017. Towards a systematic screening tool for quality assurance and semiautomatic fraud detection for images in the life sciences. *Science and Engineering Ethics* 23 (4): 1113–1128.

Kurt, Serhat. 2018. Why do authors publish in predatory journals? *Learned Publishing*. https://doi.org/10.1002/leap.1150. Accessed 6 July 2018.

Liddell, Henry George, Robert Scott, and Henry Stuart Jones, eds. 1953. *A Greek-English lexicon*. Oxford: Clarendon Press.

Lose, Gunnar. 2011. Plagiarism. *International Urogynecology Journal* 22 (8): 903–904.

Marcus, Adam. 2010. Irony alert: Shades of plagiarism undo med ethics paper on terminal care. *Retraction Watch*. December 30. http://retractionwatch.com/2010/12/30/irony-alert-shades-of-plagiarism-undo-med-ethics-paper-on-terminal-care. Accessed 6 July 2018.

Marcus, Adam, and Ivan Oransky. 2015. The plagiarism euphemism parade continues. *GMS Medizin – Bibliothek – Information* 15 (1–2): 1–3. https://doi.org/10.3205/mbi000338. Accessed 6 July 2018.

McCook, Alison. 2016. Journal charges fee to lodge allegations against papers. *Retraction Watch*. September 27. http://retractionwatch.com/2016/09/27/say-what-journal-charges-fee-to-lodge-allegations-against-papers. Accessed 6 July 2018.

———. 2018a. Medical ethicist: "I now understand that I should not have been re-using material". *Retraction Watch*, July 20. https://retractionwatch.com/2018/07/20/medical-ethicist-i-now-understand-that-i-should-not-have-been-re-using-material. Accessed 29 Aug 2018.

———. 2018b. Famous Harvard economist reused parts of 2002 paper multiple times, says journal. *Retraction watch*. June 4. https://retractionwatch.com/2018/06/04/famous-harvard-economist-reused-parts-of-2002-paper-multiple-times-says-journal. Accessed 19 July 2018.

McGill, Scott. 2012. *Plagiarism in Latin literature*. Cambridge, UK: Cambridge University Press.

Mercer, Christia. 2017. Descartes' debt to Teresa of Ávila, or why we should work on women in the history of philosophy. *Philosophical Studies* 174 (10): 2539–2555.

Morris, Sally. 2016. Foreward. In *Against plagiarism: A guide for editors and authors*, ed. Yuehong (Helen) Zhang, v–vii. Cham: Springer.

Morris, Sally, Ed Barnas, Douglas LaFrenier, and Margaret Reich. 2013. *The handbook of journal publishing*. Cambridge, UK: Cambridge University Press.

Moskovitz, Cary. 2017. Text recycling in health sciences research literature: A rhetorical perspective. *Research Integrity and Peer Review* 2 (1): 1–7. https://doi.org/10.1186/s41073-017-0025-z. Accessed 6 July 2018.

Niermeyer, J.F., C. van de Kieft, and J.W.J. Burgers. 2002. *Mediae Latinitatis lexicon minus*. Vol. 2.: M–Z. Leiden: Brill.

Norman, Ian, and Peter Griffiths. 2008. Duplicate publication and 'salami slicing': Ethical issues and practical solutions. *International Journal of Nursing Studies* 45 (9): 1257–1260.

Nylenna, Magne, and Sigmund Simonsen. 2006. Scientific misconduct: A new approach to prevention. *The Lancet* 367 (9526): 1882–1884.

Oransky, Ivan. 2013a. Who's on first? Paper on 'the ethics of being first' retracted because it was…second. *Retraction Watch*. November 15. http://retractionwatch.com/2013/11/15/whos-on-first-paper-on-the-ethics-of-being-first-retracted-because-it-was-second. Accessed 6 July 2018.

———. 2013b. P values: Scientific journals' top ten plagiarism euphemisms. *Retraction Watch*. November 29. http://retractionwatch.com/2013/11/29/p-values-scientific-journals-top-ten-plagiarism-euphemisms. Accessed 6 July 2018.

Pakaluk, Michael. 2017. Ethicist says ghostwriter's role in 'Amoris' is troubling. *Crux*. Jan 15. https://cruxnow.com/commentary/2017/01/15/ethicist-says-ghostwriters-role-amoris-troubling. Accessed 6 July 2018.

Posner, Richard A. 2007. *The little book of plagiarism*. New York: Pantheon Books.

Prasad, S. Shyam. 2017. Plagiarism for PhD thesis 'ok' as long as it's limited to 30%. *India Times*. March 23. http://bangaloremirror.indiatimes.com/bangalore/others/plagiarism-for-phd-thesis-ok-as-long-as-its-limited-to-30/articleshow/57799614.cms. Accessed 6 July 2018.

Pyne, Derek. 2017. The rewards of predatory publications at a small business school. *Journal of Scholarly Publishing* 48 (3): 137–160.

Resnik, David B., and C. Neal Stewart Jr. 2012. Misconduct versus honest error and scientific disagreement. *Accountability in Research* 19 (1): 56–63.

Resnik, David B., Elise M. Smith, Stefanie H. Chen, and Carlos Goller. 2017. What is recklessness in scientific research? The Frank Sauer case. *Accountability in Research* 24 (8): 497–502.

Richtig, G., M. Berger, B. Lange-Asschenfeldt, W. Aberer, and E. Richtig. 2018. Problems and challenges of predatory journals. *Journal of the European Academy of Dermatology and Venereology*. https://doi.org/10.1111/jdv.15039. Accessed 6 July 2018.

Rogerson Ann, M., and Grace McCarthy. 2017. Using internet based paraphrasing tools: Original work, patchwriting or facilitated plagiarism? *International Journal for Educational Integrity* 13 (2): 1–15. https://doi.org/10.1007/s40979-016-0013-y. Accessed 6 July 2018.

Roig, Miguel. 2015. *Avoiding plagiarism, self-plagiarism, and other questionable writing practices: A guide to ethical writing*, 2nd ed. U.S. Department of Health and Human Services, Office of Research Integrity. https://ori.hhs.gov/sites/default/files/plagiarism.pdf. Accessed 6 July 2018.

———. 2017a. Comment on Cary Moskovitz' 'Text recycling in health sciences literature: A rhetorical perspective'. *Research Integrity and Peer Review* 2 (2): 1–4. https://doi.org/10.1186/s41073-017-0026-y. Accessed 6 July 2018.

———. 2017b. Encouraging editorial flexibility in cases of textual reuse. *Journal of Korean Medical Science* 32 (4): 557–560. https://doi.org/10.3346/jkms.2017.32.4.557. Accessed 6 July 2018.

Rota, Gian-Carlo. 1990. Mathematics and philosophy: The story of a misunderstanding. *The Review of Metaphysics* 44 (2): 259–271.

———. 1991. The pernicious influence of mathematics upon philosophy. *Synthese* 88 (2): 165–178.

———. 1997a. Ten lessons I wish I had been taught. *Notices of the American Mathematical Society* 44 (1): 22–25.

———. 1997b. Ten lessons I wish I had been taught. In *Indiscrete thoughts*, ed. Fabrizio Palombi, 195–203. Boston: Birkhäuser.

———. 1997c. The pernicious influence of mathematics upon philosophy. In *Indiscrete thoughts*, ed. Fabrizio Palombi, 89–103. Boston: Birkhäuser.

———. 2006. The pernicious influence of mathematics upon philosophy. In *18 unconventional essays on the nature of mathematics*, ed. Reuben Hersh, 220–230. New York: Springer.

Saunders, Joss. 2010. Plagiarism and the law. *Learned Publishing* 23 (4): 279–292.

Scanlon, Patrick M. 2007. Song from myself: An anatomy of self-plagiarism. *Plagiary* 1: 57–66. http://hdl.handle.net/2027/spo.5240451.0002.007. Accessed 6 July 2018.

Schabel, Chris. 2005. *Haec ille*: Citation, quotation, and plagiarism in 14th century scholasticism. In *The origins of European scholarship: The Cyprus Millennium International Conference*, ed. Ioannis Taifacos, 163–175. Stuttgart: Franz Steiner Verlag.

Seo, J. Mira. 2009. Plagiarism and poetic identity in Martial. *American Journal of Philology* 130 (4): 567–593.

Shamoo, Adil E., and David B. Resnik. 2015. *Responsible conduct of research*. 3rd ed. New York: Oxford University Press.

Shamseer, Larissa, David Moher, Onyi Maduekwe, Lucy Turner, Virginia Barbour, Rebecca Burch, Jocalyn Clark, James Galipeau, Jason Roberts, and Beverley J. Shea. 2017. Potential predatory and legitimate biomedical journals: Can you tell the difference? A cross-sectional comparison. *BMC Medicine* 15 (28): 1–14. https://doi.org/10.1186/s12916-017-0785-9. Accessed 6 July 2018.

Smolčić, Vesna Šupak. 2013. Salami publication: definitions and examples. *Biochemia Medica* 23 (3): 137–141.
Sonfield, Matthew C. 2014. Academic plagiarism at the faculty level: Legal versus ethical issues and a case study. *Journal of Academic Ethics* 12 (2): 75–87.
St. Onge, K.R. 1988. *The melancholy anatomy of plagiarism*. Lanham: University Press of America.
Stevenson, Iain. 2004. Is open access the new vanity publishing? *Learned Publishing* 17 (2): 83–84.
Swiss National Science Foundation. 2016. Regulations on scientific misconduct (Research integrity regulations, RI regulations). http://www.snf.ch/sitecollectiondocuments/ueb_org_fehlverh_gesuchstellende_e.pdf. Accessed 6 July 2018.
Teixeira da Silva, Jaime. 2017. Copy-paste: 2-click step to success and productivity that underlies self-plagiarism. *Science and Engineering Ethics* 23 (3): 943–994.
The Office of Research Integrity. 1994. ORI policy on plagiarism. *ORI Newsletter* 3: 1. https://ori.hhs.gov/ori-policy-plagiarism. Accessed 6 July 2018.
Tsui, Anne, et al. 2014. Retraction statement for 'Ethics and integrity of the publishing process: Myths, facts, and a roadmap'. *Management and Organization Review* 10 (1): 157–162.
Umlauf, Mary Grace, and Yuki Mochizuki. 2018. Predatory publishing and cybercrime targeting academics. *International Journal of Nursing Practice* 24: S1. https://doi.org/10.1111/ijn.12656. Accessed 10 June 2018.
Vincent, Clement. 2007. The purloined bibliography. *The Chronicle of Higher Education*. July 16. http://www.chronicle.com/article/The-Purloined-Bibliography/46601. Accessed 6 July 2018.
Visker, R. 2010. Note from the editorial board/Mededeling van de redactie. *Tijdschrift voor Filosofie* 72 (1): 5–6.
Wager, Liz. 2013. Suspected plagiarism in a published manuscript. Committee on Publication Ethics. http://publicationethics.org/files/plagiarismB.pdf. Accessed 6 July 2018.
Wager, Elizabeth. 2014. Defining and responding to plagiarism. *Learned Publishing* 27 (1): 33–42.
Wager, Elizabeth, Virginia Barbour, Steven Yentis, Sabine Kleinert. 2009. Retraction guidelines. *Committee on Publication Ethics*. https://publicationethics.org/files/retraction guidelines.pdf. Accessed 16 Mar 2018.
Weber-Wulff, Debora. 2014. *False feathers: A perspective on academic plagiarism*. Heidelberg: Springer.
Weinberg, Justin. 2015. Answers from academic publishers. *Daily Nous: News for and About the Philosophy Profession*. May 28. http://dailynous.com/2015/05/28/answers-from-academic-publishers. Accessed 6 July 2018.
Yilmaz, Ihsan. 2007. Plagiarism? No, we're just borrowing better English. *Nature* 449 (7163): 658. https://doi.org/10.1038/449658a. Accessed 6 July 2018.
Zahnd, Ueli. 2015. Easy-going scholars lecturing *secundum alium*? Notes on some French Franciscan *Sentences* commentaries of the fifteenth century. In *Mediaeval commentaries on the 'Sentences' of Peter Lombard*, ed. Philipp W. Rosemann, vol. 3, 267–314. Leiden: Brill.
Zgusta, Ladislav. 1988. Copying in lexicography: Monier-Williams' Sanskrit dictionary and other cases (Dvaikośyam). *Lexicographica* 4: 145–164.
Zhang, Yuehong (Helen). 2016. *Against plagiarism: A guide for editors and authors*. Cham: Springer.

Chapter 4
A Test Case for Published Corrections: The Discipline of Philosophy

Abstract Individuals discovered to have engaged in wide-scale serial plagiarism in philosophy are relatively few, but the academic publishers falling victim to them are many. Some of the most respected publishing houses in philosophy have recently faced the issue of having published plagiarized material. The chapter uses a specific context of serial plagiarism involving 43 articles and book chapters by one author of record as a test case. The various responses by these publishers to this instance of serial plagiarism provide a real-time snapshot of the practices for correcting the scholarly record in the discipline of philosophy. I propose a new rubric for evaluating published corrections of the scholarly record for cases of demonstrated plagiarism. On this rubric, the highest-scoring corrections are those that: (1) unambiguously declare that a plagiarized work is plagiarized, (2) clearly credit the original source material misappropriated in the act of plagiarism, and (3) are easily accessible to the scholarly community without registration barriers or paywalls. This analysis yields a twofold conclusion: first, relatively little uniformity exists among publishers in philosophy for responding to plagiarism; and second, the discipline of philosophy often falls short of the accepted practices for correcting the scholarly record in contrast to the natural sciences. This chapter considers only public, documented cases of academic plagiarism in philosophy and makes no new allegations of plagiarism.

Keywords Academic publishing · Plagiarism · Retractions · Expressions of concern · Philosophy

Notable instances of serial plagiarism in the discipline of philosophy have been reported in recent years (Palus 2016; Newman 2010; Weinberg 2014; Ballor 2014; Anonymous 2015a; Paglieri 2015; Spodenkiewicz 2004; cf. Kramer 2018; Stern 2018).[1] On all accounts, plagiarism by professional philosophers is quite rare, yet the latest instances involve some of the most highly regarded university presses and commercial publishers in the English-speaking world. These publishing houses have found that they have unwittingly published plagiarized book chapters and journal

[1] This chapter considers only public, documented cases of academic plagiarism in philosophy and makes no new allegations of plagiarism. This chapter appeared previously as Dougherty 2017.

M. V. Dougherty, *Correcting the Scholarly Record for Research Integrity*, Research Ethics Forum 6, https://doi.org/10.1007/978-3-319-99435-2_4

articles. Compounding the issue are the ways in which these plagiarized works have been disseminated through the scholarly community; in most cases they were issued not just in print form but also across a range of electronic platforms. Correcting the scholarly record in a digital age presents new challenges—but also new opportunities—for maintaining the integrity of the body of published philosophical research.

This chapter considers the various public responses by publishers in philosophy to one recent instance of serial plagiarism, involving 43 plagiarized book chapters and journal articles in philosophy published in English under one name.[2] The data set of 43 cases offers a real-time snapshot of how publishers in philosophy deal with the aftermath of having published plagiarized material. As is shown below, there is relatively little uniformity in philosophy for correcting the scholarly record after the publication of plagiarized material. This inconsistency stands in contrast to the publication practices of the natural sciences, where there is a greater consensus about the importance of maintaining the integrity of the set of published scientific works as well as general agreement about the accepted methods for correction (see Hames 2007: 173–199; ICMJE 2016: 7–8; Wager et al. 2009: 1–6; Weber-Wulff 2014: 132–133).[3] One may question whether the philosophy community is justified in settling for a lower standard than the natural sciences community in the correction of the scholarly record.

4.1 Classifying Responses to Plagiarism

Academic presses in philosophy have exhibited a range of responses to the publication of plagiarized material. In assessing their responses, I use the following five categories:

1. Issuing an expression of concern
2. Issuing a statement of retraction
3. Reprinting a volume after the removal of plagiarized material
4. Halting the sale of plagiarized material
5. Taking no public action

Not all these categories are mutually exclusive. Although expressions of concern are distinguished below from statements of retraction, and taking no public action at all rules out actions falling under the remaining categories, a publisher could perform some other combination of actions drawn from these categories. A publisher can, for instance, publicize an expression of concern while halting the sale of the plagiarized material, or issue a statement of retraction while reprinting a volume after the

[2] Evidence of plagiarism for cases 1–40 has been published in Dougherty, Harsting, and Friedman 2009. For cases 41–43, see the respective statements by publishers and editors discussed below (Anonymous 2011a, 2011g; Faesen et al. 2011). I reference the 43 cases according to this established enumeration.

[3] For discussions of further improving methods of correction in the sciences, see Marcus and Oransky 2014 and Allison et al. 2016.

removal of plagiarized material. In addition, a publisher's performance of an action from one category does not necessitate the performance of an action from another. For instance, a publisher may reprint a volume after the removal of plagiarized material without ever offering either a statement of retraction or an expression of concern. Although taking no public action at all appears to be an omission rather than an action, some publishers may choose to do nothing regarding demonstrated cases of plagiarism after a lengthy process of deliberation. A publisher might choose to do nothing if, for instance, issuing a retraction were believed to be damaging to the publisher's reputation, or if halting the sale of plagiarized material were believed to entail sufficiently negative financial consequences.

In some instances, categorizing a particular public response by a publisher to a case of proven plagiarism can be challenging. For example, whether a publisher's ambiguous statement in a given case constitutes an expression of concern or a statement of retraction may not be immediately obvious. Furthermore, placing certain responses by publishers into the category of doing nothing can be difficult. When a publisher allows a volume containing plagiarized material to go out of stock silently without a reprinting, this move may appear to a charitable observer to be a gradual attempt at halting the sale of plagiarized material, but to a more critical observer to be no action at all. Despite interpretive difficulties such as these, in what follows I offer definitions and clarifications for each of the five categories, and I argue why the particular actions of publishers can be said to fall within the five categories. Finally, it should be noted that in discussing the responses by publishers to the 43 cases, I have restricted my account to publicly available material.

4.2 Issuing an Expression of Concern

The first category of responses that I wish to examine is issuing an expression of concern (hereafter *EC*). For the purposes of this chapter, an *EC* is understood to be a notice that exhibits three distinct features. The *EC*:

1. declares that a complaint has been received regarding a publication;
2. states in at least a general way the nature of the complaint; and
3. makes no substantive determination about the validity of the complaint.

In short, an *EC* is a neutral, public notification that some problem has been alleged regarding a publication. The neutrality of an *EC* is somewhat relative, however; presumably a publisher would not issue an *EC* for a received complaint that appears patently baseless. In all likelihood, publishers have differing thresholds for issuing *EC*s.

Several European publishing houses have issued *EC*s for some of the 43 cases under consideration. The commercial publishers Felix Meiner Verlag and Ashgate Publishing each issued *EC*s for plagiarized book chapters, and their respective *EC*s vary in the precise ways in which they satisfy the abovementioned three features. Meiner issued its *EC* on the website page for the book containing the plagiarized chapter under an electronic fold labeled "Mehr [i.e., More]" (Anonymous 2011a [Case 43]). This *EC*, entitled "Hinweis [i.e., Comment]," acknowledges receipt of a "Klarstellung [i.e., Clarification]" by a notifying party, and then provides a substantive extract from the received complaint describing the plagiarism. Similarly, the *EC* issued by Ashgate indicates that a notifying party has brought a complaint of plagiarism against a chapter in one of its volumes, but instead of an online notification, Ashgate sent copies of the *EC* to those who had purchased the volume directly from the publisher (Anonymous 2011b [Case 12]). These *EC*s by Meiner and Ashgate issue no judgment about the validity of the received complaints, but each simply acknowledges that a named notifying party has brought forth a complaint. The *EC*s differ significantly in accessibility, however, as the print dissemination of Ashgate's *EC* is restricted in comparison with Meiner's online availability.

For three plagiarized book chapters, Leuven University Press issued three nearly identical *EC*s, each stating that a complaint had been received and identifying by name the various notifying parties. In two instances, the notifying parties included the victims whose works had been misappropriated. Leuven University Press placed the first *EC* online on the book's home page under a table of contents tab (Anonymous 2010a [Case 39]), but two others were appended to remaining copies of the books as tipped-in plates (Anonymous 2010b [Case 15], c [Case 24]). Furthermore, all purchasers of the volumes through the press were sent labels of these respective notifications that could be inserted into copies of the volumes.

These three notifications by Leuven University Press exemplify the neutrality of *EC*s. Each is couched in legal terminology emphasizing that the publisher does not view the issuance of an *EC* to be any admission of wrongdoing: "Without giving any view on the merits of the complaint, the publisher considers it appropriate to inform the reader thereof and to refer him to the original text. The publisher reserves all rights with regard to the said complaint and the legal consequences thereof" (Anonymous 2010a, b [with slight variation], c [with slight variation]). Such indemnifying qualifiers underscore the press's public posture of detachment in publicly reporting the received complaints of plagiarism.

Undoubtedly, these *EC*s by Meiner, Ashgate, and Leuven University Press have the potential to alert readers to the problems with previously published material. Yet some difficulties remain. First, one may ask whether *EC*s should be regarded as genuine corrections of the scholarly record. It is questionable whether a publisher's public report of a received complaint, with no judgment of its validity, sufficiently addresses the harm done in cases of demonstrated violations of scholarly and publishing integrity. To be sure, *EC*s would be fitting in cases where the merits of a complaint seem quite plausible but cannot be definitively established. In such cases, an *EC* could serve as a useful warning of unverified potential problems. In cases when substantial plagiarism is demonstrated, however, one might consider that a stronger correction of the scholarly record by the publisher would be warranted.

An additional problem with *EC*s is the burden they place on named notifying parties. When publishers identify notifying parties in *EC*s, without judging the merits of the complaints, the report of misconduct has the appearance of a private allegation made by individuals. For cases where the plagiarism has been demonstrated by the notifying parties, the neutrality of *EC*s denies them the institutional support they are due. The *EC*s preserve the notifying parties as carriers of the complaint in perpetuity because there is no institutional correction of the scholarly record. In the five *EC*s noted above, the notifying parties were identified by name. Any such identification exposes them to potential legal risks, even when the merits of the complaint are beyond question. There may also be reputational risks to being identified publicly with a complaint, again even when the merits of the complaint are impeccable. Legal and reputational risks continue to be pervasive hazards for individuals who engage in academic plagiarism whistleblowing (Fox and Beall 2014: 344–345; Gunsalus 1998: 52). On this issue of repercussions for whistleblowers, see Chap. 5 below.

4.3 Issuing a Statement of Retraction

As explained above, a chief characteristic of *EC*s is their posture of neutrality. Such neutrality is not found in the second major category of public responses by publishers, the statement of retraction (hereafter *SR*). In issuing an *SR*, a publisher forgoes neutrality and recognizes publicly the merits of the complaint brought by the notifying party. In the scientific community, the *SR* is the gold standard for correcting the scholarly record; the *SR* publicly acknowledges that previously published material falls short of the principles of scholarly and publication integrity in some non-trivial way. The emphasis in *SR*s is not on the notifying parties but on the wrongs committed, and unsurprisingly the names of the notifying parties are often unmentioned in *SR*s. Even when the notifying parties are identified, the declaration that there is merit to the complaint offers some institutional support, which substantively mitigates the abovementioned legal and reputational risks. Nor is the emphasis of the *SR* primarily on those who engage in misconduct. It has been rightly observed that "the main purpose of retractions is to correct the literature and ensure its integrity rather than to punish authors who misbehave" (Wager et al. 2009: 2). The point of the *SR* is simply the correction of the scholarly record.

Retraction is not the same as removal or expurgation, however. An *SR* simply changes the status of a previously published work; the work loses the endorsement that was bestowed by the publisher through publication. Most often retracted articles are still available to the research community after undergoing a change in status. In the words of one commentator, removal after retraction "should actually be avoided unless absolutely necessary" (Hames 2007: 194).

4.3.1 Maximal and Minimal Statements of Retraction

*SR*s can correct the scholarly record with varying degrees of information. In the natural sciences, there have been debates about standardizing retraction formats, with calls for establishing a "transparency index" that will grade journals in part for their clarity in issuing retractions and corrections (Marcus and Oransky 2012). Others have proposed a rating system for evaluating the quality of retractions in the natural sciences (Bilbrey et al. 2014).[4] Some theorists have proposed that *SR*s should indicate whether the author of record agrees with the issuance of the *SR*, and further, whether the retraction extends to the whole of the article or simply a portion, such as a figure or section (Hames 2007: 195). One critic has decried what has been described as "an evasion of authorship" for many published retractions, since retractions often lack the "explicit markers—for example, salutation, sign-off, unmistakable use of first-person/third-person pronouns, references to actions by the authors of retracted papers, etc.—that clearly indicate who authored the retraction notices" (Hu 2017: 2). When a retraction includes a general apology to the reader, "it is not entirely clear who is/are apologizing" (ibid.). Even in those cases where a signed apology is offered in a published retraction, the language used often leaves the matter unclear whether the apology is merely a generic expression of unspecified regret or a genuine expression of remorse for the commission of an academic wrong.

Other critics summarize the issues involved by noting that opaque or uninformative retractions "do not serve their intended purpose, i.e., to hold all parties accountable, and to inform the scientific and wider public of the problem and reason for the paper's demise" (Teixeira da Silva and Dobránszki 2017: 521). If a retraction is to fulfill its purpose as an "essential scientific historical document" that discloses to the research community the change in status of a scholarly work, it must be informative (ibid). Although the discovery of error and the discovery of fraud are the main reasons for retractions in the natural sciences, in philosophy and other humanities disciplines retractions appear primarily for a specific kind of problem, namely, plagiarism. Therefore, a specific rubric for analyzing retractions for plagiarism in these disciplines can be useful.

In recognition that *SR*s can fall short in giving the fullest correction of the scholarly record in cases of plagiarism, I propose a new rubric for evaluating *SR*s issued in philosophy (and other humanities disciplines) for cases of plagiarism. This rubric consists of three categories:

1. *Determination*: The extent to which the *SR* exhibits support for a demonstrated claim of plagiarism set forth by notifying parties.
2. *Credit*: The extent to which the *SR* identifies the known original work(s) misappropriated by the plagiarizing article or chapter.

[4] The proposed model for the natural sciences by Bilbrey et al. (2014: 15–16) uses a value system as follows:

0—No reason for retraction can be discerned from the notice.
1—The reason for retraction can be inferred but is not stated clearly through the naming or definition of a category.
2—The reason for retraction is clearly stated, but explanation is not given as to how the rest of the article was affected by retraction.
3—The reason for retraction is clearly stated and explanation is given for if and how the entirety of the article was affected by the fault.

3. *Availability*: The extent to which the *SR* is promulgated publicly and without restriction.

On this rubric, the highest-scoring *SR*s are those that are fully explanatory. That is, they declare unequivocally that a plagiarized work is plagiarized, they clearly credit the original source material misappropriated in the plagiarism, and they are easily accessible to the scholarly community. To evaluate the *SR*s issued in philosophy on each of these three categories, I use the three scores of "maximal," "medial," and "minimal." An *SR*'s score on one category does not necessitate any particular score on another. This scoring system can also evaluate *EC*s in addition to *SR*s, on the understanding that *EC*s warrant a score of "neutral" concerning determination.

To begin, one may consider the electronic *SR* issued by the University of Notre Dame Press for a plagiarized book chapter. This *SR* states that "parts" of the chapter in question have been "subject to claims of plagiarism," and that the editor and the publisher "as a result cannot stand behind the noted material as originally contained in this volume" (Haldane 2011 [Case 14]). In light of these stated qualifications, only a reader who compares the plagiarized chapter with the original source will see that the "parts" are almost the entire chapter, including endnotes. This *SR* is therefore minimal in determination because it presents a qualified and understated acknowledgment of the plagiarism demonstrated by the notifying party. As the notification is not neutral and provides some—albeit muted—confirmation of the plagiarism, it cannot be designated as an *EC*. This Notre Dame *SR* is maximal, however, in the two remaining senses, of credit and availability: it provides a clear bibliographical reference to the original source material that was misappropriated, and it is promulgated unrestrictedly on the publisher's website. By being maximal in credit, the retraction reinstates the primary victim of the plagiarism as the original author of the material, thereby acknowledging the true history of discovery.

Another mixture of maximal and minimal scoring characterizes the *SR* issued by the Italian publisher SISMEL-Edizioni del Galluzzo for a plagiarized article that appeared in one of its annuals. The *SR* was printed in the journal 6 years after the appearance of the plagiarized article, stating that "elements of plagiarism had been identified" in the work and offering a bibliographical reference to an article listing the original sources (Anonymous 2012: vi [Case 32]). With regard to determination, this *SR* is best scored as medial, given the qualifier "elements of" in respect to the demonstrated plagiarism. Yet it is maximal in credit, as readers of the *SR* are directed to where they can find the original source materials. As to availability, the *SR* also warrants a score of medial, as it is not issued online. Notably, the article remains available for sale without a warning on the publisher's website in electronic form. These Notre Dame and SISMEL *SR*s possess a combination of various scores for determination, credit, and availability, demonstrating that notifications of plagiarism by publishers can fall short of full correction in at least three significant ways. Unfortunately, the two respective *SR*s by Notre Dame and SISMEL misspell the name of the plagiarist.

Not all *SR*s are limited to such qualified confirmations of plagiarism. The *SR* issued by De Gruyter for a plagiarized chapter consists entirely of a short online bilingual statement that simply reads, "Retracted for plagiarism/Wegen Plagiatsvorwürfen zurückgezogen" (Anon 2011c [Case 17]). The English portion leaves no room for ambiguity; yet the German that follows is somewhat weaker, stating that the article is retracted for plagiarism *allegations*. The electronic version of the article has been replaced with a document containing this same brief bilingual statement. In addition to being maximal in determination, De Gruyter's *SR* is also maximal in its unrestricted promulgation online. Yet, it is minimal in credit, as it does not identify the original source materials and therefore does not offer the fullest correction of the scholarly record.

Some *SR*s are maximal in all three senses: they offer an unqualified affirmation of plagiarism; they identify the known original source materials that were misappropriated; and they are available online without restriction. A trio of such maximal *SR*s was issued by Peeters Publishers for plagiarized articles that had previously appeared in Peeters's academic journals. With regard to determination, the *SR*s left no room for doubt. The first states: "serious plagiarism had been committed" (Visker 2010: 5 [Case 29]). The second: "such plagiarism represents a serious assault on intellectual integrity" (Editorial Board of *ETL* 2010: iii [Case 3]). The third: "evidence of plagiarism provided to the Editorial Board is overwhelming and irrefutable" (Anonymous 2009: v [Case 4]). With regard to credit, each *SR* also identifies original source materials. With regard to availability, each *SR* appeared in a later issue of the print version of the respective journals. Two *SR*s were made freely available online on the publisher's website, and two of them were issued bilingually. The correction of the scholarly record went beyond these actions, however, as the three *SR*s report that adhesive warning labels noting the plagiarism were being sent to all subscribers to the journal, with instructions that the labels should be affixed to the first and last pages of the respective articles.[5] Peeters also removed access to the electronic subscription versions of the plagiarized articles from its website, replacing the files with the relevant *SR*s. For the one article for which the *SR* was not placed online, however, there was made a modification to the freely available electronic version in the JSTOR digital library collection; it now displays the aforementioned label with the bilingual statement "plagiarized/geplagieerd" as well as a bibliographical reference to the journal's statement of retraction (Anonymous 2011d [Case 29]). Furthermore, a diagonal strikethrough line has been added over the text on each page of the article. Peeters's *SR*s in these cases are maximal in all three ways: they offer unqualified institutional recognition of demonstrated claims of plagiarism, they give clear credit to original authors, and they are easily accessible to the scholarly community.

Case 3 involves the only article from the set of 43 for which there is more than one author of record. Best practices require that statements of retraction indicate which author of record for a multi-authored work is responsible for the defects in a retracted article. The retraction statement for Case 3 was clear in assigning respon-

[5] The act of asking readers to participate in the correction of published work has an early precedent. Ann Blair has examined how in the early history of printing, errata sheets included in published volumes often had instructions asking readers to use them and correct by hand the faulty pages (see Blair 2007: 21–22).

sibility, as it states, "The plagiarism in question was related exclusively to the portion of the article for which [the first author] was responsible" (Editorial Board of ETL 2010: iii). This transparent retraction has the benefit of exonerating an innocent researcher whose work might otherwise have been viewed with suspicion, since collaborators are often harmed by academic misconduct perpetrated by colleagues (Hussinger and Pellens 2017).

4.3.2 Paywalled SRs *and Untethered* SRs

Two *SR*s issued by the commercial publisher Wiley-Blackwell are distinctive insofar as they exist behind paywalls (Anonymous 2010d: 620 [Case 35], e: 519 [Case 10]). Access to these notifications is restricted to those with private or institutional subscriptions to the respective publications, although anyone can rent access to each *SR* for a period of 24 h for $6.00 or purchase a PDF copy for $38.00. The topic of paywalling *SR*s (and also *EC*s) continues to be a much discussed issue within the scientific community. The retraction guidelines issued by the *Committee on Publication Ethics* recommend that retractions "be freely available to all readers (i.e. not behind access barriers or available only to subscribers)" (Wager et al. 2009: 2). Some researchers have alleged that paywalled corrections are straightforward instances of editorial misconduct, stating that "charging readers to learn of journal's mistakes or authors' misconduct is unethical" (Shelomi 2014: 56). Due to their placement behind paywalls on the publisher's website, these two *SR*s must be scored as minimal in availability. The first *SR* declares that the article is retracted for "overlap" with previously published material (Anonymous 2010d: 620 [Case 35]), and two sources are identified with full bibliographical material. Therefore, this *SR* should be considered maximal in both determination and credit. In addition, each page of the original article has been modified with a diagonal red watermark noting the retraction. The second *SR* by Wiley-Blackwell states that the article has been retracted for "significant overlap with previously published material" (Anonymous 2011e: 519 [Case 10]). As no references to original source materials are given, it must be graded as minimal in credit. With regard to determination, it can be scored as medial, as the reader cannot discern that the retraction for "overlap" is due to the plagiarizing misappropriation of the work of others rather than some other unspecified problem (for example, overlap with the author's own previously published material or self-plagiarism).

Not all *SR*s by Wiley-Blackwell deserve a minimal score with regard to availability, however. One unpaywalled retraction has been issued for a plagiarized book chapter "due to significant overlap with previously published material" (Anonymous 2015b [Case 13]). Since no original sources are identified for this case, and since the nature of the overlap is unspecified, this *SR* scores as minimal in credit and medial in determination. Again, a red watermark added to the original article pages further warns readers that the article has been retracted.

In correcting the scholarly record in the case of journals and annuals, best practices require that each *SR* be published on a numbered, citable page of a later issue, so that each *SR* can be easily referenced. When a journal issues a correction on a numbered, citable page of a journal, the correction itself becomes part of the scholarly record. Occasionally corrections themselves are subject to correction, as in those rare cases where an erratum is issued for an erratum, or a retraction is retracted (for example, Anonymous 1992; Patil 2015: "This is a retraction of a retraction notice"). Furthermore, for articles available in electronic form, best practices require that the original article and the *SR* be tethered electronically, so that one can easily follow a link from one to the other. These best practices are not always followed, however, and at times the *SR* appears on an unnumbered page and remains electronically untethered to the original article (e.g., Case 32).

4.3.3 SRs with University Support

Some publishers invoke the authority of other institutions in issuing their respective *SR*s. The *SR* issued by Brepols Publishers for a plagiarized article that appeared in one of its annuals is distinctive in reproducing in full a letter received from the home institution of the plagiarist. The reproduced letter discloses that a university investigation into complaints of plagiarism had concluded that the "conduct is highly questionable in terms of scientific integrity" and that the university "no longer considers this publication as being part of its scientific output, and formally retracts its affiliation with this publication" (Esposito and Porro 2009: 453 [Case 30]). The *SR* by Brepols then notes that regarding the article in question, "parts [...] present very strong similarities with, or were literally taken from," another work, which is then identified with full bibliographical information (454). As this *SR* unequivocally supports the demonstrated case of plagiarism, identifies the original source material, and is unpaywalled on the Brepols website, it must be scored as maximal on all three categories.

Likewise, the *SR* issued by the Pontifical Institute of Mediaeval Studies (PIMS) quotes from an apparently similar letter it received from the home institution of the plagiarist. This *SR* identifies five original source materials, prefaced with this statement: "Readers should be aware that much of [—]'s essay has been taken partially or verbatim from the following, largely unacknowledged, sources, though there may be more sources as well" (Newhauser and Harnum 2010 [Case 27]). Although the PIMS *SR* is thereby maximal in both its acknowledgment of the plagiarism and in its recognition of the original source materials, it is minimal with regard to availability, as it was only added to remaining unsold print copies of the book and is not issued online. The home page for the book on the PIMS website offering the volume for sale does not include the *SR* and makes no warning of the plagiarized content.

Springer too issued an *SR* that references the findings of a university inquiry. This *SR* states that a book chapter "has been retracted at the request of the employer of the author at the time of writing, because it contains passages from the previously published manuscripts without acknowledging the source" (Anonymous 2015c [Case 25]). As Springer's *SR* also gives full bibliographical references for four original source materials and also is freely available online, this correction of the scholarly record is maximal in all three senses. In addition, the word "RETRACTED" now appears as a watermark across the first two pages of the electronic version of the chapter.

4.3.4 Academic Editors and Independent SRs

The *SR*s examined above were each issued by publishers, even though some of them were signed by scholars who served as the academic editors of the books and journals in which the plagiarized works appeared. One may ask, however, whether academic editors can issue *SR*s independently. Does a public notification of plagiarism by an editor alone, without any direct affiliation with the original publisher, constitute a genuine *SR*? The issue arises in 2 of the 43 cases of plagiarism under consideration.

The first example is a notification by three editors in regard to a volume published by Peeters Publishers that contained a plagiarized chapter. The brief notification by the editors, entitled "Plagiarism," is quite comprehensive (Faesen et al. 2011 [Case 42]). It declares that the editors were made aware of the plagiarism, offers a bibliographical reference to the original source material, and appears online without any paywall or registration restrictions. In light of these three features, the requirements of a fully maximal *SR* seem to have been met unequivocally. Nevertheless, this putative *SR* appears not on the website of Peeters Publishers but on an unrelated website by a organization affiliated with the original publication of the volume. The notification looks like an *SR*, but absent the publisher's backing one may question whether the action of editors alone is sufficient to constitute a formal correction of the scholarly record.

A second, more complex example involves a case already mentioned above (Case 39), for which Leuven University Press issued an electronic *EC* (Anonymous 2010a). The original occasion of the publication of the volume was an exhibition at an academic library. Copies of the volume there were supplemented with a leaflet notification printed on the library letterhead and signed by the editor that went well beyond the above-noted neutral language of Leuven University Press's online *EC*. In the leaflet notification, the editor expresses regret that "this plagiarism has come to light" after publication and states that the chapter "contains many passages and footnotes which have been translated and quoted (without any due reference)" from a source that is then identified with full bibliographical information (Verweijj 2010 [Case 39]). For this one case of plagiarism, therefore, there are two notifications, one *EC* from the publisher, and one leaflet from the editor, differing significantly in determination. Unlike the publisher's *EC*, the editor's putative *SR* unequivocally confirms that plagiarism has been committed. This putative *SR*, while having the institutional sponsorship of the academic library, lacks the institutional support of the publisher and has very limited public circulation. Again, without the publisher's support, one may question whether the editor's notification fulfills the conditions of a genuine *SR*. Perhaps such statements can be best understood as informal retractions in the absence of any involvement by the relevant publisher.

A related issue is whether public apologies or admissions of wrongdoing by the very agents who have committed research and publishing misconduct are genuine *SR*s that correct the scholarly record. Acts of this type appear to be more common in the natural sciences, and a recent study observes that "most authors' published acknowledgments of scientific misconduct seem to minimize culpability by means of the strategic use of language" (Souder 2010: 175). Even if such published statements are not considered to be genuine *SR*s, they could be analyzed in terms of determination, credit, and availability. In recent years, the editors of several philosophy journals—*Synthese*, *Journal of Political Philosophy*, *Hypatia*, *Dialogue*, and *Philosophia Mathematica*—each have published online apologies for certain editorial failures in unrelated cases. These apologies were not published within the pages of the journals, but issued in various online fora (Goodin et al. 2017; Flaherty 2017; Sher et al. 2016; Thomas 2016; Marion 2012). Optimally, the version of record of the published item will be linked to any genuine public apology or admission of wrongdoing issued by the author of record after publication, so that researchers will be made aware if the authors of record or journals no longer fully endorse the published material.

4.3.5 *Problems with Minimal and Medial* SR*s*

*SR*s that are minimal or medial in determination, credit, or availability serve as partial corrections of the scholarly record. Undoubtedly, partial corrections are better than no corrections at all, yet they create some problems for researchers. An *SR* that is minimal or medial in determination may leave a reader in doubt about whether a serious violation of scholarly and publishing integrity has occurred. In a worst-case scenario, a researcher may continue to cite the plagiarized article on the mistaken assumption that the evidence for the plagiarism is uncertain or questionable. (For the same reasons, researchers may adopt a similar response toward articles and book chapters for which *EC*s have been issued.) Furthermore, an *SR* that is less than maximal with regard to credit may place researchers in a bind, particularly if their scholarship has relied heavily on the plagiarized work. Such researchers may feel constrained to cite the plagiarized work in recognition of their intellectual debts if they are unable to determine the original source materials themselves. Finally, those *SR*s that are less than maximal with regard to availability increase the likelihood that the correction of the scholarly record will remain unknown to the scholarly community.

4.4 Reprinting a Volume After the Removal of Plagiarized Material

The third major category of responses by presses to the publication of plagiarized material is the printing of a redacted edition. Some of the largest academic publishers in philosophy have pursued this route, and often the redacted version is accompanied by an *EC* or *SR*. Many of the notifications in these cases, however, do not score well with regard to determination, credit, or availability.

Oxford University Press provides a distinctive example falling under this third category of responses. Five years after the original publication of a *Festschrift* that appeared under the Clarendon Press imprint, Oxford reissued the volume with two significant changes: the plagiarized chapter was omitted, and a brief note appeared in the volume. The note, repeated three times (in the table of contents and at the beginning of two of the indices) states, "Chapter 20 [...] was removed from this book in 2010" (Salles 2010: ix, 517, 543 [Case 26]). Since the removed chapter had been the last in the volume, a renumbering of pages for the previous chapters was unnecessary, but the volume was shorter by 44 pages, and the back matter had to be adjusted accordingly. The note can be considered a rather marginal *EC*, as it simply describes the action of removal taken by the press without giving an overt warning or explanation; a reader cannot infer from the note that the omitted chapter was plagiarized. Oxford's response does not qualify as a full correction of the scholarly record, but it does impede additional circulation of the plagiarized material. Notably, Oxford also substituted the redacted version of the book for the original in various electronic platforms (including the book preview at Amazon.com), thereby further restricting the proliferation of the plagiarized material.

Cambridge University Press printed redacted versions of four volumes originally containing plagiarized chapters. It issued *SR*s for these four chapters and in three cases commissioned new essays to replace the plagiarized ones. Two of the four plagiarized chapters appeared in volumes from the Cambridge Companions to Philosophy series. The volume containing the earlier of the two was eventually reprinted in a second edition, and the plagiarized chapter was one of several replaced with new essays. Even though the first edition remains on the press's Cambridge Core platform, electronic access to the chapter has been replaced with the brief notification, "The contents of this chapter have been removed for rights reasons" (Anonymous 2010e [Case 11]). A reviewer of the second edition has been critical of a perceived silence by the press concerning the plagiarism: "The editors, however, fail to mention that one of the contributors to the first edition [...] has recently been exposed as a plagiarist [...]. Cynics might therefore wonder whether claims about the outdatedness of the first edition were deemed—in the interests of the Press's *bella figura*—an unfortunate requisite of the consequent and conventional *damnatio memoriae* of the offender" (Rist 2015: 444). For the second plagiarized Cambridge Companion chapter, a statement was placed on the publisher's website under the "Table of Contents" tab indicating that the chapter "has been omitted from the present printing for legal reasons," and the book is listed as unavailable for ordering

(Anonymous 2010f [Case 31]). Also in this case the chapter on the press's electronic platform is declared unavailable "for rights reasons" (Anonymous 2010g). The short online notifications declaring the chapters to be unavailable "for rights reasons" or "for legal reasons" satisfy the conditions of an *SR*, but they must be scored as minimal in both determination and credit, yet maximal in availability, as they are freely accessible online.

The two other plagiarized chapters published by Cambridge University Press appeared in a two-volume set in the Cambridge History of Philosophy series. The publisher reissued the set in a revised paperback edition 5 years after the original publication, inserting newly commissioned chapters in place of the plagiarized ones, with no interruption of prior pagination. For the electronic version of the set on the Cambridge Core platform, files for both plagiarized chapters were replaced with *SR*s similar to the ones offered in the Cambridge Companion cases, as each one states, "The contents of this chapter have been removed for legal reasons" (Anonymous 2011f [Case 37], g [Case 41]). Notably, a more detailed explanation stating that the two chapters "are now known to have been substantially plagiarized" appears on the academic editor's university webpage, with links to PDFs of the newly rewritten chapters in their non-final proof versions (Pasnau 2018). Again, one can consider whether notifications made by editors independently of a publisher constitute genuine *SR*s; in this case, significantly more information is provided by the editor than by the publisher. Table 4.1 assesses the quality of the informal *SR*s issued in these cases.

The revised paperback version of the 2-volume set makes no mention of the plagiarism and does not identify in what way the revised edition has been revised; for several years the publisher had been selling both the original, unredacted edition and the revised paperback version on its website. A reviewer of the original edition has also been critical of the press for a perceived lack of clarity regarding the response to plagiarism, stating: "One of the contributors [...] plagiarized the bulk of his contributions, which will be replaced in the forthcoming paperback version and future editions of the hardcover version (a fact Cambridge University Press should

Table 4.1 Assessing the quality of independent notifications for plagiarism issued by academic editors

	Case	Supports demonstrated claim of plagiarism (determination)	References original source (credit)	Extent of promulgation (availability)
Faesen et al. (2011)	42	+	+	+
Verweijj (2010)	39	+	+	–
Pasnau (2018)	37	+	–	+
Pasnau (2018)	41	+	–	+

Key: – Minimal, = Medial, + Maximal

do rather more to publicize)" (King 2012: 612). Such a criticism indicates a fault line regarding the origins of *SR*s and the correction of the scholarly record. Does responsibility for correction fall primarily with the editor or with the publisher?

Both editors and publishers may pursue corrections of the scholarly record. In one instance, a notification of plagiarism for a chapter in a collection published by Springer appeared in two places: on the publisher's website (Pattaro et al. 2010b [Case 34]) and in a journal published by Wiley-Blackwell (Pattaro et al. 2010a [Case 34]). The three authors of the two notifications were the editors of the volume containing the plagiarized chapter. Had the notification not also appeared on the Springer website, it likely could not be considered an *SR*, unless one judges that editors can issue *SR*s independently of the support of the original publisher. The three editors stated that they hoped Springer would reissue the volume with a newly commissioned replacement for the plagiarized chapter, one that would be written "by a scholar who [...] should be impervious to the temptations of plagiarism" (Pattaro et al. 2010a, b). Such a scenario did materialize, as Springer issued a second edition of the volume with a newly written chapter 8 years after the original edition and noted the plagiarism in a new preface (Miller Jr and Biondi 2015: xix). Furthermore, a notice of retraction was placed in the electronic version of the first edition, explaining that the retraction was warranted "because the text was not written [...], but was copied, including some passages copied verbatim" (Anonymous 2015d: E1), and a diagonal watermark "RETRACTED" was placed over each page of the article. The retraction also now appears in the Google Books online preview for the first edition of the volume.

The practice of issuing redacted editions may work well with books, but the practice does not easily translate to journals. One variation, however, occurs when an issue of a journal is republished as a stand-alone volume. This transition from journal issue to edited collection gives publishers the opportunity to omit plagiarized material. Wiley-Blackwell followed this procedure in one case (Braun and Vallance 2011 [Case 35]). As noted above, the plagiarized article had already been retracted with an *SR* printed in the journal. In the redacted printing as an edited collection, there is no reference to the omitted chapter, and the editors' introduction has been subtly modified to avoid references to it.

4.5 Halting the Sale of the Plagiarized Material

The fourth major category of responses by publishers is to stop the sale of plagiarized works. As was mentioned above, in some cases publishers have done so in conjunction with printing redacted versions or issuing notifications. Other publishers have stopped the sale without any other public actions. Routledge followed the latter procedure in 2011 for four volumes that each contained one plagiarized chapter (Cases 5, 6, 7, and 16). Also, Routledge suspended electronic access to one plagiarized encyclopedia entry (Case 2). Other publishers that have stopped sales as the sole public response to a case of plagiarism are Oxford University Press (Case 1) and Harrassowitz Verlag (Case 33).

Stopping sales undoubtedly limits the additional distribution of plagiarized material that continued sales would have brought about. Yet simply to cease the proliferation of plagiarized material is not by itself a correction of the scholarly record, as the repository of published research thereby remains damaged without amendment. Stopping sales alone does not assist researchers in knowing that the published material is corrupted by plagiarism. To most researchers, a publisher's act of stopping the sale of a problematic book without issuing an *EC* or *SR* will likely resemble the normal life cycle of a book going out of print. Not knowing that the previously published material fails to meet scholarly standards, researchers may continue to cite it in new research. Consider the following non-ideal situations involving plagiarized chapters for which sales were halted but no *EC*s or *SR*s were issued:

- One of the unretracted plagiarized Routledge chapters (Case 5) is cited approvingly in the 2014 redacted edition of the Cambridge History of Philosophy series, a redaction that itself was occasioned solely because of the inclusion of two plagiarized chapters in the original edition (Case 37, Case 41). Here a work redacted for plagiarism still commends to its readers other plagiarized work.
- The principal victim whose work was misappropriated in another unretracted plagiarized Routledge chapter (Case 7) turned to Amazon.com's "Customer Reviews" dialogue box to warn readers that the unretracted chapter "is almost totally plagiarized" from his work (Sinnema 2012). This notification by the victimized author is maximal in determination, credit, and availability, but it lacks the institutional support of the publisher.
- The plagiarism of unretracted Case 7 is now openly discussed in print by scholars. In a published collection of essays, one author writes that Case 7 "has been denounced as an extensive plagiarism" and then adds, "This piece has not, at present, been retracted by the editor and publisher" (Paganini 2016: 232, n. 41).
- The same unretracted plagiarized Routledge chapter (Case 7) continues to be cited approvingly by other scholars, including one book published by Routledge in 2017, another by Wiley-Blackwell in 2017, another by The University of Chicago Press in 2015, and two volumes published by Brill in 2013. One of the Brill volumes approvingly cites both the unretracted chapter and the original source material in the same footnote, with no mention that the former plagiarizes the latter.
- The unretracted plagiarized Harrassowitz Verlag chapter from 2007 (Case 33) was republished in part in the abovementioned collection by Felix Meiner Verlag in 2011 (Case 43), thereby resulting in the same plagiarized material being published twice by different academic publishers, with a notification issued only by the latter publisher.

One may wonder whether non-ideal situations such as these could be substantively mitigated if publishers were to issue maximal *SR*s without delay. A further problem with simply stopping sales without a published correction of the scholarly record can be the decision to reprint years later, a decision sometimes made by a different publisher that has acquired the rights to the original publication.

4.6 Taking No Public Action

In addition to the options of issuing *EC*s or *SR*s, printing redacted volumes, and halting sales, a remaining option for presses that have published plagiarized work is simply to take no public action at all. In scientific disciplines, this option has been severely criticized in discussions about maintaining the integrity of the scholarly record. In an article subtitled "The Ethics of not Correcting Errors in the Scientific Literature," one researcher has argued that "anything less than correcting an error that has been pointed out should be considered misconduct" (Teixeira da Silva 2016: 225). Another theorist has argued, persuasively my view, that "the most unethical response to known misconduct is not retracting at all," and that a refusal to retract is "editorial misconduct" (Shelomi 2014: 55, 51).

In a compendium of the best industry standards for publishing in the sciences, Irene Hames argues that an absolute and overriding "golden rule" of publishing is that "suspected or alleged misconduct must not be ignored" (2007: 173). She mentions plagiarism in the sciences as requiring action on the part of editors and publishers to protect the system of knowledge:

> It is not acceptable for Editors to take no action and ignore suspicions or accusations. If any suspicions are found to be valid, the Editor should make sure that appropriate corrective measures are put into place. Readers should be informed of any work that is found to be fraudulent, fabricated, plagiarized, or the result of serious misconduct. (162)

> When evidence is found that a paper has a serious problem, for example, that it contains fraudulent data or [...] contains plagiarized material, a "retraction" is needed. (194)

Not only is there a clear understanding of how to deal with plagiarism cases in the sciences, but there is also sophisticated theorizing about how best to ensure the integrity of the scholarly record. For example, one recent study proposes that fraudulent scientific publications should be conceptualized as viruses capable of contaminating the scientific body of knowledge. That study borrows from public health models (that seek to prevent the spread of biological viruses) to make analogous recommendations for protecting the scientific corpus of published work (Montgomery and Oliver 2017).[6] There is no reason for the discipline of philosophy not to match the best practices and sophisticated theorizing of the sciences for correcting its own scholarly record.

Table 4.2 presents the responses of the publishers for each of the 43 cases. Taking no public action at all is somewhat common among publishers in philosophy when faced with demonstrated plagiarism: in the 43 cases of demonstrated plagiarism

[6] For another appeal to a public health model for dealing with research misconduct, see Nylenna and Simonsen 2006.

Table 4.2 Responses by publishers in philosophy to an instance of serial plagiarism (Cases 1–43)

	Case	Expression of concern	Statement of retraction	Redacted edition	Sales halted	No public action
Ashgate	12	●			●	
Brepols	30		●			
Cambridge UP	11		●	●		
Cambridge UP	31		●	●	●	
Cambridge UP	37		●	●		
Cambridge UP	41		●	●		
Central European UP	23					●
De Gruyter	9					●
De Gruyter	17		●			
Felix Meiner	43	●				
Fordham UP	28					●
Four Courts	38					●
Four Courts	40					●
Harrassowitz	33				●	
Leuven UP	15	●				
Leuven UP	24	●				
Leuven UP	39	●				
Notre Dame	14		●			
Oxford UP	1				●	
Oxford UP	26	●		●		
Palgrave Macmillan	18					●
Peeters	3		●		●	
Peeters	4		●		●	
Peeters	19					●
Peeters	29		●		●	
Peeters	36					●
Peeters	42					●
PIMS	27		●			
Routledge	2				●	
Routledge	5				●	
Routledge	6				●	
Routledge	7				●	
Routledge	16				●	
Scribner	20					●
Scribner	21					●
Scribner	22					●
SISMEL	32		●			
Springer	25		●			
Springer	34		●	●		

(continued)

Table 4.2 (continued)

	Case	Expression of concern	Statement of retraction	Redacted edition	Sales halted	No public action
St. Martin's	8					●
Wiley-Blackwell	10		●			
Wiley-Blackwell	13		●		●	
Wiley-Blackwell	35		●		●	

under consideration here, 8 publishers have pursued this policy, resulting in no public response for 13 plagiarized works.

From an outsider's vantage point, it is not possible to identify with certainty the reasons for a publisher's lack of public action after being notified of a demonstrated case of plagiarism. Most striking are those instances where a publisher corrects the scholarly record decisively in some of its cases but takes no public action in others. As noted above, Peeters Publishers issued *SR*s that are maximal in determination, credit, and availability for three of its plagiarized works (Cases 3, 4, and 29), yet it took no public action for three others (Cases 19, 36, and 42). The only obvious feature differentiating the six plagiarized works is that the former group consists of journal articles, while the latter group consists of book chapters. Peeters appears to distinguish between publication genres when issuing corrections of the scholarly record. No such distinction, however, explains why De Gruyter issued the above-mentioned *SR* for one plagiarized book chapter (Case 17) but took no action at all for another plagiarized chapter (Case 9). In the digital age, when publishers have website pages for their books, and when many publishers offer chapters electronically on various platforms, the rationale for treating book chapters differently from articles in the correction of the scholarly record is unclear.

In addition to damaging the repository of published research, unretracted plagiarized articles and book chapters can lead to secondary corruptions of the scholarly enterprise. An unretracted plagiarized Fordham University Press article from 2006 (Case 28) was republished before the plagiarism was discovered in part in a collection by Leuven University Press in 2009 (Case 39), thereby again resulting in a situation where the same plagiarized material is added twice to the repository of published research by different academic publishers. The republication of unretracted plagiarized material is perhaps unsurprising, as, in the considered judgment of some observers, "those who engage in plagiarism tend to do so in multiple publications; plagiarism tends not to be an isolated one-off event" (Fox and Beall 2014: 346). Another corruption can occur when a volume is reviewed in academic journals. Late-appearing academic book reviews of volumes containing unretracted plagiarized chapters published by Four Courts Press (Case 38, Case 40), for example, made no reference to the plagiarism, thereby resulting in reviewers who thereby commended plagiarized material to their readers.

Since both academic publishers and scholars have a strong self-interest in a trustworthy repository of published research, one might be puzzled at finding instances of unretracted articles and book chapters for which plagiarism has been demonstrated.

Without insider information, however, one cannot know why a publisher fails to issue a notification in a given case. The issuance of a notification generally involves the work of a press editor employed by the publisher and an academic editor (usually a university researcher) who is responsible for editing a journal or a volume of collected chapters. Others may be involved, such as a journal's editorial board, a book series editor, or the legal department of a publisher. The involvement of many individuals may create impediments to the issuance of a notification in a given case. Consider the following possible scenarios:

- A press editor expresses privately to a notifying party that the evidence of plagiarism is unassailable, but the press's legal department simply advises that no action be taken.
- An academic editor tells a notifying party that a close personal relationship with the plagiarist precludes involvement in a given case, as the editor believes that maintaining ties of friendship overrides any obligation to participate in the correction of the scholarly record.
- A press editor responds to a notifying party's request for an *SR* by saying that a given case of plagiarism is sufficiently well known in the discipline and that pursuing the matter with a formal retraction would be cruel to the plagiarist.
- An academic editor defers all consideration of an *SR* to the publisher, but the press editor does not want to act without the participation of the academic editor.
- A press editor believes that corrections of the scholarly record should occur in publications by academics in the field, rather than through *SR*s issued by the publisher.
- A press editor tells a notifying party that the issuing of an *SR* would simply be too damaging to the press's reputation.

In addition, a press editor or an academic editor may assume that there are bad motives on the part of the notifying parties, and therefore no action is taken. Such a scenario is discussed in whistleblowing literature in the natural sciences, where one finds the following response: "if the facts reported are true, the motive of the whistleblower should not matter" (Gunsalus 1998: 53). Even if the various editors (rightly or wrongly) assume bad motives on the part of notifying parties, such an assumption should be irrelevant to the decision to correct the scholarly record. In the end, without insider information one is bound to speculate in cases where publishers have been presented with demonstrated cases of plagiarism but have not responded publicly with a correction of the scholarly record.[7]

[7] Some academic editors in biomedical disciplines explicitly encourage readers to disclose evidence of suspected wrongdoing. In an article titled, "Plagiarism and other Scientific Misconducts," editors K. Höffken and H. Gabbert, write "Please support us with our efforts. Do not hesitate to inform us about any irregularity, violation or infringement" (2009: 328).

4.7 The Varying Quality of Corrections in Philosophy

Table 4.3 summarizes the quality of *SR*s and *EC*s that were issued for 23 of the 43 cases examined here. As I have indicated, there is relatively little uniformity among publishers in philosophy in responding to the publication of plagiarized articles and book chapters. In comparison with the natural sciences, the discipline of philosophy often falls short of the standard practices for correcting the scholarly record. Many

Table 4.3 Assessing the quality of notifications for plagiarism issued by publishers in philosophy

	Case	Supports demonstrated claim of plagiarism (determination)	References original source material (credit)	Extent of promulgation (availability)	Types of notification
Ashgate	12	ø	+	–	□
Brepols	30	+	+	+	■
Cambridge UP	11	–	–	+	■
Cambridge UP	31	–	–	+	■
Cambridge UP	37	–	–	+	■
Cambridge UP	41	–	–	+	■
De Gruyter	17	+	–	+	■
Felix Meiner	43	ø	+	+	□
Leuven UP	15	ø	+	=	□
Leuven UP	24	ø	+	=	□
Leuven UP	39	ø	+	+	□
Notre Dame	14	–	+	+	■
Oxford UP	26	ø	–	–	□
Peeters	3	+	+	+	■
Peeters	4	+	+	+	■
Peeters	29	+	+	+	■
PIMS	27	+	+	–	■
SISMEL	32	=	+	=	■
Springer	25	+	+	+	■
Springer	34	+	+	+	■
Wiley-Blackwell	10	=	–	–	■
Wiley-Blackwell	13	=	–	+	■
Wiley-Blackwell	35	+	+	–	■

Key: ø Neutral, – Minimal, = Medial, + Maximal, □ Expression of concern, ■ Statement of retraction of the corrections in philosophy fail to be maximal in determination, credit, or availability.

To sum up, fully maximal *SR*s can be viewed as valuable instances of post-publication peer review that strengthen a publisher's relationship to the body of published research. The rubric proposed in this chapter for analyzing *SR*s could be used by editors and publishers for crafting *SR*s for new cases of plagiarism, thereby ensuring a high standard for future corrections of the scholarly record and bringing the publication practices and standards of the discipline of philosophy in line with those of the natural sciences.

References

Allison, David B., Andrew W. Brown, Brandon J. George, and Kathryn A. Kaiser. 2016. A tragedy of errors. *Nature* 530: 27–29. https://doi.org/10.1038/530027a. Accessed 6 July 2018.

Anonymous. 1992. Erratum. *Faith and Philosophy* 9(3): 409.

———. 2009. A note from the editorial board. *Recherches de Théologie et Philosophie médiévales* 76(2): v–vi. http://www.peeters-leuven.be/pdf/RTPM7602009.pdf. Accessed 6 July 2018. (Case 4).

———. 2010a. *Notification*. Leuven: Leuven University Press. http://upers.kuleuven.be/en/book/9789058677761. Accessed 6 July 2018. (Case 39).

———. 2010b. *Notification*. Leuven: Leuven University Press. (Case 15).

———. 2010c. *Notification*. Leuven: Leuven University Press. (Case 24).

———. 2010d. Retraction. *Renaissance Studies* 24 (5): 620. https://doi.org/10.1111/j.1477-4658.2010.00694.x. Accessed 6 July 2018 (Case 35).

———. 2010e. *Cambridge Core. The Cambridge companion to Augustine*. https://www.cambridge.org/core/books/the-cambridge-companion-to-augustine/augustine-and-medieval-philosophy/210A2841B1D0E452F6B7D41557898A18. Accessed 6 July 2018. (Case 11).

———. 2010f. Table of contents. *The Cambridge companion to early modern philosophy*. http://www.cambridge.org/us/academic/subjects/philosophy/early-modern-philosophy/cambridge-companion-early-modern-philosophy?format=HB. Accessed 6 July 2018. (Case 31).

———. 2010g. *Cambridge Core. The Cambridge companion to early modern philosophy*. https://www.cambridge.org/core/books/the-cambridge-companion-to-early-modern-philosophy/scholastic-schools-and-early-modern-philosophy/F8FC655C5880B4A1E956965836811D05. Accessed 6 July 2018. (Case 31).

———. 2011a. *Hinweis*. Hamburg: Meiner Verlag für Philosophie. https://meiner.de/departure-for-modern-europe-aufbruch-in-das-moderne-europa-9498.html. Accessed 6 July 2018. (Case 43).

———. 2011b. *Notification*. Aldershot: Ashgate. (Case 12).

———. 2011c. *Retracted for plagiarism/Wegen Plagiatsvorwürfen zurückgezogen*. Berlin: De Gruyter. https://doi.org/10.1515/9783110204544.6.277. Accessed 6 July 2018. (Case 17).

———. 2011d. Plagiarized/geplagieerd. *JSTOR*. http://www.jstor.org/stable/40889945. Accessed 6 July 2018. (Case 29).

———. 2011e. Retraction. *Proceedings of the Aristotelian Society* 111 (3.3): 519. https://doi.org/10.1111/j.1467-9264.2011.00322.x/epdf. Accessed 6 July 2018. (Case 10).

———. 2011f. *Cambridge Core. The Cambridge history of medieval philosophy*. https://www.cambridge.org/core/books/the-cambridge-history-of-medieval-philosophy/the-care-of-souls-and-practical-ethics/163AABF637061563F0F4E9E6BE8CD3A5. Accessed 6 July 2018. (Case 37).

———. 2011g. *Cambridge Core. The Cambridge history of medieval philosophy*. https://www.cambridge.org/core/books/the-cambridge-history-of-medieval-philosophy/philosophy-and-theology/5F444AD024722833B7D0B8DC1F6BF719. Accessed 6 July 2018. (Case 41).

———. 2012. We wish to inform our readership [...]. *Documenti e Studi sulla Tradizione Filosofica Medieval* 23: vi. (Case 32).

———. 2015a. *It has been brought to our attention [...]*. Leiden: Brill. https://brill.com/view/title/159. Accessed 6 July 2018.

———. 2015b. This chapter has been retracted. *Wiley Online Library*. https://doi.org/10.1002/9780470998847.ch2. Accessed 6 July 2018. (Case 13).

———. 2015c. *Retraction notice*. Cham: Springer. https://doi.org/10.1007/1-4020-3001-0_4. Accessed 6 July 2018. (Case 25).

———. 2015d. *Retraction note*. Cham: Springer. (Case 34).

Ballor, Jordan J. 2014. Plagiarism in a digital age. *Journal of Markets and Morality* 17 (2): 349–352.

Bilbrey, Emma, Natalie O'Dell, and Jonathan Creamer. 2014. A novel rubric for rating the quality of retraction notices. *Publications* 2 (1): 14–26. https://doi.org/10.3390/publications2010014. Accessed 6 July 2018.

Blair, Ann. 2007. Errata lists and the reader as corrector. In *Agent of change: Print culture studies after Elizabeth L. Eisenstein*, ed. Sabrina Alcorn Baron, Eric N. Lindquist, and Eleanor F. Shevlin, 21–41. Amherst: University of Massachusetts Press.

Braun, Harald E., and Edward Vallance, eds. 2011. *The Renaissance conscience*. Malden: Wiley-Blackwell.

Dougherty, M.V. 2017. Correcting the scholarly record in the aftermath of plagiarism: A snapshot of current-day publishing practices in philosophy. *Metaphilosophy* 48 (3): 258–283.

Dougherty, M.V., Pernille Harsting, and Russell L. Friedman. 2009. 40 cases of plagiarism. *Bulletin de Philosophie médiévale* 51: 350–391.

Editorial Board of ETL. 2010. Note from the editorial board/Note de la redaction. *Ephemerides Theologicae Lovanienses* 86 (1): iii. http://www.peeters-leuven.be/pdf/ETL8601EditNote.pdf. Accessed 6 July 2018. (Case 3).

Esposito, Costantino, and Pasquale Porro. 2009. A note from the editors. *Quaestio* 9: 453–454. https://doi.org/10.5555/J.QUAESTIO.6.780188810102040000008030507080778. Accessed 6 July 2018. (Case 30).

Faesen, Rob, Bernard Deprez, and Leo Kenis. 2011. Plagiarism. *Jesuitica*, November 3. https://www.jesuitica.be/info-item/30. Accessed 6 July 2018. (Case 42).

Flaherty, Colleen. 2017. *Hypatia* disavows the *Hypatia* disavowal. *Inside Higher Ed*, May 19. https://www.insidehighered.com/news/2017/05/19/journals-board-directors-offers-its-support-editors-decision-publish-article. Accessed 6 July 2018.

Fox, Mark, and Jeffrey Beall. 2014. Advice for plagiarism whistleblowers. *Ethics and Behavior* 24 (5): 341–349.

Goodin, Robert E., Lea Ypi, Nicholas Southwood, and Christian Barry. 2017. Open letter from the editors of the *Journal of Political Philosophy. Philosopher*. May 25. https://politicalphilosopher.net/2017/05/25/open-letter-from-the-editors-of-the-journal-of-political-philosophy. Accessed 6 July 2018.

Gunsalus, C.K. 1998. How to blow the whistle and still have a career afterwards. *Science and Engineering Ethics* 4 (1): 51–64.

Haldane, John. 2011. Editor and publisher's note. University of Notre Dame Press. http://undpress.nd.edu/books/P00799#description. Accessed 6 July 2018. (Case 14).

Hames, Irene. 2007. *Peer review and manuscript management in scientific journals: Guidelines for good practice*. Malden: Blackwell Publishing.

Höffken, K., and H. Gabbert. 2009. Plagiarism and other scientific misconducts. *Journal of Cancer Research and Clinical Oncology* 135 (3): 327–328.

Hu, Guangwei. 2017. Authorship of retraction notices: 'If names are not rectified, then language will not be in accord with truth'. *Publications* 5 (2): 1–3. https://doi.org/10.3390/publications5020010. Accessed 6 July 2018.

Hussinger, Katrin, and Maikel Pellens. 2017. Guilt by association: How scientific misconduct harms prior collaborators. *Zentrum für Europäische Wirtschaftsforschung Discussion Papers* 17–051. http://ftp.zew.de/pub/zew-docs/dp/dp17051.pdf. Accessed 6 July 2018.

ICMJE. 2016. Recommendations for the conduct, reporting, editing, and publication of scholarly work in medical journals. *International Committee of Medical Journal Editors*. http://www.icmje.org/icmje-recommendations.pdf. Accessed 6 July 2018.

King, Peter. 2012. Review of *The Cambridge history of medieval philosophy*, rev. ed. *Journal of the History of Philosophy* 50(4): 612–613.

Kramer, Bernd. 2018. Universität Göttingen: Aufstieg und Fall des Dr. M. *Die Zeit*. January 5. http://www.zeit.de/campus/2018/01/universitaet-goettingen-betrug-dozent-professor. Accessed 6 July 2018.

Marcus, Adam, and Ivan Oransky. 2012. Bring on the transparency index. *Scientist*. August 1. http://www.the-scientist.com/?articles.view/articleNo/32427/title/Bring-On-the-Transparency-Index. Accessed 6 July 2018.

———. 2014. What studies of retractions tell us. *Journal of Microbiology and Biology Education* 15 (2): 151–154. https://doi.org/10.1128/jmbe.v15i2.855. Accessed 6 July 2018.

Marion, Mathieu. 2012. Dialogue co-editor apologizes. *Feminist Philosophers*. November 4. https://feministphilosophers.wordpress.com/2012/11/04/dialogue-co-editor-apologizes. Accessed 6 July 2018.

Miller, Fred D., Jr., and Carrie-Ann Biondi. 2015. Preface to volume 6 (second edition). In *A history of the philosophy of law from the ancient Greeks to the scholastics*, ed. Fred D. Miller Jr. and Carrie-Ann Biondi, 2nd ed., xix. Dordrecht: Springer. (Case 34).

Montgomery, Kathleen, and Amalya L. Oliver. 2017. Conceptualizing fraudulent studies as viruses: New models for handling retractions. *Minerva* 55 (1): 49–64.

Newhauser, Richard, and Bill Harnum. 2010. *With deep regret [...]*. Toronto: Pontifical Institute of Mediaeval Studies. (Case 27).

Newman, Melanie. 2010. Plagiarism probe sees UK scholar quit Belgian post. *Times Higher Education*, March 11. http://www.timeshighereducation.co.uk/410670.article. Accessed 6 July 2018.

Nylenna, Magne, and Sigmund Simonsen. 2006. Scientific misconduct: A new approach to prevention. *The Lancet* 367 (9526): 1882–1884.

Paganini, Gianni. 2016. Thomas Hobbes against the Aristotelian account of the virtues and his Renaissance source Lorenzo Valla. In *Early modern philosophers and the Renaissance legacy*, ed. Cecilia Muratori and Gianni Paganini, 221–237. Cham: Springer.

Paglieri, Fabio. 2015. Reflections on plagiarism. *Topoi* 34 (1): 1–5.

Palus, Shannon. 2016. Philosopher earns 14th retraction for plagiarism. *Retraction Watch*, June 8. http://retractionwatch.com/2016/06/08/philosopher-earns-13th-retraction-for-plagiarism. Accessed 6 July 2018.

Pasnau, Robert. 2018. Pasnau in print. http://spot.colorado.edu/~pasnau/inprint. Accessed 6 July 2018. (Case 37, Case 41).

Patil, Madhuri. 2015. Retraction: Retraction. *Journal of Human Reproductive Sciences*: 187. https://doi.org/10.4103/0974-1208.170594. Accessed 6 July 2018.

Pattaro, Enrico, Fred Miller Jr., and Carrie-Ann Biondi. 2010a. Plagiarism. *Ratio Juris* 23 (3): 435–436. (Case 34).

———. 2010b. *Plagiarism*. Dordrecht: Springer. http://www.goo.gl/1kjfXy. Accessed 6 July 2018. (Case 34).

Rist, John. 2015. Review of *The Cambridge companion to Augustine*, 2nd ed. *Journal of Theological Studies 66* (1): 443–446.

Salles, Ricardo. 2010. Chapter 20. In *Metaphysics, soul, and ethics in ancient thought: Themes from the work of Richard Sorabji*, ed. Ricardo Salles, 517–543. Oxford: Clarendon Press. (Case 26).

Shelomi, Matan. 2014. Editorial misconduct—definition, cases, and causes. *Publications* 2: 51–60. https://doi.org/10.3390/publications2020051. Accessed 6 July 2018.

Sher, Gila, Otávio Bueno, and Wiebe van der Hoek. 2016. Statement from *Synthese* editors. *Daily Nous: News for and About the Philosophy Profession*. January 20. http://dailynous.com/2016/01/27/statement-from-synthese-editors-moratorium-on-special-issues. Accessed 6 July 2018.

Sinnema, Donald. 2012. Article is plagiarism. *Amazon.com*. November 5. http://www.amazon.com/Humanism-Modern-Philosophy-Studies-History/dp/0415186161. Accessed 6 July 2018. (Case 7).

Souder, Lawrence. 2010. A rhetorical analysis of apologies for scientific misconduct: Do they really mean it? *Science and Engineering Ethics* 16 (1): 175–184.

Spodenkiewicz, Paweł. 2004. Przykra afera plagiatowa na Uniwersytecie Łódzkim. *Naszemiasto*, December 18. http://lodz.naszemiasto.pl/archiwum/przykra-afera-plagiatowa-na-uniwersytecie-lodzkim,428478,art,t,id,tm.html. Accesssed 6 July 2018.

Stern, Victoria. 2018. University defends researcher accused of plagiarizing former Pope. *Retraction Watch*, January 31. https://retractionwatch.com/2018/01/31/university-defends-researcher-accused-plagiarizing-former-pope. Accessed 6 July 2018.

Teixeira da Silva, Jaime A. 2016. An error is an error… is an erratum: The ethics of not correcting errors in the science literature. *Publishing Research Quarterly* 32 (3): 220–226.

Teixeira da Silva, Jaime A., and Judit Dobránszki. 2017. Notices and policies for retractions, expressions of concern, errata and corrigenda: Their importance, content, and context. *Science and Engineering Ethics* 23 (2): 521–554.

Thomas, Robert. 2016. Pronouns and *Philosophia Mathematica*. *Phlox*, March 6. https://phlox-group.wordpress.com/2016/03/06/pronouns-and-philosophia-mathematica. Accessed 6 July 2018.

Verweijj, Michiel. 2010. *The article […]*. Brussels: Koninklijke Bibliotheek van België. (Case 39).

Visker, R. 2010. Note from the editorial board/Mededeling van de redactie. *Tijdschrift voor Filosofie* 72 (1): 5–6. (Case 29).

Wager, Elizabeth, Virginia Barbour, Steven Yentis, Sabine Kleinert. 2009. Retraction guidelines. *Committee on Publication Ethics*. http://publicationethics.org/files/retraction guidelines.pdf. Accessed 6 July 2018.

Weber-Wulff, Debora. 2014. *False feathers: A perspective on academic plagiarism*. Heidelberg: Springer.

Weinberg, Justin. 2014. A case of extensive plagiarism. *Daily Nous: News for and About the Philosophy Profession*, November 5. http://dailynous.com/2014/11/05/a-case-of-extensive-plagiarism-guest-post. Accessed 6 July 2018.

Chapter 5
Academic Whistleblowing

Abstract Despite the essential role that academic whistleblowers serve in initiating the oftentimes lengthy process of correcting the scholarly record, individuals who disclose evidence of suspected plagiarism are often subject to considerable backlash. To be sure, the evidence they provide, even when impeccable, can create a significant workload of verification for editors and publishers, as well as for research integrity officers at the institutional homes of the suspected plagiarists. I examine the benefits and hazards of multi-targeted whistleblowing and discuss the harassment and witness intimidation typically experienced by those who blow the academic whistle in good faith. The increasing awareness among researchers and institutional authorities that to harass whistleblowers is itself a form of misconduct reflects an important recent shift in academic culture. On the other hand, academic whistleblowers in recent times have been described as post-publication vigilantes for their efforts in securing corrections of the scholarly record, so the professional dangers of academic whistleblowing should not be understated.

Keywords Whistleblowing · Retaliation · Harassment · Reprisals

While preparing to write an invited book review in early 2012 of a recently published collection of papers on Renaissance philosophy, I noticed that one of the chapters contained a short passage that was strikingly familiar. Initially I suspected that the author of record for the chapter was simply re-using material from previously-published work. Was this a case of so-called "self-plagiarism" or text recycling? The name of the author of record for the chapter was unfamiliar to me, however. And then I remembered: I had met the passage in a specialized monograph on Renaissance philosophy published 35 years earlier that I had read recently, and the author of that monograph was not identical with the person listed as the author of record for the book chapter. Retrieving the monograph and comparing it with the chapter, I matched small portions of text that seemed to have similarities. I wondered whether the degree of overlap was a case of mere sloppiness or outright academic plagiarism. I slowly started to suspect that the author of record for the chapter had acted deficiently in appropriating the words of another scholar. Do seemingly established scholars really do this? Due to my previous work with serial plagiarism

M. V. Dougherty, *Correcting the Scholarly Record for Research Integrity*, Research Ethics Forum 6, https://doi.org/10.1007/978-3-319-99435-2_5

cases, I had already formed the general view that people who misappropriate the words of others tend to do so on multiple occasions and across many publications, a view confirmed by some theorists (Fox and Beall 2014: 346). I eventually decided to test this view again by examining some of the other publications appearing under the name of the author of record for the chapter, beginning with two articles in a philosophy journal in which I had also happened to publish. After careful examination, I confirmed that the two journal articles by the author of record had similar deficiencies in the liberal use of the words of others, but to a much more serious degree than in the initial case of the book chapter. I then looked at a few more articles and book chapters by the same author of record, both within and outside my field of expertise, and the cases multiplied.

The articles and book chapters exhibited various degrees of gravity in the omission of sources for sentences and paragraphs. Some of the apparent violations of standard academic practices appeared to me to be quite serious, but others were admittedly minor. One theorist has stated—rightly in my view—that plagiarism is "a continuous spectrum of text manipulations and not just one particular method of using other people's words" (Weber-Wulff 2014: 114). I noticed that some of the apparently faulty works that I was examining had co-authors of record, and I wondered what responsibility these other individuals might have for any of the deficiencies present in the works. Furthermore, I was puzzled that a single individual could publish so many articles and chapters in more than one discipline on such a vast array of topics. In my experience, polymaths on such a scale are anomalous in today's academic culture, which is generally marked by intense professional specialization. After a careful consideration of the evidence, I concluded that several of the journal articles and book chapters by the author of record fulfilled the particular heuristic proposed in Chap. 3 of the present book; that is, they exhibited *a non-trivial appropriation of words, with inadequate credit, generating the appearance of original authorship, in discrete items belonging to the scholarly record.*

But was it plagiarism? There are competing conceptions of plagiarism. As discussed in Chap. 3.2, some popular definitions require a *mens rea*, and others do not. Some conceptions are restrictive regarding the quantity and quality of misappropriated material, and some are permissive. I consulted the definitions of plagiarism endorsed by the home institutions of the author and co-authors of record. Those definitions, however, were issued by the institutions for cases of undergraduate and graduate student plagiarism, and they were not explicitly offered to cover academic plagiarism by professors. One definition—at the home institution of the author of record—declared unequivocally that a student is guilty of plagiarism when submitted work contains "one or more portions copied or closely paraphrased from an unidentified original, without acknowledging the source or formally quoting" (USI 2008: 13). The university policies also reflected a strong stand on research integrity; all students are required to sign an anti-plagiarism pledge where they affirm the principle of academic honesty, promise that they will undertake to cite exhaustively all sources, and warrant that no portion of any scholarly work will be copied. These statements made no mention of a *mens rea*, and they appeared to be restrictive in terms of quantity and quality, as even "portions" of unacknowledged or uncited material were said to constitute plagiarism. It seemed to me that the suspected deficiencies I had discovered in several published works by one author of record violated not only the research standards of the world of learning, but also the local standards of the author of record's home institution.

5.1 The Whistleblowing Dilemma

Researchers who find themselves in situations like this one will face the question of whether to report what they have found. That is, they will have to decide whether to become academic whistleblowers. Potential whistleblowers face a dilemma: do nothing, or enter into the long and uncertain path of declaring to relevant authorities that they have discovered evidence of suspected wrongdoing. There are tremendous advantages to the first option; by doing nothing their valuable research time will not be expended in the matter. They can demur on the grounds that someone more established can discover and report the suspected problem, or they may tell themselves that the apparent violations of scholarly standards on such a scale must already be well-known to relevant parties. Many studies of whistleblowing in the corporate sector warn that whistleblowing generally does not benefit the whistleblower; most often the whistleblower suffers significantly from a host of repercussions, which include stress, loss of reputation, professional retaliation, among others (Devine and Maassarani 2011: 19–40). Academic whistleblowers fare no better. As some commentators have noted, "being a plagiarism whistleblower is inherently stressful and can leave one vulnerable to criticism and retaliation by colleagues and others" (Fox and Beall 2014: 341). Even if a whistleblower succeeds in getting external validation by an institution or authority, this confirmation may come at a high cost, leaving one "changed and often damaged" at the end of a lengthy process (Gunsalus 1998: 52). Some theorists have concluded that "bringing forward concerns of fraud imposes reputational risks, enormous time commitments, and no tangible reward" (Nelson et al. 2018: 526).

First-person accounts of academic whistleblowing are filled with descriptions of adverse consequences that include harassment, intimidation, emotional trauma, and financial expense. These problems even afflict whistleblowerers who are firmly established at research universities as tenured and accomplished academics (Sprague 1993). A study of the effects of whistleblowing by the United States Office of Research Integrity identified a wide range of negative consequences for whistleblowers, including the loss of employment, the denial of salary (or promotion or tenure), the reduction in research and travel funds, ostracism from peers, legal threats from lawyers, and delays in having submitted manuscripts and grant applications reviewed (Office of Research Integrity 1995: 18–19). A recent account of the fallout of whistleblowing activity concluded that "potential whistleblowers might be disinclined to make their concerns public" as a basic self-protective measure to safeguard their careers and reputations (Hussinger and Pellens 2017: i).

Whistleblowers who point out problems in the published literature may also be labeled by detractors as "bullies." A whistleblower who writes to editors and publishers to request published corrections of the scholarly record may be accused by detractors of carrying out "personal attacks." In the present-day academic culture, some individuals fail to distinguish between criticism of a published work and criticism of a person; whistleblowers should expect that some parties will attempt to focus attention away from the submitted evidence and toward some assumed nefarious motivations of the whistleblower.

Potential whistleblowers should be aware that preparing evidence of suspected plagiarism can be extremely time-consuming, and there is no guarantee that one's efforts will result in any published correction of the scholarly record. Even if a whistleblower is successful in eliciting a published correction (such as a retraction, erratum, or corrigendum), the evidence meticulously prepared by the whistleblower may later be attributed to a source other than the whistleblower in the published correction. A whistleblower should not be surprised if a published correction for plagiarism gives the impression that the missing citations and quotation marks were discovered and brought forward by the author of record, rather than by a third-party whistleblower. What is more, all of the evidence compiled by the whistleblower may be published with the correction, but presented as the discovery of an investigating research integrity office or the discovery of the editors of the original publication. In short: whistleblowers should not be under the illusion that they will be thanked or acknowledged for their work, and even if they are successful in getting the scholarly record corrected their work may be credited publicly to another party altogether. As noted with examples in Chap. 4.2, a whistleblower will not likely be named in retractions, but may be named in expressions of concern, which may publicize the whistleblower as a complainant without the benefit of any institutional confirmation of the quality of evidence offered.

Despite these potential problems, there are strong reasons to blow the academic whistle. The primary one is getting the scholarly record corrected for future students and researchers. Without academic whistleblowers, fewer corrections of the scholarly record would be published. The issuance of a published correction by an editor or publisher is typically the culmination of a lengthy chain of events that first began when a third party discovered evidence of a suspected violation of research integrity and then reported it. Despite the personal costs of whistleblowing, such researchers remain the "guardians of research integrity" and whistleblowing remains "a means of enforcing a community's standards" (Anderson et al. 2013: 222, 253).

Good-faith whistleblowers may have subsidiary reasons beyond securing a published correction of the scholarly literature. They may desire to protect any unsuspecting and innocent co-authors from being involved in further deficient publications. Whistleblowers may also wish to shield unwary graduate students who otherwise would be interested in working with the author of record for the deficient publications. Early-career co-authors and doctoral students of those professors who commit research misconduct often face professional difficulties in establishing their own academic careers when the misdeeds of their senior co-authors or their advisors become known. In worst-case scenarios, misconduct by advisors can completely derail a young academic's career and prevent the awarding of a degree (Marcus and Oransky 2016). One might also be motivated to blow the whistle to stem the flood of future compilations by suspected plagiarists from entering the scholarly record. A potential whistleblower might consider how the proliferation of plagiarized works in journals and in published collections displaces the work of genuine researchers. Perhaps a whistleblower finds it to be unfair to see suspected plagiarists enjoying professional success, while highly-talented, non-plagiarizing colleagues continue to receive rejection letters from journals, grant agencies, and search committees. The life of research is competitive, and knowledge of unfair practices can elicit strong condemnations from those abiding by the accepted moral codes of academia and the ethical practices of the larger research culture.

Some whistleblowers may have unwittingly cited the deficient work of suspected plagiarists in previous publications and now be frustrated to have added to the success of the author of record by commending in print the deficient work to others and by adding to the suspected plagiarist's citation count. Students and scholars who read and incorporate the work of plagiarists in their research are harmed, since readers are left with a false view of the history of discovery and additional undeserved authority may be bestowed onto plagiarists.

Potential whistleblowers might be aware that those who engage in serial research misconduct may be counting on the negative repercussions of whistleblowing to keep their misdeeds from being known. Such potential whistleblowers might consider, for example, that "a plagiarist might be emboldened by knowing that the sole responsibility for exposing a plagiarist falls to an original author or whistle-blower, rather than an impartial sanctioning body" (Cox et al. 2018: 925). That plagiarists might be hoping that "the high financial and emotional costs borne by an individual in exposing a plagiarist" will suffice to keep academic misdeeds under cover might lead potential whistleblowers to become actual ones (ibid).

Having surveyed all the harmful consequences and inefficiencies that academic plagiarism injects into the world of learning, researchers who contemplate whistleblowing may conclude that they ought to disclose their findings, even when faced with many forms of retaliation, intimidation, and harassment. In discussing academic plagiarism, some journal editors have expressed the view that academic whistleblowing is not just permissible but obligatory. One counsels, "where an individual does have good cause to suspect plagiarism or other research misconduct, he or she has a responsibility to raise it with the relevant authority" (Martin 2007: 908). In light of considerations as these, a prospective whistleblower may concur with the view that "unwillingness to respond with indignation to acts of plagiarism is itself a moral failing" (Scanlon 2007: 62).

A prospective academic whistleblower, wondering what to do after finding evidence of suspected research misconduct, will likely have two questions:

1. To whom can one blow the ethical whistle in an academic setting?
2. How could such an action affect the academic whistleblower?

This chapter addresses these two questions. While there is much literature on whistleblowing in governmental and business sectors, less has been said about whistleblowing in an academic context (Fox and Beall 2014; Gunsalus 1998; Doran 2016; Bouter and Hendrix 2017; see also Yong et al. 2013; Devine and Reaves 2016). The ramifications of a particular act of academic whistleblowing cannot always be predicted accurately, yet the range of possible responses can be outlined in general.

5.2 To Whom Can One Blow the Academic Whistle?

Researchers who discover strong evidence of suspected research misconduct, and who are committed to reporting it, will face the question of the best recipient of their findings. Perhaps they have discovered what they believe to be a situation of serial plagiarism. They can blow the ethical whistle to all or some of the following persons, institutions, or venues:

1. The suspected plagiarist
2. A post-publication review venue
3. The relevant journal, volume, or series editors
4. The relevant publisher
5. The research integrity office of the suspected plagiarist's university
6. The sponsoring grant agency
7. A journalist
8. The genuine author(s) victimized by the suspected plagiarist

There are risks and benefits associated with disclosing evidence of suspected plagiarism to each of these potential recipients.

5.2.1 Whistleblowing Directly to the Suspected Plagiarist

Some claim that a whistleblower should first contact the party believed to have committed wrongdoing prior to reporting to anyone else. On this view, disclosing evidence of suspected wrongdoing to the apparent wrongdoer provides the opportunity to clear up confusions or to find an unknown exculpatory explanation prior to involving any institutional authorities. To be sure, what appears to be an egregious violation of scientific integrity can turn out to be a misunderstanding that is easily clarified with additional information. An apparent case of academic plagiarism, for example, can turn out to be a situation where an author legally changed a first name, a surname, or both, and then re-used some previously-published material in a later publication under the new name, perhaps even with a different institutional affiliation.

Another reason generally offered in support of first whistleblowing to suspected plagiarists is to allow them the option of initiating the process of correcting the scholarly record themselves. If error or negligence, rather than outright fraud, is the cause of the deficiencies in the plagiarized item, the authors of record themselves might be willing to "do the right thing" and request the publication of a statement of retraction or correction. The journalists at *Retraction Watch* categorize examples of author-initiated retractions and corrections under the tag "Doing the Right Thing" (Retraction Watch 2010–), and the expression now appears in articles that discuss the benefits of these "self-retractions" (Hosseini et al. 2018). Author-initiated retractions are rare but not unheard of, and some commentators have urged a greater acceptance for such a mechanism of correction. At least one study has argued that researchers do not suffer a "citation penalty" for prior work over the course of a career for self-initiated corrections of the scholarly record, and such researchers may enjoy "possibly positive citation benefits among prior work" (Lu et al. 2013: 4). Furthermore, author-initiated retractions generally appear in print more quickly than retractions requested by third parties, as the former do not typically involve as many procedural hurdles, administrative delays, or lengthy verification investigations.

Despite these potential benefits, the view that one should contact the author of record prior to whistleblowing elsewhere has been subject to severe criticism. An early disclosure of evidence of suspected wrongdoing to the suspected wrongdoers may give unscrupulous researchers the opportunity to cover up or even destroy other evidence of misconduct. In short, such direct whistleblowing may give academic wrongdoers a head start before any investigating agencies become involved:

> Contacting authors before anyone else knows about potential issues in their work, only serves to give unethical scientists time to hide their tracks [...]. That will make it much more difficult for universities and oversight agencies to investigate cases properly (Marcus and Oransky 2015: 37).

Other commentators have similarly warned that if researchers blow the whistle first to suspected plagiarists, they may be giving them "time to attempt to thwart [...] making the allegations known to others or to preemptively retaliate" (Fox and Beall 2014: 344). Not all plagiarists appreciate being presented with evidence of suspected plagiarism. In an article titled, "Responses of Authors Accused of Plagiarism by Journal Editors," two medical journal editors report that in their experience "many plagiarists tried to ignore the situation or act as [if] there is nothing to worry about" when approached about potentially deficient manuscripts (Wiwanitkit and Wiwanitkit 2017: 310). Occasionally suspected plagiarists have threatened whistleblowers with legal action when approached with concerns about plagiarism (Fox and Beall: 344–345).

Direct involvement with a suspected wrongdoer may also introduce unforeseen complications much later during various investigative processes. As outside parties may believe (sometimes rightly, perhaps) that a whistleblower is acting out of a grudge or a vendetta, it may be helpful for a whistleblower to be in a position to declare throughout the entire process that she or he has never met, interacted with, or corresponded with the alleged wrongdoer. As personality conflicts have been known to generate dubious allegations of misconduct, "institutional officials who receive charges will thus almost always probe (consciously or unconsciously) for evidence of personal animosity in the initial stages" (Gunsalus 1998: 53). In light of this possibility, a whistleblower can benefit from maintaining distance from the suspected wrongdoer.

A lack of personal interaction with the alleged wrongdoer may also make life easier for the whistleblower later, as it minimizes any personal element of the situation, particularly if the alleged wrongdoer attempts to plead, threaten, bribe, cajole, or otherwise try to influence the whistleblower from taking any further action. Even if a whistleblower desists from contacting the alleged wrongdoer and reports to other parties instead, later in the process the whistleblower may hear personally from the accused or from associates of the accused and may be told how the whistleblowing may have caused personal difficulties for the accused. Keeping some distance will likely assist a whistleblower in maintaining a clear separation between the required actions for correcting the scholarly record and any irrelevant personal or emotional appeals.

5.2.2 Whistleblowing in a Post-publication Review Venue

Having resolved not to contact a suspected wrongdoer directly, a whistleblower might decide to publicize the findings of suspected wrongdoing to the greater academic community for the sake of generating support for a published correction of the scholarly record. There are a variety of post-publication opportunities for such disclosures, including online or print, as well as anonymous or named venues. The manner of discovery of the suspected misconduct may determine the venue through which a whistleblower discusses it. If (to continue the scenario that opened this chapter) the discovery of suspected plagiarism is found by a book reviewer, the published book review itself can be used to disclose the problems found. Even though book reviews are generally the place to mention scholarly deficiencies and criticisms of a book under review, evidence of violations of research integrity are generally considered to be problems of a different order, and journal editors are not always willing to approve a book review that alleges research misconduct. Book reviews that contains such disclosures will generally require the clearance of editors at the journal, and they might also be vetted with the editor of the volume in question and the author of record in question. On rare occasions, the immediate victim of plagiarism will happen to be the reviewer herself or himself, and such reviews can be outright unsparing. For example, Kathleen L. Riley reviewed a volume in *The Catholic Historical Review* and stated, "a more honest and accurate title for this slight volume would be *An Unauthorized, Thinly Disguised and Abbreviated Version of the Dissertation of Kathleen Riley*" (2000: 712). The book was withdrawn by the publisher after Riley's review was published. On other occasions, the reviewer may choose to be silent in the text of the review about plagiarism yet write about the matter in detail elsewhere (see, for example, Vincent 2007).

The website *PubPeer* has quickly become established as an influential venue for post-publication review. Originally the website allowed only those possessing both a publication record and an affiliation with a research institution to post anonymous comments on published articles, but now the ability to post anonymous comments is unrestricted. Most of the postings on the site concern articles that report experimental results in empirical sciences, but some postings concern philosophy and other humanities disciplines. *PubPeer*'s stated founding purpose is "to foster a scientific environment where robust, high-quality research is valued, while providing a forum to discuss the problems of unreproducible, misleading, misconceived or fraudulent work" (PubPeer 2014a). Despite this purpose, the website guidelines ask participants to desist from making any accusations of fraud, and instead encourage participants simply to present public, factual information that readers can independently verify (PubPeer 2014b). One researcher, who has earned (to date) 21 retractions and 10 corrections largely prompted by the extensive scrutiny of his works on *PubPeer*, filed a lawsuit in an attempt to force *PubPeer* to disclose the identities of the whistleblowing scientists who had revealed the problems with his work on the website (Abritis 2018). The researcher lost in court and *PubPeer* prevailed because of the anonymity protections offered by the case law about the freedom of the press protected by the First Amendment of the Constitution of the United States (State of Michigan Court of Appeals 2016).

As authors of comments on *PubPeer* can conceal their identities, the website allows whistleblowers (who might otherwise feel vulnerable in providing evidence of suspected misconduct) to reveal their findings with the hope that someone with greater institutional or professional security might take up the case. A similar post-publication venue to *PubPeer* was *PubMed Commons* (2013–2018), which was hosted by the United States National Library of Medicine at the National Institutes of Health. *PubMed Commons* had allowed registered researchers to post critiques of any articles indexed in the *MEDLINE* database, but the service was discontinued in early 2018 due to "the low level of participation" (NCBI Insights 2018). According to some observers, the reason for the lack of interest in *PubMed Commons* was the inability to comment anonymously (McCook 2018).

The practice of post-publication review in anonymous online venues has been subject to heated criticism. In a much-debated episode, psychologist Susan T. Fiske wrote that those who participate in such activities are "online vigilantes" engaging in "methodological terrorism" (2016a: 1). These words appeared in an early-release in-press editorial for the *Observer*, the flagship publication of the American Psychological Association. Critics responded quickly (Gelman 2016; Singal 2016), and the final published version of Fiske's article was substantively changed to omit such language (Fiske 2016b). Nevertheless, the views expressed by Fiske in the first version of her editorial has support in some quarters. The editor-in-chief of *Plant Physiology* had previously published an editorial that criticized what he saw as "vigilante science currently facilitated by PubPeer" (Blatt 2015: 909).[1] Fiske and

[1] Jaime A. Teixeira da Silva similarly warns of an "aggressive post-publication science watchdog vigilante movement" that is characterized by "a highly impositional form of post-publication vigilantism" (2017: 610, 611).

Blatt defend an approach where critics are counseled to raise questions privately, either with authors of record or through authoritative channels, rather than placing evidence of suspected research misconduct online for open consideration by all members of a research field. PubPeer founders have responded that anonymous post-publication review promotes vigilance, rather than vigilantism (PubPeer 2015), and PubPeer has been defended as a bulwark against scientific misconduct and "fake news" in science (Forest 2018). The criticism of anonymous online post-publication review mirrors the criticism that has been directed toward the Vroniplag Wiki, which is an online forum where members under pen-names scientifically analyze German dissertations and other works for plagiarism. As Debora Weber-Wulff has chronicled, many European universities have neglected to withdraw the degrees involving such plagiarized dissertations, despite the overwhelming evidence of plagiarism publicly documented by the Vroniplag Wiki members (2014: 31–36, 94–107; see also Dannemann 2018).

The weakness of whistleblowing in a public post-publication review venue is that the activity may be entirely ignored by those parties able to initiate a correction of the scholarly record. Furthermore, the relevant parties of the scientific or academic community may be ignorant of any claims made on independent post-publication review venues. Since post-publication review venues are autonomous and untethered either to an investigating body of an institution or to the editorial offices of a publisher, whistleblowing in such venues may be insufficient to trigger an investigation that could lead to a retraction. Even though new *PubPeer* postings generate a notification to all authors of record for articles subject to criticism, the authors of record may not wish to enter a public debate and may simply avoid the online discussion. Even published book reviews in journals that indicate deficiencies in the works under review may remain unknown or may be disregarded. Unless one reports the evidence offered in a post-publication review to an institutional authority, a formal process of correcting the scholarly record may not be initiated.

There have been some individuals who have posted evidence of suspected research misconduct on their personal websites or on their privately maintained blogs. Disclosures in such venues lack the institutional support found with other outlets of whistleblowing, and whistleblowing in this way can more easily make one a target of legal threats. Furthermore, universities may not fully support whistleblowing employees who disclose evidence of suspected plagiarism on personal websites outside of normal institutional channels. Nevertheless, some recent high-profile retractions have been occasioned by whistleblowing on personal websites (see Fosmoe 2017; Singal 2015).

5.2.3 Whistleblowing to a Journal or Volume Editor

The typical recipients of evidence of suspected plagiarism by whistleblowers are journal editors and academic editors of published collections of papers. Both kinds of editors serve a crucial and well-defined pre-publication role in vetting works through a peer-review process, which now generally involves text-matching and image manipulation software to screen for possible violations of scholarly integrity.[2] The role of these editors in dealing with post-publication problems in articles and chapters is not often acknowledged, however. Verifying evidence of suspected plagiarism can be extremely time-consuming, even when the evidence provided is impeccable and irrefutable. Most journal editors and academic editors hold teaching or research positions in addition to their editorial ones, and editorial work is often uncompensated service to the profession. Editors will occasionally have their time wasted with bad-faith or ill-conceived requests for retraction (Lewis 2016). Some editors have sought to restrict their responsibility for adjudicating cases of suspected misconduct by imposing a self-made statute of limitations for claims brought against published articles. When a journal editor recently promulgated a 6-year time limit for considering any claims of research integrity violations concerning articles appearing in the journal, some readers concluded that the journal no longer vouches for the integrity of any of its articles after 6 years (McCook 2017). Since the presumed integrity of articles would then have a 6-year expiration date, some suggested that it would be best not to cite any older articles from the journal (ibid.). In contrast, other journal editors have adopted a more forensic role in maintaining research integrity by undertaking an internal systematic review of all previously published work appearing in the journals (Grens 2017).

The responsibility of both editors and readers for identifying academic plagiarism is emphasized by some editors. Ben R. Martin, editor of the journal *Research Policy*, observes:

> The task of policing the problem of plagiarism falls in part on the editors of journals [...]. Very importantly, they also need the help of readers. If readers spot a paper where a significant element appears to have been plagiarised, they should bring this immediately to the attention of the editor of the journal concerned (Martin 2007: 906).

Journal editors who encourage readers to submit any tips on suspected breaches of research integrity for articles published at any time demonstrate a greater sensitivity to their stewardship over their journals and to the contributions of these journals to the scholarly record.

In the words of one theorist, editors "have a duty to try to keep the literature as sound and uncontaminated as possible" and must act when credible evidence of academic violations are made known to them (Hames 2007: 197). The principle "suspected or alleged misconduct must not be ignored" has been touted as a golden rule for journal editors (173). Occasionally one still finds editors who seem genuinely surprised to learn that their editorial accountability extends beyond getting articles to print. Whistleblowers may have to persuade some editors that their ambit of responsibility includes work already in print and that whistleblowing is essential to the correction of the scholarly record. Journal editors are ultimately the gatekeepers of published corrections that appear within the pages of their journals, so they may need to be reminded that editorial responsibility does not cease after articles

[2] Data analysis software may soon also be added to an editor's standard arsenal for preventing the publication of deficient manuscripts (Simonsohn 2014).

come into print. One study found that journal editors suffer from a "lack of agreement in plagiarism definitions and in the threshold of acceptable text re-cycling" (Roig 2014: 59). Some editors, unfortunately, are ill-equipped to consider evidence of suspected research misconduct. Editors who fail to issue retractions for demonstrably defective work leave to their editorial successors the problem of cleaning up for their failures.

Whistleblowers need not be discouraged; if the evidence is strong, an eventual change in editorial leadership at a journal may occasion a new consideration of a request for retraction. Editors who hesitate in issuing the necessary corrections—on the mistaken belief that such retractions will reflect badly on their editorial tenure—might do well to consider how those same retractions would appear after they leave office and are issued by their successors. If the evidence for retraction is strong, the problem of the uncorrected literature does not simply go away; the defects in published articles may be rediscovered by other researchers who themselves will be motivated to become whistleblowers.

5.2.4 Whistleblowing to a Publisher

In some disciplines, items are published under the stewardship of an acquisitions editor of a publishing house rather than though the work of an academic serving as an editor of a journal or collection of essays. In these situations, a whistleblower may need to approach an acquisitions editor to request a published correction for a deficient work. If an acquisitions editor decides to issue a statement of retraction, usually after some vetting of the retraction statement through the legal department of a publisher, the acquisitions editor will have to decide upon the manner of distribution. The statement might be inserted in remaining unsold copies of the volume, or an electronic statement might be posted to the webpage of the volume on the publisher's website. The retraction statement might also be appended to the electronic versions of the volume sold through the publishers' subscription platform or added to the e-book version.

Whistleblowing to a publisher can be more challenging when the acquisitions editor who guided the deficient item through the publication process is no longer employed at the publishing house. Smaller publishers are often acquired by larger ones, and therefore a new publisher with a different name may come to own the rights to a deficient volume published years earlier. Sometimes the same volume is published by a different publisher overseas, or in translation by still another publisher, so that more than one publisher has been involved in bringing the deficient material into the scholarly record and distributing it to the global academic community. In cases such as these, a whistleblower may find it difficult to identify an editor at a publishing house who feels sufficiently responsible to address the situation originating during the stewardship of another acquisitions editor.

5.2.5 Whistleblowing to a University Research Integrity Office

Major universities typically have research integrity offices that are tasked with investigating credible evidence of misconduct for work produced under that university's institutional affiliation. These offices have formal procedures that whistleblowers can follow for filing a complaint and requesting an investigation. The process of engaging with a research integrity office can be time-consuming, as one may be expected to respond to a series of questions, testify in a hearing, or even be subject to questioning (perhaps even by the accused). Research integrity offices have significant authority at their disposal. They may, for example, contact journals to request retractions for articles appearing in print with the university's institutional affiliation; instruct an author of record to request retractions from editors and publishers as a condition of continued employment; or mandate that offenders complete an external research integrity rehabilitation program.[3] At universities that require the highest standards of professional conduct by their researchers, a research integrity office may even start the process of employment termination. A whistleblower who acquires the support of a research integrity office in making retraction requests to journals and publishers is likely to succeed in getting the scholarly record corrected.

In some cases, a whistleblower may not be entirely clear about which university's research integrity office should be contacted. Authors of record may have published deficient articles with various university affiliations over the course of a career, and the presence of common co-authors from other institutions may further complicate the matter. When reporting suspected misconduct, some whistleblowers have contacted both the university listed as the institutional affiliation in print and the current employing institution.

Another complicating matter is that a university's research integrity office may not have the requisite objectivity to investigate evidence of suspected plagiarism by its own employees. In a worst-case scenario, a research integrity office might seek primarily to protect the image of the institution by minimizing any appearance of wrongdoing, or even by attempting to impede the issuance of corrections of the scholarly record by publishers. When those who commit research misconduct have been successful in securing large grants for an institution, a university's research integrity office may have a financial incentive to absolve the guilty. A research integrity office of this kind functions surreptitiously as a *de facto* public relations firm or damage-control committee, issuing carefully-worded press releases meant to shield the reputation of the institution rather than support corrections of the scholarly record. In the judgment of one commentator, "when the stakes are high and large sums of research money are involved from outside grants and contracts, some universities still have problems adequately investigating and protecting young

---

[3]One such program is The P. I. Program at the Center for Clinical and Research Ethics at Washington University in St. Louis (USA) directed by philosopher James Dubois. The program was formerly called Restoring Professionalism and Integrity in Research (RePAIR). For an account, see Cressey 2013.

whistleblowers from pressure, intimidation, job loss, and even worse consequences" (Sprague 1993: 130). In short, not all institutions are equally committed to maintaining a reliable scholarly record; not all institutions hold their professors to the same basic standards of academic honesty that a reputable university would expect even undergraduate students to follow (see Dannemann and Weber-Wulff 2015). For investigations into suspected plagiarism, the "worst possible outcome is unalloyed exoneration of the guilty" (Hauptman 2011: 8). Nevertheless, well-managed research integrity committees have significant influence in supporting corrections of the scholarly record, and they can offer substantial support to whistleblowers.

5.2.6 Whistleblowing to a Grant Agency

When deficient articles appear in print with acknowledgment of funding by an external grant agency, a whistleblower may wish to disclose the evidence of suspected wrongdoing to the named grant agency. Some grant agencies have internal research integrity offices that investigate received evidence of suspected grant misuse, and some have formal procedures for submitting claims. The grant agency can assist in the correction of the scholarly record by sending requests for retraction of deficient articles that involve research conducted under its sponsorship, so a whistleblower may be greatly advantaged to obtain the support of a grant agency.

5.2.7 Whistleblowing to a Journalist

A whistleblower may consider divulging evidence of suspected research misconduct to the press if whistleblowing elsewhere has failed to get the scholarly record corrected. Local and national newspapers at times are interested in reporting of academic misdeeds—especially when the misconduct has been committed by an individual employed in a publicly-funded institution—but sometimes reporting occurs only after some finding of research misconduct has been made by a university authority. Specialized news outlets (e.g., *Times Higher Education*, *The Chronicle of Higher Education*, *Inside Higher Ed*) sometimes report on the misdeeds of researchers, and publicity can motivate institutions to carry out more comprehensive investigations. University-run student newspapers are also often interested in carrying stories about academic misdeeds, and reporting on them can motivate seemingly lethargic investigating committees to complete their work (see, for example, Broeksteeg 2010; cf. Vandevelde 2010).

The involvement of journalists can generate pressure for the home institution of the accused and for the publication houses that have issued the publications in question. One possible good effect of involving the press is that wider public attention makes it difficult for institutions to hide alleged wrongdoing or to delay indefinitely any corrections of the scholarly record. For these reasons, some commentators on plagiarism recommend the involvement of the media. In contrast, some whistleblowers have noted that the act of speaking to the press can cause difficulty within the academic community. One whistleblower has described "an unwritten code in science that a person is not supposed to complain outside the confines of the scientific community," so that for individuals to speak to the press is to do so "at their own peril" (Sprague 1993: 116). In a variety of disciplines, however, disclosures of evidence of misconduct to news outlets are often credited for overcoming delays or reluctance to correcting the published literature (Parrish 1999; Schechner et al. 2009).

5.2.8 *Whistleblowing to the Genuine Authors*

Finally, a whistleblower may wish to let the immediate victims of plagiarism know that their works have been misappropriated. At times a whistleblower will encounter the assertion that a request for retraction of plagiarized work requires the participation of the genuine authors whose works have been misappropriated. That view is mistaken; academic plagiarism harmfully affects the larger research community and needs to be rectified independently of the views of the original authors whose works have been misappropriated. A lack of involvement on the part of the immediate victims should not be seen as an impediment to correcting the scholarly record. If such were the case, retractions could not be issued if the immediate victims are unable to be located or are deceased. Furthermore, at times the victims of plagiarism do not wish to be involved due to a personal relationship with the plagiarist (as when the victim is the plagiarist's colleague), or due to institutional power structures (as when the victim is the plagiarist's graduate student). Not all individuals have the same response upon learning that they have been victimized by a plagiarist; some are quite uninterested in supporting a published correction of the scholarly record. But some are not: Michael Dansinger's open letter titled "Dear Plagiarist: A Letter to a Peer Reviewer Who Stole and Published Our Manuscript as His Own" provides an unadorned first-hand account of being victimized, as it explains:

> It took 5 years from conceptualization of the study to publication of the primary analysis [...]. In all, this body of research represents at least 4000 hours of work. When you published our work as your own, you were falsely claiming credit for all of this work and for the expertise gained by doing it (2017: 143).

The need to correct the scholarly record should be considered independently of the views of the immediate victims; victim participation is a possibly helpful, but not essential, feature of retraction requests.

Even though a claim of suspected plagiarism can be evaluated without involvement of the immediate victims of plagiarism, the authority of victims in requesting a correction of the scholarly record should not be underestimated. Editors and publishers generally recognize that victims are often in the best position to take legal action for the misappropriation of their work, and a victim's persuasive account of the experience of having her or his research taken may provide a compelling justification to prompt an otherwise uninterested editor or publisher to investigate further. Nevertheless, a whistleblower may request a correction of the scholarly record for suspected plagiarism prior to knowing anything about the reaction of the immediate victim of the plagiarism. Some whistleblowers have copied the immediate victims of plagiarism when electronically submitting evidence of suspected plagiarism, so that editors and immediate victims learn of the request at the same time. Doing so underscores that a correction of the record for the larger research community is the overriding goal of a retraction request, rather than assuaging any personal feelings that may attend victimization. A whistleblower may also wish to notify the publisher of the original source material about a misappropriation by a suspected plagiarist. In doing so, the whistleblower can make clear that he or she will be going forward with a request to have the scholarly record corrected without tying such a request to any participation by either of the additional parties. In such a situation, the victim and original publisher may come to support the request, but such participation is not necessary.

Sometimes academic plagiarism is discovered by the original author whose work has been misappropriated rather than by a third party. Whistleblowing by immediate victims, rather than by third parties, can be especially draining, and the publication of retractions does not always assuage the "sense of violation and anger" that victims-turned-whistleblowers may have throughout the process (Bowers 1994: 549). One recent first-person account by a victim of academic plagiarism recounts that "I was stunned because I couldn't believe a full professor of high global standing—a respected leader in their field—would do this [...]. But I was also worried, because I now had to prove the originality of my work [...]. I avoided my institution while the plagiarist recently visited" (Anonymous Academic 2017). In this case, the whistleblowing plagiarism victim attained some success in correcting the scholarly record yet still had some doubts about the affair:

> The editors also printed a corrigendum in the next issue. But not once in the process did the author or the editors concede that the issue was, in fact, plagiarism. Retraction should have been discussed. Despite this having happened three years ago, something still rankles: the plagiarist got away scot-free (ibid).

In some disciplines, it has been claimed that plagiarism allegations "are most frequently made by the original author whose material was plagiarized," since the primary victim is likely most acquainted with the latest publications in a research subfield (Woolf 1988: 82; Woolf 2016: 85). It may be that victims are most likely to discern acts of academic plagiarism, but correcting the scholarly record need not be seen as the prerogative of the victims, and in some cases a third party may be in a better position to report the evidence of suspected misconduct.

5.3 Multi-Targeted Whistleblowing

As outlined above, there are at least eight major potential recipients for whistleblowers to turn to with evidence of suspected research misconduct. However, no consensus is found in the whistleblowing literature regarding which of these eight a whistleblower should contact first. This lack of consensus is likely a function of the unpredictable ways in which the various receiving parties might respond to evidence of suspected wrongdoing. Some journal editors prefer not to issue retractions until a finding of wrongdoing has been established by a university's research integrity office. On the other hand, the presence of newly published retractions can provide strong evidence to a university's research integrity office that a larger investigation is warranted. A stalemate can occur when one party waits for another party to issue findings as a condition of taking any further action. For example, standoffs can occur between grant agencies and research integrity offices. Likewise, many journalists may not want to go forward with a story—even with overwhelming evidence—without some external validation of the evidence by an institution. Published advice about academic whistleblowing is often contradictory. Some experts recommend that whistleblowers should submit "to the highest authority" and find that journals and funding agencies are "best equipped to evaluate the significance of errors within their publications and have the power to retract papers" (Doran 2016: 405). Others provide the opposite advice, recommending that one should blow the whistle to journal editors only after a research integrity board has been notified and has rendered a decision (Bouter and Hendrix 2017). Given the unpredictability of how each of the eight major potential whistleblowing recipients might respond when presented with evidence of suspected research misconduct, many whistleblowers opt for multi-targeted whistleblowing by disclosing evidence to several persons or institutions simultaneously.

Another reason in favor of multi-targeted whistleblowing is that whistleblowers are often simply ignored by parties who have received claims from them. Miguel Roig has observed that "some journal editors lack familiarity with scientific publication issues" and "other editors are not concerned with publication ethics" (2014: 58). Certain book editors appear to have learned that an easy way to make whistleblowers go away is to make vague promises to the effect that all deficiencies in a plagiarized book will be corrected in an otherwise unspecified "future revised edition" that never in fact appears. Not all editors observe the golden rule of editors mentioned above: "suspected or alleged misconduct must not be ignored" (Hames 2007:173), so whistleblowing to more than one party may be necessary in order to increase the likelihood of success in securing a published correction.

Some individuals accused of plagiarism have claimed that it is unfair to have to defend themselves on more than one front for the same act of potential academic wrongdoing. Consider a situation where a whistleblower has reported evidence of suspected plagiarism at the same time to a journal editor, a university research integrity office, a grant agency, and the genuine author whose work has been apparently misappropriated. To be sure, it can be time-consuming for the accused to have to react to accusations in more than one venue or forum. On the other hand, whistleblowing to more than one person or institution lessens the likelihood that any of the recipients of complaints will be tempted to minimize the offenses or engage in cover-ups to avoid responsibility.

Potential whistleblowers should also consider that even though there are eight major recipients of whistleblowing, there may be some overlap between them in certain cases. Some editors of biomedical journals allow whistleblowers to publish signed accusations of plagiarism within the pages of their journals, along with supporting evidence, followed by published responses by the accused and a final statement by the editors (Brentlinger et al. 2009; Jacobs et al. 2014). This practice, almost unheard of in humanities disciplines, supports the correction of the scholarly record and serves as a strong deterrent to would-be plagiarists. In submitting evidence of suspected plagiarism for publication in this way, a whistleblower simultaneously alerts a post-publication review venue and a journal editor.

In her article, "How to Blow the Whistle and Still Have a Career Afterwards," C. K. Gunsalus advises potential whistleblowers never to present their disclosures as charges or accusations, but instead to frame their disclosures as questions. In substituting interrogatives for declaratives, a whistleblower facilitates a conversation with the receiving party. Gunsalus explains, "Your questions should proceed on the implicit premise that there is something you do not understand and thus that you are seeking help to improve your own comprehension" (Gunsalus 1998: 56). If one possesses seemingly irrefutable evidence of suspected wrongdoing, framing the evidence in the form an inquiry can be challenging, but a calm and open approach may increase the likelihood of success in securing a published correction of the scholarly record.

5.4 How Will the Whistleblowing Impact the Academic Whistleblower?

Today's world of learning is a cosmopolitan enterprise. The discovery and transmission of knowledge, as well as its preservation through the maintenance of a reliable scholarly record, involve scholars and researchers throughout the world committed to a common good. Even though the high ideals of the world of learning transcend borders, regional factors can be relevant when it comes to how whistleblowers will be treated upon the disclosure of evidence of potential academic wrongdoing. Due to historical and cultural reasons, some sectors of the academic world may not view acts of whistleblowing as favorably as others. In many academic communities, specialists and practitioners in a subfield form a small close-knit group, often with long histories of collaboration, co-authorship, mentoring, and friendship. When one member of this group is accused of potential wrongdoing, especially by an outsider, the first reaction to defend the accused may be aggression toward the whistleblower. The intensity of the blowback experienced by first-time whistleblowers may come as a surprise, particularly to those who have an idealized view of the academic enterprise. The political and sociological factors that guide the research culture are not always apparent to early-career individuals, so whistleblowers are often unprepared for the resistance and pressures they quickly encounter.

The risks of being harassed are many and the harassment may happen over a long period of time. If one discloses strong evidence of suspected wrongdoing to an institutional research integrity board or to a grant agency, it is likely that such an action will initiate investigative proceedings by these bodies, and these investigations will likely require further involvement that can last many months if not years. Additional involvement by a whistleblower may take the form of testifying in a hearing, providing written answers to a committee's questions, responding to a rebuttal by the accused, or preparing additional documentation of the suspected wrongdoing. These often lengthy institutional proceedings can cause a whistleblower's degree of participation to evolve from a simple notification of suspected wrongdoing to serving in the ongoing capacity of a witness. As proceedings drag on, a whistleblower may encounter external pressure to stop participating or to cease testifying in an institution's investigation. A particular form of whistleblower harassment can be described as academic witness intimidation. Both harassment and witness intimidation are forms of retaliation, and they can affect whistleblowers in various ways. Acts of whistleblower harassment and academic witness intimidation can impede successful corrections of the scholarly record.

5.5 What Might Whistleblower Harassment and Academic Witness Intimidation Look Like?

Whistleblower harassment and witness intimidation can take many forms in an academic context, especially when one has engaged in multi-targeted whistleblowing by providing evidence of suspected wrongdoing to more than one person or agency. Unfortunately, there is no way to predict what precisely one will experience in any particular situation of academic whistleblowing. Part of the difficulty in assessing the likelihood of harassment and intimidation based on the experience of others is that the various acts of retaliation are largely invisible to an outside observer, even in those cases where a whistleblower has been relatively successful in obtaining corrections of the scholarly record.

The increased recognition within the scholarly community of the problem of harassment and intimidation toward whistleblowers has prompted declarations that "interference with a misconduct investigation" is itself a form of misconduct (Resnik 2003: 123, 132). In 2012, the Universities UK organization published a "concordat" on research integrity that states, "Research misconduct can take many forms, including [...] reprisals against whistleblowers" (Universities UK 2012:17). This approach, which now appears to be a mainstream view, if not a consensus, is a positive development. Despite an increased recognition of the rights of whistleblowers and their crucial role in maintaining research integrity, the path of academic whistleblowers nevertheless remains quite challenging.

To illustrate this, I will turn to the real-life experience described at the beginning of this chapter. Whistleblowing in those specific cases of suspected plagiarism resulted in 12 published corrections of the scholarly record, but also involved repercussions that are not evident from the published corrections.

Starting in late 2015, I collected evidence of suspected plagiarism that I had come across in a number of English-language articles and book chapters by an author of record working in the fields of philosophy and communications. I sent the relevant portions of the evidence to the following persons and institutions:

1. The journal editors in which the articles had appeared;
2. The acquisitions and academic editors of the publishers of the book chapters;
3. The commission on research integrity of the national grant agency acknowledged for funding in three of the publications;
4. The research integrity office of the author of record's home university;
5. The research integrity offices of two co-authors who had each published at least three articles with the author of record.

In my applications, I requested that the relevant journal and acquisitions editors (1, 2) correct the scholarly record by issuing retractions or corrigenda for the articles and chapters in question, and I asked to the national grant agency and the research integrity offices (3, 4, 5) to lend their institutional support for published corrections of the scholarly record. After 36 months, journals and publishers had issued corrections of the scholarly record for 12 articles and book chapters. The corrections consisted of: 5 retractions (van Eemeren 2015a; Anonymous 2016a, b, 2017c; Martinelli 2018a), 3 errata (van Eemeren 2015b; Anonymous 2017a, 2018a), and 4 corrigenda (Anonymous 2017b, d, 2018b, c).

The published corrections used a variety of expressions to explain the deficiencies of the articles and book chapters in question. Two retractions from one publisher for a pair of book chapters stated that "direct reference and citation of the works of other scholars is often inconsistent and in some cases totally lacking" and that the chapters do "not meet standards currently expected of an academic publication" (Anonymous 2016a, b). Two other publishers used the expression "severe shortcomings in the references" to characterize the deficiencies (van Eemeren 2015b: 481; Anonymous 2017b). One correction expressed the issue as one of "erroneous citations" (Anonymous 2017a), and another explained that an article "does not appropriately acknowledge [...] the primary source of the comments made in the article" (van Eemeren 2015a: 493). Only one of the published corrections of the scholarly record by editors and publishers used the explicit term *plagiarism*, declaring that a chapter is "plagiarized from various sources" (Martinelli 2018a) and further noting in an addendum that the retraction concerned "an evident example of academic misconduct" (Martinelli 2018b).

The three research integrity offices responded in quite different ways. The first—at the home institution of the author of record—completed its investigation but did not make its final report public. Instead, the university issued a press release stating that a commission had found only "minor violations of copyright [violazioni minori del diritto d'autore]" and "incorrect methods of referring to sources" which however were considered to be "of minor importance" [modalità scorrette di citazione delle fonti di minor gravità] (USI 2016), later re-affirming this view after a new case came to light (Illarietti 2018). Absent in the press releases was any acknowledgment of the published retractions, errata, and corrigenda issued by editors and publishers.

The research integrity officer at the home institution of one of the co-authors, after examining three of the co-authored articles, concluded:

> I agree with the complainant that substantial plagiarism has occurred that extends beyond reasonable repetition of common statements, that there is misrepresentation of who has done the research stated in these articles, there is a lack of proper citations, and there is extensive use of others' statements verbatim (Virginia Tech Research Integrity Office 2016: 1).

This unequivocal finding of "substantial plagiarism" by the co-author's institution contrasted with the finding of minor copyright violation and minor citation errors by the research integrity office of the home institution the author of record.

The third research integrity office, at the home institution of another co-author, conducted a lengthy investigation into three articles and subsequently issued a judgment that mandated published corrections for two of three co-authored articles. The members of the investigating board were not able to arrive at judgment regarding the remaining case, and they noted that a witness (who was the suspected victim of plagiarism and a former student of the co-author) had declined to offer a statement to the investigators. The conclusion of the investigation was the mandated correction of two of the three co-authored articles, noting that the complaint "is found to be justified" [gegrond wordt verklaard] but on the understanding that the co-author "himself has not violated the principles for scientific research" [zelf de principes van wetenschappelijk onderzoek niet heeft geschonden] (College van Bestuur van de Universiteit van Amsterdam 2017: 2). The research integrity office directed the co-author to pursue two published corrections, and, if possible, to do so in consultation with the first author of record.

An outsider examining the public record might think that the whistleblowing in these cases was somewhat successful, given the 12 published corrections of the scholarly record as well as the explicit affirmation of "substantial plagiarism" and "academic misconduct." Several of the published corrections were maximal in determination, credit, and availability, insofar as they authoritatively confirmed the problems with the publications, identified the original undocumented sources that had been misappropriated, and appeared in print on paginated citable pages of the relevant journals or online without paywalls. The public record rarely mirrors the experience of whistleblowing activity, however. During the lengthy process of seeking and attaining published corrections of the scholarly literature, I experienced quite a bit of pressure from several parties.

The following three sections of this chapter offer an account of these repercussions. I have no reason to think that my experiences are in any way atypical for academic whistleblowers; in fact, they appear to be consistent with what is reported elsewhere in the academic whistleblowing literature (Sundin and Jutfelt 2018; Sonfield 2014; Zilberberg 2012; Trevino 1996; Bowers 1994; Vincent 2007;

Dansinger 2017). My account is thus one specific example of what is sadly enough a general tendency. By reporting on our experiences as whistleblowers it is hoped that we can contribute to bringing about a needed change. Reprisals against whistleblowers are too common, despite a growing awareness of the essential role of whistleblowers in maintaining a reliable scholarly record.

5.5.1 The National Grant Agency

I submitted an application requesting a review of suspected research violations to the research integrity commission of a national grant agency. In total the national grant agency had funded three works from the author of record (two articles and one book chapter). In response to my application, a lawyer from the national grant agency stated that the allegations of suspected misconduct would be investigated within a reasonable time in accordance with the grant agency's policies. Furthermore, the lawyer stated that those who submit applications will never be informed of the results of the investigation or any actions undertaken by the agency in the matter. In short, I would not have access to the judgment or final report of the agency; I was left to trust the process from afar. I was not completely in the dark, however. Since I had also sent the evidence of suspected plagiarism to the three relevant journal and book editors, I knew that one of these three works had received a published retraction and came to know that the two others would receive published errata.

In light of the stated parameters of secrecy touted by the national grant agency's lawyer, I was quite surprised to receive an email from a journalist at a European newspaper 11 months after submitting my documentation to the national grant agency. The journalist stated that a *Forschungsrat* (research councilor) at the national grant agency had told him that my actions constituted an offense according to the country's penal code for slander, and that the *Forschungsrat* had provided to the reporter references to the statutes and penalties that concerned slander. The reporter asked me to respond to the position of the *Forschungsrat* on this point. I stated in an email response that I worried about the future of science if the national grant agency now considered the reporting of evidence of suspected scientific misconduct to be illegal. Shortly thereafter, the journalist published a lengthy article about the case with the title, "Witch-Hunt in the Realm of Reason" [Hexenjagd im Reich der Vernunft] (Hafner 2017). Without using my name, the article presented me in unflattering terms, described my academic position and research interests, and stated in an unqualified way—apparently based on the claims of the *Forschungsrat*—that I had violated the local law in reporting the accusations of suspected plagiarism. Playing on my publication history in my field of medieval philosophy, the newspaper portrayed me as an overzealous crusader of the middle ages looking to immolate innocents at the stake. Throughout the article, the author of record was portrayed as the victim of a so-called "witch-hunt."

In reporting these cases of suspected plagiarism, I had followed the policies of the grant agency for reporting concerns of research misconduct. To have one of its officials talking with a journalist to allege "illegal activity" on the record about the matter, while denying to me a copy of the grant agency's final report, seemed punitive and unfair. The newspaper article failed to mention explicitly that corrections of the scholarly record had appeared for two of the three cases that I had reported to the national grant agency: a book chapter had been retracted for inadequate citation and a journal article had received a detailed erratum. I wondered why the grant official had apparently not mentioned these two relevant corrections of the scholarly record to the journalist, as their issuance seemed to be a confirmation of the quality of the evidence I had supplied with my application covering three articles. At the time of my application, the author of record for the three articles was holding a position of authority at the national grant agency, and I wondered how this context might have affected the actions of the *Forschungsrat*. As I do not live in the country of the national grant agency, I was grateful that I would never need to apply for grant funds there, but I worried about any future whistleblowers who might submit similar applications in order to correct the scholarly record.

5.5.2 Journal Editors

While examining the articles by the author of record in question, I had encountered what appeared to be varying degrees of inadequate citation in the apparent appropriation of verbatim and near-verbatim texts from many sources. These variances determined the kinds of action I decided to take. For what seemed to be the more severe cases, I requested straightforward published statements of retraction. For some of the less severe cases, I sometimes softened the requests by asking for a correction. For some of the minor cases, I sent no requests. After sending a request for retraction to a journal published by the large international publisher Taylor and Francis, I engaged in lengthy correspondence with two editors of the journal, complying with their follow-up queries for scans of three original sources. The editors thanked me for the additional information, stating it was helpful, and they noted that they would be discussing with the publisher the adoption of a text-matching software check for all articles submitted in the future. They explained that this new process would allow them to detect problems with manuscripts immediately. In the last communication that I received from the editors, they stated that the journal would re-issue the article in a corrected form and that they would let me know once the revised article was placed online. These cordial and professional communications with the editors, the promise to publish a corrected version of the article, and the plan of the editors to adopt text-matching software to check future submissions, jointly led me to conclude that my whistleblowing had been somewhat successful at the journal.

I considered the matter closed and awaited the promised notification by the editors about the correction. It never arrived, however, and the article remained uncorrected. Instead, 3 months later, I was unexpectedly asked to meet with a senior administrator at my university to discuss a letter that had been sent by the two editors of the journal. The lengthy letter, on official letterhead of the Taylor and Francis journal, began, “We write in our capacity as editors of the journal.” The letter declared that its purpose was to inform my university of “questionable ethical conduct on the part of one of your employees.” The editors noted that they had received a request for retraction, and that they had learned of other retraction requests that had been sent by me to other journals. Although they conceded that my actions were “not illegal,” they nevertheless declared that “the ethical basis for those actions is highly questionable.” The letter concluded with the query:

> Given the amount of our precious time that Prof. Dougherty has wasted in his distasteful campaign, we would ask whether you are aware that he devotes what is presumably a huge amount of paid time from Ohio Dominican University to these activities and represents your university in this way.

Accompanying the letter were copies of our email correspondence and a copy of my original retraction request. The letter was signed by both journal editors.

Rather than correcting the scholarly record, the editors had written to my institution alleging vague and unspecified claims of unethical activity. The letter did not mention that three retractions and one erratum had already been issued for other works by the author of record by the time their letter had been sent; the letter also neglected to state that the editors had previously promised to publish a revised version of the paper with source documentation. (This background was discernable through the supplied email correspondence, omitted in the letter itself.) The editors had apparently changed their minds about the matter of a correction at some point. These facts, when brought to the attention of my university, allowed the letter to be seen for what it was.

I was left to wonder about the apparent change of position on the part of the editors, but I was at a loss. As I learned at a later point, one of the editors had a long history of collaboration with the author of record (who was, as noted above, slowly acquiring published corrections for other articles and book chapters). In contrast, I have myself never met, corresponded with, or had any contact in any way with the author of record, having only encountered the articles and book chapters appearing in print under the name. I later learned also that the same editor had published the author of record’s work in another journal while serving in the capacity of editor there, and he was also co-editing a book series with the author of record. I had not yet sent a retraction request to that journal with the evidence of suspected plagiarism for that new case.

The following month I received an email from a different editor from a different journal also published by Taylor and Francis. This editor explained that she had “been urged by other editors” to contact my university to complain about “this campaign in which you appear to be engaging.” To me, it seemed that the two editors were attempting to suborn a third editor into pressuring me to desist from requesting additional published corrections of the scholarly record. At the time, not only was I providing evidence of suspected plagiarism to various editors and publishers, but was participating as a witness in the on-going investigations being conducted by several research integrity offices. Not least as a whistleblower, I was grateful to work at a university led by genuine scholars committed to research integrity, who supported whistleblowing activity as a necessary part of maintaining a reliable body of published literature, and who could properly interpret the attempted interference by the journal editors. Potential whistleblowers should be aware that not all members of the world of learning will separate scholarly criticism of published articles from criticism of persons; that is, not all will be willing to distinguish the professional from the personal.

5.5.3 Research Integrity Offices

Some decisions made by chairs of research integrity offices can present challenges to whistleblowers. A research integrity office may, for example, tell a whistleblower that no new cases will be considered if they are discovered and reported once an investigation has begun. The imposition of what is functionally an *ad hoc* statute of limitations for reporting any additional cases of suspected plagiarism may catch a whistleblower by surprise, especially when new cases are coming to light after the initial cases. Furthermore, if a research integrity office is in a country with more than one national language, its research integrity officer may suddenly switch languages in corresponding with a whistleblower in a later part of an investigation, even though a common language between whistleblower and the research integrity officer has already been used for many months. In the cases mentioned above, English correspondence mysteriously turned to Italian and Dutch midway through various proceedings.

Furthermore, a research integrity office may tell a whistleblower that any retractions, errata, or corrigenda issued by publishers while an investigation is underway one will be considered as merely third-party opinions. Additionally, a research integrity office may change the nature of the investigation to a different focus altogether midway through an investigation: what begins as an investigation into suspected academic plagiarism may be changed to an inquiry regarding whether copyright laws of a certain jurisdiction may have been breached. These tactics may disadvantage a whistleblower who attempts to secure institutional support for published corrections of the scholarly record.

In early 2018, the home institution of the author of record issued a new press release about the situation, occasioned by a journalist's queries about a retraction for plagiarism that had just appeared in print. The very brief press release asserted that the university would not initiate a new investigation and that the university stood by the conclusions of the earlier investigation. The press release then discussed the whistleblower, saying that the university

> wants to underscore how the source of these reports is always the same: the fury with which this person proceeds obliges a certain caution in acting and an accurate evaluation of the good foundation of the continuous 'charges' (tiene poi a sottolineare come la fonte di queste segnalazioni sia sempre la stessa: l'accanimento con il quale questa persona procede impone una certa cautela nell'agire e un'accurata valutazione del buon fondamento delle continue 'denunce') (Illarietti 2018).

With this statement, the university seemed to be trying to direct attention away the confirmations of research misconduct by other authoritative research stakeholders (e.g., editors, publishers, research integrity offices) and instead to focus attention on the whistleblower. The press release implied that bad motivations by the whistleblower (i.e., "fury") was the real issue at hand, rather than any suspected research misconduct. In issuing such a press release, the university appeared to oppose a basic principle of misconduct inquiries, expressed by C. K. Gunsalus as:

> if the facts reported are true, the motive of the whistleblower should not matter. Even where the whistleblower delights in the problems of the wrongdoer, if the objective evidence reveals that important professional standards were violated, the motives of the person raising the question should be irrelevant (Gunsalus 1998: 53).

Several news outlets focused on the university's attempt to direct attention away from academic plagiarism by insinuating a bad motive by the whistleblower (Weinberg 2018a; Stern 2018; Illarietti 2018; Oggiscienza 2018). When asked by a journalist about my views on the matter, I responded:

> Even if my motives were somehow untoward, they would be immaterial to whether misconduct has been committed [...]. In truth, I have never had any dealings with the author of record in these cases, and my interest is professional: establishing a reliable scholarly literature in cases of demonstrated academic plagiarism, particularly in my field of philosophy and related disciplines (Stern 2018).

In light of the university's position in this case, the philosophy news website *Daily Nous* ran a story titled, "Plagiarist's University Issues Criticism ... of the Whistleblower," which ended with the statement that the university's "attempt to smear him is bizarre and unbefitting an institution of higher education" (Weinberg 2018b).

Even if a whistleblower reports evidence of suspected serial scientific misconduct to a university's official research integrity board—evidence that generates many published corrections of the scholarly record by editors and publishers—the response of a university may still be to attempt to impugn the integrity of the whistleblower, for example, through an official press release. Whistleblowers should expect that some academic stakeholders—even ones holding positions of authority—will cast aspersions on their motivations, even though the motivations of whistleblowers are immaterial to whether or not the reported suspected misconduct has been committed.

There are larger implications here. The way a university treats whistleblowers in a given case can have a profound effect upon future potential whistleblowers who will consider whether to disclose evidence of potential research misconduct. Successful attempts to besmirch whistleblowers may be an efficient way for institutions to minimize the chances that potential whistleblowers will use their official reporting structure in the future. As noted above, however, many institutions have come to affirm in recent years that whistleblower harassment and academic witness intimidation is itself a form of misconduct, but that position is not yet universally held in the world of learning. Prospective whistleblowers should be prepared to have their motivations publicly called into question, even though such motivations are irrelevant to the quality of the evidence they provide for suspected research misconduct.

5.6 Conclusion

In sum, potential whistleblowers should expect to receive pushback in the submission of any evidence of potential research misconduct. The sobering first-person narratives offered by some academic whistleblowers can provide valuable insights for any researchers faced with the choice of blowing the whistle who might wonder what kind of retaliation or harassment might be in store for them (Sundin and Jutfelt 2018; Sonfield 2014; Zilberberg 2012; Trevino 1996; Bowers 1994; Vincent 2007; Dansinger 2017). In describing the various forms of retaliation that he experienced, whistleblower Robert Sprague has set forth advice for others who are deliberating whether to disclose evidence of suspected research misconduct. Despite difficulties that included emotional trauma, financial expense, lack of support from colleagues, and many hours lost, Sprague counsels, "Do the right thing even though such action may place you at considerable risk" (1993: 131). According to a study of whistleblower consequences by the United States Office of Research Integrity, of those who suffered one or more negative consequences for their whistleblowing activities, 60% still reported that they would *definitely* blow the whistle again and another 15% would *probably* blow the whistle again (Office of Research Integrity 1995: 48). In short, potential whistleblowers must predict whether the harms likely to be experienced will be worth the good of having contributed to the integrity of research and the reliability of the scholarly record.

Appendix: Sample Letters for Requesting Corrections of the Scholarly Record

Requests for corrections of the scholarly record can take many forms. Four sample letters are offered here. The first (Fig. 5.1) seeks a retraction on the basis of suspected duplicate publication. The second (Fig. 5.2) concerns suspected plagiarism and is signed by several parties. The third (Fig. 5.3) is a request for a corrigendum for the use of an undisclosed pseudonym, and the fourth (Fig. 5.4) seeks a published clarification of authorship.

From: Dougherty, M. V.
Sent: Thursday, May 4, 2017 6:02 AM
Subject: Business Ethics Quarterly / Duplicate and redundant publication inquiry

Dear Professor:

I am writing to inquire whether the publication of an article in *Business Ethics Quarterly* is consistent with the journal's policy on duplicate and redundant publication. The article is:

[—] "Value Maximalization, Stakeholder Theory and the Corporate Objective Function," *Business Ethics Quarterly* 12.2 (2002): 235-256.

The reason for my inquiry is that the text of this article has already appeared in print twice, so its re-appearance in *BEQ* marks at least the third verbatim publication of the same text. The two previous unreferenced instances of publication are:

"Value Maximalization, Stakeholder Theory, and the Corporate Objective Function," *European Financial Management* 7.3 (2001): 297-317.

"Value Maximalization, Stakeholder Theory, and the Corporate Objective Function," *Journal of Applied Corporate Finance* 14.3 (2001): 8-21.

Although Prof. [—] states in the first footnote that the paper first appeared in 2000 in a volume published by the business school of his home institution, I find no indication to the reader that the article has already been published, completely, twice before, in the two above-listed journals. The article's re-appearance in BEQ is likely to lead readers to believe that it presents new, original work in the field, rather than the third verbatim re-presentation of the same text.

I respectfully request that if you find the article to be in violation of the BEQ policy on duplicate and redundant publication, and the norms of scientific publishing, that BEQ and Cambridge University Press correct the scholarly record by publishing a statement of retraction for [—] 2002 within the pages of BEQ.

I have consulted the Cambridge University Press policy on duplicate publication, which states that authors must "acknowledge and cite those sources" when "portions of the content overlap with published or submitted content." The 2002 BEQ version is identical in text, figure, quotations, and footnotes with the previously-published versions in *European Financial Management* and in *Journal of Applied Corporate Finance*, except for a different abstract.

I should mention that I have never met, corresponded with, nor had any communication with Prof. [—], about this or any other matter.

Thank you for considering my confidential request, and I look forward to hearing from you.

Fig. 5.1 Sample letter requesting a retraction for suspected duplicate publication

April 12, 2017

Dear Professor:

We, whose signatures are below, are the current instructor and students with the Honors Program at Ohio Dominican University (USA) in a course titled *HON 379: Critical Research and Writing*. The purpose of this upper-level Honors course is to consider scholarly research from the point of view of justice.

We have observed that a chapter appearing in a volume published by the International Semiotics Institute appears to fall short of adequate citation practices. It is:

> [—], "Subjectivity from a Semiotic Point of View," in *Nordic-Baltic Summer Institute for Semiotic and Structural Studies, Part IV. Ecosemiotics,* ed. Eero Tarasti, et al. (International Semiotics Institute, 2001): 149–159.

The chapter appears to consist substantively of texts pieced together from various authors without quotation marks, either with inadequate attribution or no attribution at all. The document accompanying this letter highlights select passages from the article that are taken verbatim or near verbatim from works by other authors.

As the document makes evident, the fundamental problem is that readers of the chapter have no way of knowing that sentences and paragraphs that appear to be written by Prof. [—] are in fact verbatim and near-verbatim extracts from other authors. A range of citation problems appear to plague the chapter; even when at times the original sources are listed in the bibliography and referenced with an in-text citation, in the absence of quotation marks the reader has no way of knowing that the sentences are verbatim the work of authors other than Prof. [—].

For significant portions of the article, the writings of Pope John Paul II, Anthony Kenny, and Calvin Schrag appear in the article, and no reference to their work is given anywhere in the chapter. We believe that these three undocumented sources in particular constitute the core of the article.

We ask you to consider whether the conditions of academic plagiarism have been met on the basis of this evidence of suspected plagiarism.

We believe that the publication of this chapter may constitute a serious breach of publication integrity. Its inclusion within the repository of published works creates a situation where credit may be due to those original authors whose books and articles have apparently been misappropriated.

For these reasons, we are writing to request that the International Semiotics Institute issue an online statement of retraction of the chapter for plagiarism, if the enclosed evidence is considered to be sufficient. In doing so, ISI will be following the best practices for correcting the scholarly record in this digital age. As we understand it, other corrections of the scholarly record for 9 other works by Prof. [—] have appeared (or will shortly appear) by editors and publishers on the basis of inadequate citation practices.

While we understand that this is a delicate matter, we are convinced that publication integrity and high reputation of the International Semiotics Institute in the world of learning require a correction of the scholarly record whenever plagiarism has been demonstrated.

We should note that we have no relationship with Prof. [—]; we have never met, corresponded with, nor seen him. There is nothing personal in our request for the correction of the scholarly record. In our view, any unretracted plagiarized articles contaminate the scholarly record in various ways: they deny genuine authors the credit they deserve for their discoveries, they distort the genealogy of ideas, and they corrupt the bequest of human knowledge passed on to the next generation of students and researchers.

Thank you for taking up this matter, and we look forward to hearing your reply.

Fig. 5.2 Sample letter requesting a retraction for suspected plagiarism

From: Dougherty, M. V.
Sent: Thursday, July 6, 2017 2:46 PM
Subject: Australasian Journal of Philosophy / Request for Corrigendum

Dear Professor:

I am writing a book on research integrity in the discipline of philosophy, and I came across the issue of Prof. David Lewis' use of a pseudonym ("Bruce Le Catt") in publishing the following article in *Australasian Journal of Philosophy*:

Bruce Le Catt, "Censored Vision," *Australasian Journal of Philosophy* 60.2 (1982): 158–162

In the article, David Lewis apparently uses the pseudonym to critique work published under his own name in the journal two years earlier. The online version of the article in questionon the Taylor and Francis website does not indicate the true authorship of the article. Occasionally, but not always, the pseudonymously-published article is attributed to Prof. Lewis by others publishing on the topic discussed. Not all philosophers are aware of the identity between Lewis and Le Catt, and it is conceivable that many younger members of the profession could read the 1982 article without knowing that Lewis is providing a critique of his own work.

I am writing to request a correction of the scholarly record. Would you kindly publish a brief corrigendum to the article in question, both in the pages of the journal and tethered electronically to the online version of the article, that declares the true authorship of the article? In doing so, *Australasian Journal of Philosophy* would be following the best practices for maintaining publishing and research integrity.

Thank you for considering my request for a correction of the scholarly record.

Fig. 5.3 Sample letter requesting a corrigendum for an undisclosed pseudonym

From: Dougherty, M. V.
Sent: Tuesday, November 07, 2017 3:26 PM
Subject: Request for published clarification of authorship / Philosophical Studies

Dear Editors:

I am writing to request a published clarification of authorship for an article in *Philosophical Studies* that appeared under a pseudonym. The article is:

M. Lisagor, "On Harman's Theory of Knowledge," *Philosophical Studies* 29.6 (1976): 433-439.

I am requesting the publication of a short note, within the pages of the journal, stating that "M. Lisagor" is a pseudonym for Prof. Joseph Margolis, Temple University, USA. I have spoken with Prof. Margolis to confirm that he is "M. Lisagor," and he supports this request for a published clarification. Shortly after the publication of the article, Prof. Margolis confirmed his use of the pseudonym with Prof. David Shatz of Yeshiva University, who discovered the use of the pseudonym. (Both Prof. Margolis and Prof. Shatz are copied on this email.)

There are precedents for my request. In recent months, *Australasian Journal of Philosophy*, *The Journal of Philosophy*, and *The University of California Press* have each issued clarifications that identify the true authorship of articles published under pseudonyms for separate cases involving philosophers David Lewis, Amélie Rorty, and Neven Sesardić.

Thank you for considering this request for a published clarification of the scholarly record. Such a clarification would properly identify the article as belonging to Prof. Margolis's philosophical works.

Fig. 5.4 Sample letter requesting a clarification of authorship

References

Abritis, Alison. 2018. Caught our notice: Researcher who sued PubPeer commenter up to 21 retractions. *Retraction Watch*, April 5. https://retractionwatch.com/2018/04/05/caught-our-notice-researcher-who-sued-pubpeer-commenter-up-to-21-retractions. Accessed 10 July 2018.

Anderson, Melissa S., Marta A. Shaw, Nicholas H. Steneck, Erin Konkle, and Takehito Kamata. 2013. Research integrity and misconduct in the academic profession. In *Higher education: Handbook of theory and research*, ed. Michael B. Paulsen, 217–261. Dordrecht: Springer.

Anonymous. 2016a. It has been brought to our attention [...]. Leiden: Brill. https://brill.com/abstract/title/27457. Accessed 6 July 2018.

———. 2016b. It has been brought to our attention [...]. Leiden: Brill. https://brill.com/view/title/27478. Accessed 6 July 2018.

———. 2017a. Erratum. *Swiss Medical Weekly* 14406 (1): 147. https://smw.ch/en/resource/jf/journal/file/download/article/smw/en/smw.2017.14406/SMW14406.pdf. Accessed 6 July 2018.

———. 2017b. It was brought to our attention [...]. Amsterdam: John Benjamins Publishing Company. https://doi.org/10.1075/cvs.1.05sch. Accessed 6 July 2018.

———. 2017c. Retraction. *Studies in Communication Sciences* 6 (2): 215. http://www.hope.uzh.ch/scoms/article/view/1603. Accessed 6 July 2018.

———. 2017d. Corrigendum/Scoms correction note. *Studies in Communication Sciences* 9 (1): 248A. http://www.hope.uzh.ch/scoms/article/view/1616. Accessed 6 July 2018.

———. 2018a. Erratum. Cham: Springer. https://link.springer.com/chapter/10.1007/978-90-481-3949-1_11. Accessed 6 July 2018.

———. 2018b. Corrigendum. *Journal of Communication in Healthcare* 11 (1): 81.

———. 2018c. Corrigendum. *Patient Education and Counseling*. https://doi.org/10.1016/j.pec.2018.07.014. Accessed 29 August 2018.

Anonymous Academic. 2017. Plagiarism is rife in academia, so why is it rarely acknowledged? *The Guardian*, October 27. https://www.theguardian.com/higher-education-network/2017/oct/27/plagiarism-is-rife-in-academia-so-why-is-it-rarely-acknowledged. Accessed 6 July 2018.

Blatt, Michael R. 2015. Vigilante science. *Plant Physiology* 169: 907–909. https://doi.org/10.1104/pp.15.01443 Accessed 6 July 2018.

Bouter, Lex M., and Sven Hendrix. 2017. Both whistle blowers and the scientists they accuse are vulnerable and deserve protection. *Accountability in Research* 24 (6): 359–366.

Bowers, Neal. 1994. A loss for words: Plagiarism and silence. *The American Scholar* 63 (4): 545–555.

Brentlinger, Paula E., et al. 2009. Plagiarism. Plagiarism: Authors' response. Plagiarism: Editor's note. *Transactions of the Royal Society of Tropical Medicine and Hygiene* 103 (8): 855–857.

Broeksteeg, Ielse. 2010. K.U. Leuven involved in new plagiarism scandal? Did the Institute of Philosophy cover up a fraudulent professor? *The Voice*, February 16.

College van Bestuur van de Universiteit van Amsterdam, 2017. Besluit: Definitief oordeel op klacht. April 20. N. 2017cb0099. https://retractionwatch.com/wp-content/uploads/2018/01/PDF_2_University_of_Amsterdam.pdf. Accessed 6 July 2018.

Cox, Adam, Russell Craig, and Dennis Tourish. 2018. Retraction statements and research malpractice in economics. *Research Policy* 47 (5): 924–935.

Cressey, Daniel. 2013. 'Rehab' helps errant researchers return to the lab: As cases of misconduct rise, ethicists test an approach for reforming offenders. *Nature* 493: 147. https://doi.org/10.1038/493147a Accessed 6 July 2018.

Dannemann, Gerhard. 2018. Crowd-based documentation of plagiarism: The Vroniplag Wiki experience. In *Research ethics in the digital age: Ethics for the social sciences and humanities in times of mediatization and digitization*, ed. Farina Madita Dobrick, Jana Fischer, and Lutz M. Hagen, 45–67. Wiesbaden: Springer.

Dannemann, Gerhard, and Debora Weber-Wulff. 2015. Viel Licht und noch mehr Schatten: Wie Universitäten auf Plagiatsdokumentationen reagieren. *Forschung & Lehre*, April: 278–280. http://www.wissenschaftsmanagement-online.de/beitrag/viel-licht-und-noch-mehr-schatten-wie-universit-ten-auf-plagiatsdokumentationen-reagieren5216. Accessed 6 July 2018.

Dansinger, Michael. 2017. Dear plagiarist: A letter to a peer reviewer who stole and published our manuscript as his own. *Annals of Internal Medicine* 166 (2): 143. https://doi.org/10.7326/M16-2551 Accessed 6 July 2018.

Devine, Tom, and Tarek F. Maassarani. 2011. *The corporate whistleblower's survival guide*. San Francisco: Berrett-Koehler Publishers.

Devine, Tom, and Alicia Reaves. 2016. Whistleblowing and research integrity: Making a difference through scientific freedom. In *Handbook of academic integrity*, ed. Tracey Bretag, 957–972. Singapore: Springer.

Doran, Michael. 2016. How to survive as a whistle-blower. *Nature* 532: 405. https://doi.org/10.1038/nj7599-405a Accessed 6 July 2018.

Fiske Susan T. 2016a. Mob rule or wisdom of crowds? *APS Observer.*, [in Press]: 1–2. http://datacolada.org/wp-content/uploads/2016/09/Fiskepresidential-guest-column_APS-Observer_copy-edited.pdf. Accessed 6 July 2018.

———. 2016b. A call to change science's culture of shaming. *APS Observer*. https://www.psychologicalscience.org/observer/a-call-to-change-sciences-culture-of-shaming. Accessed 6 July 2018.

Forest, C. 2018. PubPeer contre "fake news" en sciences? *Ethics, Medicine and Public Health.* https://doi.org/10.1016/j.jemep.2018.01.009 Accessed 6 July 2018.

Fosmoe, Margaret. 2017. Court says IU South Bend professor who filed suit wasn't defamed. *South Bend Tribune*, January 4. http://www.southbendtribune.com/news/education/court-says-iu-south-bend-professor-who-filed-suit-wasn/article_d4109bd1-18c2-5893-b3bc-5388c2cdf574.html. Accessed 6 July 2018.

Fox, Mark, and Jeffrey Beall. 2014. Advice for plagiarism whistleblowers. *Ethics and Behavior* 24 (5): 341–349.

Gelman, Andrew. 2016. What has happened down here is the winds have changed. *Statistical Modeling, Causal Inference, and Social Science*, September 21. http://andrewgelman.com/2016/09/21/what-has-happened-down-here-is-the-winds-have-changed. Accessed 6 July 2018.

Grens, Kerry. 2017. Journal cleans up image archives. *The Scientist*, June 12. http://www.the-scientist.com/?articles.view/articleNo/49642/title/Journal-Cleans-Up-Image-Archives. Accessed 6 July 2018.

Gunsalus, C.K. 1998. How to blow the whistle and still have a career afterwards. *Science and Engineering Ethics* 4 (1): 51–64.

Hafner, Urs. 2017. Hexenjagd im Reich der Vernunft. *Neue Zürcher Zeitung*, January 21. https://www.nzz.ch/feuilleton/plagiat-hexenjagd-im-reich-der-vernunft-ld.142092. Accessed 6 July 2018.

Hames, Irene. 2007. *Peer review and manuscript management in scientific journals: Guidelines for good practice*. Malden: Blackwell Publishing.

Hauptman, Robert. 2011. *Authorial ethics: How writers abuse their calling*. Lanham: Lexington Books.

Hosseini, Mohammad, Medard Hilhorst, Inez de Beaufort, and Daniele Fanelli. 2018. Doing the right thing: A qualitative investigation of retractions due to unintentional error. *Science and Engineering Ethics* 24 (1): 189–206.

Hussinger, Katrin, and Maikel Pellens. 2017. Guilt by association: How scientific misconduct harms prior collaborators. *Zentrum für Europäische Wirtschaftsforschung Discussion Papers* 17–051. http://ftp.zew.de/pub/zew-docs/dp/dp17051.pdf. Accessed 6 July 2018.

Illarietti, Davide. 2018. Il docente dell'Usi ha copiato anche il Papa. *Ticino Online*, January 18. https://www.tio.ch/ticino/attualita/1236159/il-docente-dell-usi-ha-copiato-anche-il-papa. Accessed 6 July 2018.

Jacobs, Brian, et al. 2014. Plagiarism identified in *JOMS* article. *Journal of Oral and Maxillofacial Surgery* 72 (11): 2098.

Lewis, James R. 2016. Sucking the '*De*' out of me: How an esoteric theory of persecution and martyrdom fuels falun gong's assault on intellectual freedom. *Alternative Spirituality and Religion Review* 7 (1): 93–109.

Lu, Susan Feng, Ginger Zhe Jin, Brian Uzzi, and Benjamin Jones. 2013. The retraction penalty: Evidence from the Web of Science. *Scientific Reports* 3 (3146): 1–5. https://doi.org/10.1038/srep03146 Accessed 6 July 2018.

Marcus, Adam, and Ivan Oransky. 2015. We gave bad advice: Better not contact authors first. *Lab Times*, November 22. http://www.labtimes.org/labtimes/ranking/dont/2015_06.lasso. Accessed 6 July 2018.

———. 2016. For young scientists, a supervisor's fraud can derail a career. *Stat*, November 25. https://www.statnews.com/2016/11/25/postdocs-grad-students-fraud. Accessed 6 July 2018.

Martin, Ben R. 2007. Keeping plagiarism at bay—A salutary tale. *Research Policy* 36 (7): 905–911.

Martinelli, Dario. 2018a. Statement of retraction. International Semiotics Institute, January 8. https://www.facebook.com/isisemiotics/posts/880483692121348. Accessed 6 July 2018.

———. 2018b. Concerning our statement of retraction. International Semiotics Institute, January 18. https://www.facebook.com/isisemiotics/posts/883179261851791. Accessed 6 July 2018.

McCook, Alison. 2017. Journal won't look at allegations about papers more than six years old. *Retraction Watch*, June 9. http://retractionwatch.com/2017/06/09/journal-wont-look-allegations-papers-six-years-old-public-websites. Accessed 6 July 2018.

———. 2018. PubMed shuts down its comments feature, PubMed Commons. *Retraction Watch*, February 2. http://retractionwatch.com/2018/02/02/pubmed-shuts-comments-feature-pubmed-commons. Accessed 6 July 2018.

NCBI Insights. 2018. PubMed Commons to be discontinued. February 1. https://ncbiinsights.ncbi.nlm.nih.gov/2018/02/01/pubmed-commons-to-be-discontinued. Accessed 6 July 2018.

Nelson, Leif D., Joseph Simmons, and Uri Simonsohn. 2018. Psychology's Renaissance. *Annual Review of Psychology* 69: 511–534.

Office of Research Integrity. 1995. *Consequences of whistleblowing for the whistleblower in misconduct in science cases: Final report*. https://ori.hhs.gov/sites/default/files/final.pdf. Accessed 6 July 2018.

Oggiscienza. 2018. In filosofia, caccia alle bufale = caccia alle streghe. *Oggiscienza: La ricerca e i suoi protagonisti*, March 9. https://oggiscienza.it/2018/03/09/filosofia-caccia-bufale-caccia-streghe. Accessed 6 July 2018.

Parrish, Debra M. 1999. Scientific misconduct and correcting the scientific literature. *Academic Medicine* 74 (3): 221–230.

PubMed Commons. 2013–2018. https://www.ncbi.nlm.nih.gov/pubmedcommons. Accessed 6 July 2018.

PubPeer. 2014a. FAQ. *PubPeer: The Online Journal Club*. https://web.archive.org/web/20140203061445/https://pubpeer.com/faq. Accessed 6 July 2018.

———. 2014b. Suspected research misconduct and poor research practices. *PubPeer: The Online Journal Club*. https://pubpeer.com/misconduct. Accessed 6 July 2018.

———. 2015. Vigilant Scientists. *PubPeer: The Online Journal Club*, October 5. https://blog.pubpeer.com/publications/B6CF3DB974A8ECC64B1A0303BBCD6F. Accessed 6 July 2018.

PubPeer: The Online Journal Club (n.d.). https://pubpeer.com. Accessed 6 July 2018.

Resnik, David B. 2003. From Baltimore to Bell Labs: Reflections on two decades of debate about scientific misconduct. *Accountability in Research* 10 (2): 123–135.

Retraction Watch (n.d.). http://retractionwatch.com. Accessed 6 July 2018.

Riley, Kathleen L. 2000. Review of: The life and times of Archbishop Fulton J. Sheen. *The Catholic Historical Review* 86 (4): 712–713.

Roig, Miguel. 2014. Journal editorials on plagiarism: What is the message? *European Science Editing* 40 (3): 58–59. http://europeanscienceediting.eu/articles/journal-editorials-on-plagiarism-what-is-the-message. Accessed 6 July 2018.

Scanlon, Patrick M. 2007. Song from myself: An anatomy of self-plagiarism. *Plagiary* 1: 57–66. http://hdl.handle.net/2027/spo.5240451.0002.007. Accessed 6 July 2018.

Schechner, Richard, Talia Rodgers, Claire L'Enfant, Judith Butler, Marvin Carlson, Tracy C. Davis, David Savran, Shannon Jackson, Branislav Jakovljevic, Jill Dolan, Zarrilli Phillip, W.B. Worthen, Joseph Roach, and Peggy Phelan. 2009. Concerning theory for performance studies. *TDR: The Drama Review* 53 (1): 7–46.

Simonsohn, Uri. 2014. Just post it: The lesson from two cases of fabricated data detected by statistics alone. *Psychological Science* 24 (10): 1875–1888.

Singal, Jesse. 2015. The case of the amazing gay-marriage data: How a graduate student reluctantly uncovered a huge scientific fraud. *New York Magazine*, May 29. http://nymag.com/scienceofus/2015/05/how-a-grad-student-uncovered-a-huge-fraud.html. Accessed 6 July 2018.

———. 2016. Inside psychology's 'methodological terrorism' debate. *New York Magazine*, October 12. http://nymag.com/scienceofus/2016/10/inside-psychologys-methodological-terrorism-debate.html. Accessed 6 July 2018.

Sonfield, Matthew C. 2014. Academic plagiarism at the faculty level: Legal versus ethical issues and a case study. *Journal of Academic Ethics* 12 (2): 75–87.

Sprague, Robert L. 1993. Whistleblowing: A very unpleasant avocation. *Ethics and Behavior* 3 (1): 103–133.

State of Michigan Court of Appeals. 2016. Fazlul Sarkar, Plaintiff-Appellant, v John Doe, Defendant-Appellee, and Jane Doe, Defendant, and PubPeer Foundation, Appellee. Fazlul Sarkar, Plaintiff-Appellee, v John Doe and Jane Doe, Defendants, and PubPeer Foundation, Appellant. *State of Michigan Court of Appeals*, December 6. No. 326667, No. 326691. Wayne Circuit Court, LC No. 14–013099-CZ. https://www.aclu.org/sites/default/files/field_document/20161206_c326667_125_326667.opn_.pdf. Accessed 6 July 2018.

Stern, Victoria. 2018. University defends researcher accused of plagiarizing former Pope. *Retraction Watch*, January 31. https://retractionwatch.com/2018/01/31/university-defends-researcher-accused-plagiarizing-former-pope. Accessed 6 July 2018. Accessed 6 July 2018.

Sundin, Josefin, and Fredrik Jutfelt. 2018. Keeping science honest. *Science* 359 (638330): 1443. https://doi.org/10.1126/science.aat3473 Accessed 6 July 2018.

Teixeira da Silva, Jaime A. 2017. Ethical exceptionalism: Can publishing rules be manipulated to give the impression of ethical publishing? *Bangladesh Journal of Medical Science* 16 (4): 610–614. https://doi.org/10.3329/bjms.v16i4.33623 Accessed 6 July 2018.

Trevino, Linda Klebe. 1996. Whistleblowing in the academic community: Personal reflections of a victim of plagiarism. *Proceedings of the International Association for Business and Society* 7: 883–890.

Universities UK. 2012. *The Concordat to support research integrity*. London: Universities UK. http://www.universitiesuk.ac.uk/policy-and-analysis/reports/Pages/research-concordat.aspx. Accessed 6 July 2018.

USI. 2008. Faculty of communication sciences study regulations. http://www.com.usi.ch/regolamento-studi-com-2008-en-39555.pdf. Accessed 6 July 2018.

———. 2016. Violazioni del diritto d'autore ma nessun dolo. Conclusa l'inchiesta per comportamento scientifico scorretto, August 29. http://www.usi.ch/en/comunicati-stampa/1894. Accessed 6 July 2018.

van Eemeren, Frans H. 2015a. Retraction note. *Argumentation* 29 (4): 493.

———. 2015b. Erratum. *Argumentation* 29 (4): 481–491.

Vandevelde, Toon. 2010. *Reaction of the Dean of the Institute of Philosophy*. Leuven: K. U. https://web.archive.org/web/20170103135518/http://hiw.kuleuven.be/ned/nieuws/persknipsels/ReactionDeanResignationStone.pdf. Accessed 16 March 2017.

Vincent, Clement. 2007. The purloined bibliography. *The Chronicle of Higher Education*, July 16. http://www.chronicle.com/article/The-Purloined-Bibliography/46601. Accessed 6 July 2018.

Virginia Tech Research Integrity Office. 2016. Response to complaint regarding potential plagiarism. http://retractionwatch.com/wp-content/uploads/2018/01/PDF_1_Virginia_Tech-1.pdf. Accessed 6 July 2018.

Weber-Wulff, Debora. 2014. *False feathers: A perspective on academic plagiarism*. Heidelberg: Springer.

Weinberg, Justin. 2018a. Plagiarizes again—And is caught by philosophy prof.'s class (updated). *Daily Nous: News for and About the Philosophy Profession*, January 15, http://dailynous.com/2018/01/15. Accessed 6 July 2018.

———. 2018b. Plagiarist's university issues criticism …of the whistleblower. *Daily Nous: News for and About the Philosophy Profession*, February 1. http://dailynous.com/2018/02/01/plagiarists-university-issues-criticism-whistleblower. Accessed 6 July 2018.

Wiwanitkit, Somsri, and Viroj Wiwanitkit. 2017. Responses of authors accused of plagiarism by journal editors. *Science and Engineering Ethics* 23 (1): 309–311.

Woolf, Patricia K. 1988. Deception in scientific research. *Jurimetrics* 29 (1): 67–95.

———. 2016. Deception in scientific research. In *Research Ethics*, ed. Kenneth D. Pimple, 71–98. London: Routledge.

Yong, Ed, Heidi Ledford, and Richard Van Noorden. 2013. 3 ways to blow the whistle. *Nature* 503: 454–457. https://doi.org/10.1038/503454a Accessed 6 July 2018.

Zilberberg, Marya. 2012. How does it feel to have your scientific paper plagiarized? And what can you do about it? *Retraction Watch*, March 12. http://retractionwatch.com/2012/03/12/how-does-it-feel-to-have-your-scientific-paper-plagiarized-and-what-can-you-do-about-it. Accessed 6 July 2018.

Chapter 6
Publishing Corrections of the Scholarly Record: Some Test Cases

Abstract This chapter examines the responses of editors and publishers who were presented with evidence of suspected plagiarism in a series of 14 articles and book chapters for one author of record. As the publications divide into those in philosophy and those in health communication, a comparison is possible between the manner in which different disciplines respond to evidence of suspected plagiarism. Using news reports as well as publicly issued statements from the home institutions of the author of record and of co-authors as a backdrop, I examine various textual parallels in light of the 12 published retractions, errata, and corrigenda that have been issued in these cases. Attention is given to the variety of text manipulations present in the articles and book chapters that have been subject to corrections by editors and publishers.

Keywords Retractions · Errata · Corrigenda · Publishing · Plagiarism

The whistleblowing activity recounted in the previous chapter generated 12 corrections of the published literature for one author of record. To date, five retractions, three errata, and four corrigenda have been issued by editors and publishers. The home institution of the author of record had initially issued a public statement declaring that the researcher was the subject of an investigation for plagiarism, but in the end the university judged that no articles or book chapters were plagiarized and that eight publications were deficient in only minor respects. As first announced by the university and then reported in the press, two of these eight were deemed to constitute "minor violations of copyright" [violazioni minori del diritto d'autore] and the remaining six were deemed to exhibit "incorrect methods of referring to sources" which however were considered to be "of minor importance" [modalità scorrette di citazione delle fonti di minor gravità] (USI 2016b; Illarietti 2016; see also Illarietti 2018). The university was unequivocal in its ruling that no plagiarism had occurred in the articles examined by its research integrity committee.

The 12 published corrections by editors and publishers thus exceeded the university's public count of deficient publications. This difference in number was not the only discrepancy, however. The language employed in the corrections suggested that a more significant violation of research norms had occurred than the minor

M. V. Dougherty, *Correcting the Scholarly Record for Research Integrity*, Research Ethics Forum 6, https://doi.org/10.1007/978-3-319-99435-2_6

breaches conceded by the university. The published retractions, errata, and corrigenda stated, for example, that there were "severe shortcomings in the references" (Anonymous 2015b: 481, 2017b), that "direct reference and citation of the works of other scholars is often inconsistent and in some cases totally lacking" (Anonymous 2016a, b) and that the works do "not meet standards currently expected of an academic publication" (ibid.). One retraction stated that "the essay in question plagiarized from various sources" (Martinelli 2018a). The difference in tone (if not in substance) between the university's public statements and the published corrections by editors and publishers should perhaps be unsurprising; studies on the rhetoric of apology for academic lapses has suggested that wrongdoers consistently understate their wrongdoing in their acts of apologizing (Souder 2010).

In what follows, I do not offer any new accusations or judgments of academic plagiarism. It should be noted that the home institution of the researcher was the first to publicly use the term *plagiarism* in this specific context by declaring through press releases and in on-the-record statements to journalists that it was conducting "an investigation in plagiarism" (USI 2016a; Palus 2016b). On the contrary, I only consider the publicly available evidence, namely, the actions of editors and publishers in the issuance of retractions, corrigenda, and errata, as well as the statements by three investigating institutional integrity offices.

The examination here of the public evidence provides a test-case for considering the efficacy of the traditional methods for correcting the scholarly record in philosophy and other disciplines. I contextualize the published corrections in light of the evidence that prompted them, paying particular attention to the various types of corrections that were used for the articles and book chapters. The evidence offered in this chapter suggests that philosophy and related disciplines fall short in maintaining the integrity of the scholarly record. The body of published literature is damaged not just by faulty publications, but also by the issuance of corrections that are ambiguous and by the absence of corrections for work that is unreliable. In short, editorial and institutional failures can compound the failures of authors of record.

6.1 Disambiguating Retractions, Errata, and Corrigenda

The published corrections of the scholarly literature for these 12 cases belong to three classes: retractions, errata, and corrigenda. Retractions are considered the gold-standard for corrections of the scholarly record; they are the most serious of all classes of published corrections and have been characterized as the "nuclear option" for editors and publishers (Marcus and Oransky 2017: 119). In issuing a retraction, a publisher changes the status of a published article; the published retraction provides an authoritative warning to readers that the evidence for an article's unreliability is conclusive. Retractions are the strongest external or *ad extra* corrections for mitigating the harm occasioned by deficient published literature.

There is less clarity, however, about the precise scope and meaning of errata and corrigenda. In a presentation of the best practices of scientific publishing, Irene

Hames counsels that errata should be issued "to correct an error made by the journal or publisher during the publishing process," whereas corrigenda should be issued "to correct an error made by the authors" (2007: 193). Despite having these distinct meanings, the two types of corrections are "sometimes confused and used incorrectly" (ibid.). Hames's proposed taxonomy of published corrections should be understood as prescriptive rather than descriptive; indeed, many editors and publishers—particularly in philosophy and in humanities disciplines—use the terms *erratum* and *corrigendum* interchangeably. Some major scientific publishers (e.g., Elsevier) and some journals (e.g., those under the Nature imprint) explicitly endorse this distinction to manifest to readers the exact origin of fault for errors that have occasioned published corrections (Elsevier 2016; Nature n.d.). Historical support for restricting the term *erratum* to errors attributable to the publisher alone is found in the field of book publishing. Early modern presses often issued lists of printer's errors—generally titled "*Errata*"—which would be inserted into books after printing but prior to distribution (Blair 2007).

Certain journals in philosophy follow this long-standing convention of using the term *erratum* for the correction of errors caused by the publisher during the publication process. Some published errata in philosophy journals state explicitly that the problems generating the corrections reside exclusively with the publisher and are "no fault of the author" (Anonymous 2000a: 158) or they explain that the mistakes were "introduced during the publishing process" (Anonymous 2017j). Errata have been issued in philosophy journals in cases where a line of printed text has been mistakenly deleted (Anonymous 1984: 154), a figure misplaced (Anonymous 1993a: 377), or even where a paper's entire bibliography has been inadvertently omitted in production (Anonymous 1998: 613–615). They have been published to correct errantly presented figures (Anonymous 2013: 856), flow charts (Anonymous 1999a: 324), formulas (Anonymous 1999b: 791), proofs (Anonymous 2000b: 127), and indices (Anonymous 1989: 171–172). Publishers in philosophy have also issued errata when errant numbering has made a paper's footnotes unreliable (Anonymous 2003: 237, 1993b: 346, 1968: 158), when figures have become unintelligible by being printed in black-and-white rather than color (Anonymous 2009a: 107–111), or when a key definition has been given incorrectly (Anonymous 2014: ii). At times errata correct other kinds of production errors in philosophy journals, as when the word "not" is omitted in a crucial sentence (Anonymous 2012a: 2), an author's name is misspelled (Anonymous 2017i: 433, 2017g: 607, 2009b: i; 1953: 96, 2012b: 505), or an article is published with the wrong person credited as the author of record (2017h: 319; Dutton 2001).

Not all editors and publishers in philosophy restrict the use of published errata to the correction of production errors, however; often the term is used for correcting errors introduced by authors. Thus, Gilbert Ryle published an erratum in the journal *Mind* to correct the mistaken assertion that a certain philosopher had died. In the erratum, Ryle expressed the hope that none of the philosopher's "many friends and admirers have believed the false tidings" (1932: 138). In August 2017, the online philosophy journal *Ergo* added an undated erratum to a 2-year-old article explaining that what was being cited and criticized in the original article as an "unpublished manuscript" by philosopher Alan Soble was really not written by him. The erratum stated soberly that the cited manuscript was "by an anonymous author who is not Alan Soble" and it concluded with an apology (Anonymous 2017f: 175). In an unusual erratum titled "Editorial Apology," *The Heythrop Journal* acknowledged a failure in "editorial oversight, specifically in not correcting certain expressions that are offensive in nature and hold no place in professional academic discourse" (Anonymous 2017k: 1). The journal had published more than a dozen book reviews in previous years in which a single reviewer had consistently referred to all female writers as "lady authors." In this case, the reviewer had recently passed away, and the erratum couched the matter as an editorial failure rather than an authorial one.

Some philosophy journal editors ignore the distinction between errata and corrigenda by correcting production errors through corrigenda (e.g., Anonymous 1990: 223, 2016c: 699) or by publishing errata for apparent plagiarism (e.g., Michalos 2010: 495). To be sure, class-bending corrections are not only found in philosophy. Some scientific editors and publishers have issued statements that blend the three types of correction, as in the unusual hybrid "Erratum/Corrigendum: Expression of Concern" (e.g., Anonymous 2016c: 5031) or when an expression of concern states that an article is being withdrawn for plagiarism (e.g., Rydzyński 2006: 149–150).

In addition to class-transgressing, some corrections in philosophy publications can also be obscure. In 2009, Cambridge University Press appended an erratum to the paperback version of a 2008 bioethics monograph stating that in earlier published works from 1997 to 2000 another researcher had "independently developed an argument along the lines" offered in one chapter of the book (Lee and George 2009: x). This erratum is puzzling, as it appears to be defending against a charge rather than admitting or correcting any error, and there is no indication that the publisher or the two authors of record are at fault for any issue.

Further complicating attempts to disambiguate the classes of published corrections is the practice of some philosophy journals to allow authors to publish short follow-up notes that update their previously published articles, especially when those notes appear with such titles as "Erratum," "Corrigendum," or even "Retraction." Despite the similarity in title, these author-initiated supplementary notes are different in kind from the corrections issued by editors and publishers (e.g., Morton 2016: 445; Deutch 2014: 629; Bailin 2011: 1096; Cohen 1967: 352). The notes are published under the name of the author of record, whereas corrections by editors and publishers typically make no mention of an author of record and simply appear under the aegis of the journal or publisher.[1]

If adopted more widely, Hames's abovementioned prescriptions to reserve the term *erratum* for any correction when the publisher is at fault and the term *corrigendum* for when the author of record is at fault would secure a greater transparency in published corrections of the scholarly literature. Furthermore, some editors and publishers have suggested that since the term *retraction* generally carries a negative connotation for authors, it should be used to correct errors only when the author is at fault, and never when the publisher has made an error during the production

[1] This widespread practice of issuing corrections without an identifiable author has been subject to increased criticism in recent years (see Hu 2017; Xu and Hu 2018).

process (Etkin and Oransky 2016/2017). The inconsistent use of the terms for correction (e.g., retraction, errata, corrigenda, or expressions of concern) has led some theorists to suggest that the entire nomenclature should be abandoned. On one proposed model, corrections would all be designated as "amendments" that would indicate the degree of unreliability of an article (Barbour et al. 2017:7; see also Fanelli et al. 2018). If adopted, this proposal would be a complete overhaul of the longstanding practices of corrections issued by editors and publishers: there would no longer be issued any retractions, and articles could be amended multiple times.

Recent years have seen an expansion, rather than a reduction, of the types of corrections appearing in print, however. In the discipline of philosophy, corrections now occasionally appear under an additional form, titled "addenda" by some publishers (e.g., Anonymous 2017e: 127). Some fields have seen an increase of "retract and replace" situations, where a revised article is substituted for a defective one at the time of retraction (McCook 2016). Other correction variations include "partial retractions" where only a portion of a published article (such as a figure or a paper subsection) is removed. These alternate forms of corrections are not always well received. In the words of one critic, for example, "the category of a partial retraction is invalid (i.e., nonsensical) and does not correspond to the practical reality of what constitutes a 'whole' scientific paper" (Teixeira da Silva 2016a: 45).

In light of Hames's prescriptions, it is not clear why any of the 12 published corrections by one author of record discussed at the beginning of this chapter should have warranted the designation of *errata*, except perhaps if the term was being used imprecisely by some editors and publishers. The deficiencies generating the 12 corrections appear to have originated in the submitted manuscripts, rather having been introduced inadvertently by editors and publishers. The articles and chapters had incorporated the work of others with inadequate or no references to the original publications. The select cases that follow below exemplify the variegated ways in which editors and publishers responded to requests for published corrections of the scholarly record in philosophy and in other disciplines.

6.2 Case 1: Proceedings from the International Semiotics Institute

Case 1 is a chapter published in the proceedings of an annual semiotics conference sponsored by the International Semiotics Institute. Titled "Subjectivity from a Semiotic Point of View," this 2001 chapter appears to set forth new reflections on the philosophical notion of the self from the perspective of semiotics, an interdisciplinary field involving philosophy, linguistics, and communications, among other disciplines. The chapter in question exhibits substantive textual parallels with earlier published work by several authors, most often with little or no reference to the earlier texts. Table 6.1 offers a typical example, with the text from the 2001 chapter in the first column and the unreferenced 1993 antecedent in the second. The verbatim parallels between the two texts are highlighted.

Table 6.1 Text parallels with an earlier work by Pope John Paul II (Karol Wojtyła)

[—], "Subjectivity from a Semiotic Point of View," in *Ecosemiotics: Studies in Environmental Semiosis*, Nordic-Baltic Summer Institute for Semiotic and Structural Studies, Part IV, ed. Eero Tarasti et al. (Imatra: International Semiotics Institute, 2001): 149–159, at 156–157.	Pope John Paul II / Karol Wojtyła, "Subjectivity and the Irreducible in the Human Being," in *Person and Community: Selected Essays*, trans. Theresa Sandok (New York: Peter Lang, 1993), 209–217.	
the primordial uniqueness of the human being, and thus for the basic irreducibility of the human being to the natural world. This assumption forms the basis of understanding the human being as a person. Traditional Aristotelian anthropology was based on the definition *o anthropos zoon noetikon (homo est animal rationale)*. This definition fulfills Aristotle's requirements for defining the species (human being) through its proximate genus (living being) and the feature that distinguishes the given species in that genus (endowed with reason). At the same time, however, the definition is constructed in such a way that it excludes – at least at first glance – the possibility of accentuating the irreducible in the human being. In this definition the human being is mainly an object, one of the objects in the world to which the human being visibly and physically belongs. In this perspective, objectivity was connected with the general assumption of the reducibility of the human being. The term subjectivity, on the other hand, proclaims that the human being's proper essence cannot be reduced and explained by the proximate genus and specific difference. In other words: Subjectivity is a synonym for the irreducible in the human being.	the primordial uniqueness of the human being, and thus in the basic irreducibility of the human being to the natural world [...]. This belief stands at the basis of understanding the human being as a person [...]. // Traditional Aristotelian anthropology was based, as we know, on the definition *o anthropos zoon noetikon, homo est animal rationale*. This definition fulfills Aristotle's requirements for defining the species (human being) through its proximate genus (living being) and the feature that distinguishes the given species in that genus (endowed with reason). At the same time, however, the definition is constructed in such a way that it excludes—when taken simply and directly—the possibility of accentuating the irreducible in the human being. // In [...] the definition *homo est animal rationale*, the human being was mainly an object, one of the objects in the world to which the human being visibly and physically belongs. Objectivity in this sense was connected with the general assumption of the reducibility of the human being. Subjectivity, on the other hand, is, as it were, a term proclaiming that the human being's proper essence cannot be totally reduced to and explained by the proximate genus and specific difference. Subjectivity is, then, a kind of synonym for the irreducible in the human being.	Wojtyła 211 Wojtyła 210-211

The earlier text is a short published essay by Pope John Paul II. First written in Polish under the pre-papal name Karol Wojtyła for a philosophy conference in 1975, when Wojtyła was the Cardinal Archbishop of Kraków, it was translated twice into English, first in 1978 (prior to the papal election) in the published proceedings of the conference and again in 1993 (Wojtyła 1978; John Paul II/Wojtyła 1993). The text here is from the 1993 translation that appeared during John Paul II's papacy. Titled "Subjectivity and the Irreducible in the Human Being," Pope John Paul II's essay expands upon the themes that were explored in *The Acting Person*, the pope's pre-papal *magnum opus* (John Paul II/Wojtyła 1979). The portion that re-appears unreferenced and without quotation marks in the researcher's 2001 chapter comes from the second part of John Paul II's essay, where he presents a history of the concept of subjectivity from the time of Aristotle to the early medieval period. There is no reference in the 2001 chapter by the researcher to any work by Pope John Paul II/Karol Wojtyła. The verbatim parallels are extensive; even the non-verbatim portion of the text not highlighted in Table 6.1 constitutes simply a paraphrase or synonym substitution for the pope's words. For example, the expression "in this sense" is elongated to "in this perspective," "this belief stands" grows to "this assumption forms," "when taken simply and directly" substitutes for "at least at first glance," and the preposition "in" becomes "for." On the basis of the text of the 2001 chapter alone, the reader has no way of knowing that the material is found verbatim and near-verbatim in the philosophical writings of the man who would become pope. Words are displayed in the 2001 chapter in a way that generates for the reader every appearance of being authored by the researcher. There are no quotation marks, footnotes, or other signals that direct the reader to the earlier papal text.

The textual parallels in Table 6.1 might appear to some observers to fulfill the four conditions presented above in Chap. 2 as: (1) a non-trivial appropriation of words, (2) with inadequate credit, (3) that generates an appearance of original authorship, (4) in a discrete item belonging to the scholarly record. Perhaps one might object that the first condition of *non-triviality* fails to be met because the extent of the parallel passages between the 2001 paper and the 1993 chapter is not sufficiently lengthy. In the researcher's 2001 chapter, the textual parallel with Pope John Paul II's words forms only a little more than half of one printed page's worth of material. One response to this objection is to point out that there are other even lengthier examples of unreferenced material in the 2001 chapter. In analyzing the chapter as a whole, the selection from the work of the Pope should not be discarded. The textual parallels that form the basis for what is arguably the central thesis of the 2001 chapter can be found in the writings of Oxford philosopher Sir Anthony Kenny, who in 1988 authored a short book titled *The Self*. Table 6.2 displays an example of verbatim and near-verbatim text passages shared between the 2001 chapter and the earlier 1988 book by Kenny. Neither Kenny's book, nor any other work by Kenny, is referenced in the chapter by the researcher. Table 6.2 does not exhibit exhaustively all the unattributed overlap between the two texts; in total the parallel with Kenny's earlier work accounts for approximately 19% of the researcher's 2001 chapter.

There are minor word substitutions that distinguish the 2001 and the 1988 texts in the Table 6.2 excerpt. Some are very slight; to mention but a few examples, the pronouns "this" and "you" become "that" and "someone" in the later text; the conjunction "for" becomes "because." What is most notable about the parallel exhibited in Table 6.2, however—beyond the extensiveness of the text common to both—is a striking modification that has been made near the beginning of the 2001 version in Table 6.2. In the original text, Kenny presents an account of his autobiographical thought experiment that features himself in the act of thinking about himself, and Kenny's words record the experiment in first-person Cartesian fashion. In the undocumented appearance of the same text in 2001 chapter, the name of the researcher appears in the place of Anthony Kenny's name, thereby creating the impression for the reader that the researcher is the originator of the autobiographical thought experiment. The re-use of Kenny's self-referential introspective example, without reference, and with the name switch, creates the appearance for the reader that the researcher himself is doing the original philosophical work:

Table 6.2 Text parallels with an earlier work by Sir Anthony Kenny

[—], "Subjectivity from a Semiotic Point of View," in *Ecosemiotics: Studies in Environmental Semiosis*, Nordic-Baltic Summer Institute for Semiotic and Structural Studies, Part IV, ed. Eero Tarasti et al. (Imatra: International Semiotics Institute, 2001): 149–159, at 153–154.	Anthony Kenny, *The Self. The Aquinas Lecture, 1988* (Milwaukee, WI: Marquette University Press, 1988), 4–7.
in a misunderstanding of the reflexive pronoun. To ask what kind of substance my "self" is, is like asking what the characteristic of "ownness" is, an attribute which my own property has in addition to being mine. When, outside philosophical reflections, I talk about myself, I am simply talking about the human being, [name of researcher], and my self is nothing other than myself. In some way it is a philosophical muddle to allow the space which differentiates "my self" from "myself" to generate the illusion of a mysterious metaphysical entity distinct from, but obscurely linked to, the human being who is talking to someone. The grammatical error which is the essence of the theory of the self may seem obvious when it is pointed out. But it is by no means easy to give an accurate account of the logic, or deep grammar, of the words "I" and "myself". It will not do, for instance, to say simply that "I" is the word each of us uses to refer to himself, a pronoun which, when it occurs in sentences, is synonymous with the name of the utterer of the sentence. This is not difficult to show. Julius Caesar, in his *Commentaries*, regularly described his own actions in the third person, using the name "Caesar". Let us imagine a language in which there were no first person pronouns, and in which everyone talked about themselves by using their own names. We can ask whether everything we can say in English can also be said in this language. The answer is clearly no. If Caesar wishes to deny that he is Caesar, then in English he can tell the lie, "I am not Caesar". In the special language no similar option is open to him. "Caesar is not Caesar" doesn't work, and neither does "the person who is speaking to you is not Caesar", because in the special language, that sentence if spoken by Caesar is equivalent to the English sentence, "The person who is speaking to you is not I". The truth is that "I" does not refer to the person who utters it in the same way in which a proper name refers to its bearer, and neither does "myself". (That does not mean that these words refer to something else, say, myself.) I shall not pursue the grammatical issues further here. Certainly, the belief in a self is in one sense a grammatical error, which has different roots. One of these roots is the notion of the self in Cartesian scepticism. Descartes, in his *Meditationes*, convinces himself that he can doubt whether he has a body. He then goes on to argue, "I can doubt whether I have a body; but I cannot doubt whether I exist; for what is this I which is doubting?" The "I" must refer to something of which his body is not part, and hence to something which is not a part of the human being Descartes. The Cartesian ego is a substance whose essence is pure thought; it is the mind, or *res cogitans*. This is the self in the second of the philosophical senses identified by the *Oxford English Dictionary*, "the ego identified with the soul or mind as opposed to the body".	in a misunderstanding of the reflexive pronoun. To ask what kind of substance my *self* is is like asking what the characteristic of *ownness* is which my own property has in addition to being mine. When, outside philosophy, I talk about myself, I am simply talking about the human being, Anthony Kenny, and my self is nothing other than myself. It is a philosophical muddle to allow the space which differentiates "my self" from "myself" to generate the illusion of a mysterious metaphysical entity distinct from, but obscurely linked to, the human being who is talking to you. The grammatical error which is the essence of the theory of the self is in a manner obvious when it is pointed out. But it is […] by no means easy to give an accurate account of the logic, or deep grammar, of the words "I" and "myself." It will not do, for instance, to say simply that "I" is the word each of us uses to refer to himself, a pronoun which when it occurs in sentences, is synonymous with the name of the utterer of the sentence. This is not difficult to show. Julius Caesar, in his *Commentaries*, regularly described his own actions in the third person, using the name "Caesar" [...]. There might be a language, call it Caesarian, in which there were no first person pronouns, and in which everyone talked about themselves by using their own names. We can ask whether everything we can say in English can also be said in Caesarian. The answer is clearly no. If Caesar wishes to deny that he is Caesar [...], then in English he can tell the lie, "I am not Caesar." In Caesarian no similar option is open to him. "Caesar is not Caesar" will not do the trick. Nor will "the person who is speaking to you is not Caesar." For in Caesarian that sentence in Caesar's mouth is equivalent to the English sentence, "The person who is speaking to you is not I." The truth is that "I" does not refer to the person who utters it in the way in which a proper name refers to its bearer. Neither does "myself." This does not mean that these words refer to something else, say, my self [...]. I do not intend to pursue further the grammatical issues explored earlier. For though the belief in a self is in one sense a grammatical error […]. The error has a number of different roots: [...] root of the notion of the self is Cartesian scepticism. Descartes, in his *Meditationes*, convinces himself that he can doubt […] whether he has a body. He then goes on to argue, "I can doubt whether I have a body; but I cannot doubt whether I exist; for what is this I which is doubting?" The "I" must refer to something of which his body is no part, and hence to something which is no more than a part of the human being Descartes. The Cartesian ego is a substance whose essence is pure thought, the mind, or *res cogitans*. This is the self in the second of the philosophical senses identified by the O.E.D. "the ego identified with the soul or mind as opposed to the body."

1988 chapter by Kenny: When, outside philosophy, I talk about myself, I am simply talking about the human being, Anthony Kenny, and my self is nothing other than myself.

2001 chapter by researcher: When, outside philosophical reflections, I talk about myself, I am simply talking about the human being, [name of researcher], and my self is nothing other than myself.

Again, on the basis of the 2001 text alone, and in the absence of the normal conventions and explicit markers for manifesting to the reader what is original and what is not (e.g., quotation marks), the reader has no way of knowing that this 2001 text has been published before, with Anthony Kenny as the earlier author of record.

The textual parallel exhibited in Table 6.2 is not limited to the presentation of the highly original autobiographical thought experiment with a name substitution. A second unusual thought experiment, one premised on an imaginary language in which there are no first-person pronouns, appears in both texts. In a subtle homage to a medieval logical puzzle, Anthony Kenny calls this hypothetical language "Caesarian" (see Buckner and Zupko 2014: 53, 83). In the 2001 version by the researcher, however, the example appears blandly as "the special language." Furthermore, the type-sentences offered by Kenny as examples to illustrate the limitations of languages that lack first-person pronouns appear as well in the 2001 chapter. Both texts state that the utterances, "Caesar is not Caesar" and "the person who is speaking to you is not Caesar" do not equate to the sentence "The person who is speaking to you is not I."

The verbatim and near-verbatim parallels with Kenny's previously published words go beyond the re-presentation of Kenny's autobiographical thought experiment, his technical discussion of a hypothetical language, and examples of type-sentences of such a language. Kenny is a highly acclaimed English translator of the works of the philosopher René Descartes, and his translations of Descartes's letters were included in the three-volume Cambridge University Press edition of the philosophical works of Descartes (1984–1991). The 2001 chapter in Table 6.2 also includes, without attribution, Kenny's English rendering of a Latin passage from Descartes's *Meditations on First Philosophy*.

The concluding textual parallel in Table 6.2 between Kenny and the researcher is a shared quotation from the entry in the *Oxford English Dictionary* for the word "self", and the discussion of the definition appears verbatim in the 2001 chapter of the researcher with the exception that Kenny's abbreviation "O. E. D." has been fully expanded. It is indisputable that there are textual parallels between Kenny's 1988 book and the researcher's 2001 chapter in their discussions of the *Oxford English Dictionary* definition, but can it be further demonstrated that the 1988 and the 2001 texts are intimately related? In his 1988 book, Kenny has slightly misquoted the definition as found in the *Oxford English Dictionary*: he has left out one word and he has removed the parentheses that are used in the dictionary entry. As printed in the dictionary, the text in question appears as: "the ego (often identified with soul or mind as opposed to the body)" (Oxford English Dictionary 1971: 410c). Kenny's idiosyncratic omission of the word "often" and the removal of the parentheses are inexplicably preserved in the 2001 chapter by the researcher.

The identification of repeated idiosyncrasies or errors has been long recognized as a valuable technique in textual criticism to identify dependencies among texts. When considering the relation of medieval manuscripts, for instance, text editors use the repetition of errors or "common variations" across manuscripts to determine the relationships of manuscripts to each other, since idiosyncrasies tend to be preserved by medieval scribes copying an earlier source to produce new manuscripts (Boyle 1976). Thus, manuscripts that repeat the idiosyncrasies can be judged to be genetically related to each other or each derived from the same earlier source. To apply the same principle here: the repetition in the 2001 chapter of Kenny's idiosyncratic rendering, 13 years earlier, of the *Oxford English Dictionary* definition further strengthens the appearance that the 2001 chapter bears some relation to the 1988 book by Kenny.

As noted in the Chap. 1, the *Authorship Condition* expresses the present-day requirement of academic publishing that the name of a work's author must appear along with the published work. Declarations of authorship are typically found on the first or last page of a published article or chapter, and on the title page in monographs. The appearance of a name identifies an individual (or a group of individuals) as the author of record, who is thereby considered responsible for the content of the work. In the 2001 chapter, the name of the researcher is promoted as the author of record on the first page, just above the title of the chapter. As has been seen in Tables 6.1 and 6.2, however, the text contains substantive verbatim and near-verbatim parallels—in these examples—with the words of Pope John Paul II and Sir Anthony Kenny, whose works are not referenced in the chapter. The presence of the researcher's name a second time in the chapter, this time substituting for Anthony Kenny's name in the parallel 1988 passage, further serves to create in the reader the impression of authentic authorship. One is left to ask what portions of the chapter are the most original or unprecedented, or in other words: what portions of the chapter do not have any textual parallels with the undocumented, previously published work of others?

In answer to this question, I propose that the most original or unprecedented portions of the chapter are those sentences where the author of record presents meta-narrative commentary to explain, purportedly, what is being accomplished in the chapter. Meta-narrative comments are generously strewn throughout the chapter, appearing at the beginning of each of the subheadings that divide the work into three main sections. The opening sentence of the chapter (which is also the first subsection of the chapter) begins in the mode of meta-narrative, as it states, "I would like to say something here [...]" (149). Such explicit first-person meta-narrative sentences commonly found in academic writing dispose the reader to think that the author of record is the real author of what follows, which is an impression already sustained with the appearance of an author of record's name on the first page of most articles. When authors use first-person meta-narrative in non-fiction scholarly works, they address the reader directly, further strengthening the appearance that the author of record and the author of the work are one and the same.

A similar pattern of meta-narrative is found in each of the two subsequent subsections of the researcher's chapter, as each begins with a meta-narrative assertion that again explicitly lays claim to authorship for the words that follow. The first sentence of section two of the chapter (which contains the lengthy unreferenced textual parallels with the words of Anthony Kenny), begins, "Let me now address [...]" (151). To introduce the third section of the chapter (which contains the unreferenced textual parallel with the words of Pope John Paul II), the researcher writes, "Let me now come to the closing part of my reflections, which will be less analytical and more constructive" (155). As if the presence of the researcher's name as the author of record at the beginning of the chapter were insufficient, the words "my reflections" serve as an implicit assertion of authorship vis-à-vis the reader. To mention another instance of the similar use of meta-narrative elsewhere in the 2001 chapter: the reader encounters the assertion, "I would argue that [...]" inserted within the section that is parallel with Kenny's words (153).

The parts of the 2001 chapter that parallel unreferenced works by Pope John Paul II, Sir Anthony Kenny, and others, constitute the substance of the chapter. The meta-narrative claims found throughout the chapter provide explicit additional assertions of authorship for published text, and they strengthen the appearance of authorship for the reader regarding sentences that parallel unreferenced and deficiently documented texts by others. These meta-narrative claims arguably constitute the most original part of the 2001 chapter, yet they do not report research findings or express new insights into the notion of the self, which is the professed topic of the chapter. Rather, the original meta-narrative claims support a reader's default assumption that an identity obtains between (1) the author of record and (2) the original (i.e., genuine) author.

In spring 2017, I was teaching an upper-level course titled "Critical Writing and Research" for a group of honors students at Ohio Dominican University, and as a class project we worked on identifying the various apparent overlaps between this 2001 chapter and the previously published works of Pope John Paul II, Sir Anthony Kenny, and several others. We organized the evidence of suspected plagiarism and sent a retraction request to the International Semiotics Institute in April 2017. After many months without a judgment, I notified the copyright holder of Kenny's work of the situation, and Marquette University Press issued a statement supporting the request for retraction due to the "severe plagiarism of a copyrighted work" (South 2018: 1). In January 2018, the International Semiotics Institute responded by issuing a retraction for plagiarism. The lengthy retraction statement noted that the claim of suspected plagiarism

> was supported by ample evidence, and was followed by a "request for retraction" from our part. To verify the claim, we have appointed a committee to read carefully through Prof. [—]'s text, and cross-check with the sources indicated by the colleagues at Ohio Dominican University. After the examination, we regret to say that the committee has fully confirmed the claim, and therefore has declared the essay in question plagiarized from various sources. Particularly the following two: – Anthony Kenny. *The Self*. The Aquinas Lecture, 1988 (Milwaukee, WI: Marquette University Press, 1988). – John Paul II/Karol Wojtyla. "Subjectivity and the Irreducible in the Human Being," in *Person and Community: Select Essays*, trans. Theresa Sandok (New York: Peter Lang, 1993). To this effect, the present letter constitutes an official statement of retraction of that essay (Martinelli 2018a).[2]

This detailed retraction is maximal in determination, credit, and availability, since it explicitly states that the chapter is plagiarized; identifies the principle texts by Pope John Paul II and Anthony Kenny that were misappropriated; and appears online without a paywall. After appearing online, the retraction generated some news coverage that emphasized the appropriation of the work of Pope John Paul II and Sir Anthony Kenny, at times highlighting the student involvement in securing the

[2] I have anonymized the retraction statement here; as published it identified the researcher by name.

correction of the scholarly record (Stern 2018a, b; Weinberg 2018a, b; Illarietti 2018; Oggiscienza 2018). Several days after the retraction appeared, the International Semiotics Institute issued an addendum to the retraction objecting to what it considered to be a "sensationalistic" reporting of the plagiarism, but yet affirming that, "We do not have second thoughts concerning the statement of retraction" and that the article is "an evident example of academic misconduct" (Martinelli 2018b). The University of Lugano issued a brief press release on January 18, 2018, stating that this new case would not occasion a new inquiry at the university (Illarietti 2018).

6.3 Case 2: *Subject Matters: A Journal of Communications and the Self*

Case 2 provides a second instance of substantive parallels with the previously published work of others, but additional precisions are necessary to describe the nature of some of the parallels. Published in the journal *Subject Matters: A Journal of Communications and the Self*, the 2004 article is presented under the name of the researcher and appears to offer a novel analysis of the history of the concept of subjectivity. Focusing mostly on predecessors to the French philosopher Alain Badiou, the article considers in part the philosophical account of human existence found in Martin Heidegger's major work, *Being and Time*. One portion of the 2004 journal article parallels, without quotation marks or citation conventions of any kind, a 1995 article authored by Thomas K. Carr. This instance of overlap is presented in Table 6.3.

The fundamental difficulty is that readers of the 2004 journal article have no way of knowing—at least from the text itself—that many of the sentences exhibit verbatim and near-verbatim parallels with the text by Carr published 9 years earlier. In the original 1995 article, Carr had critically analyzed John Macquarrie's then-recent book, *Heidegger and Christianity*. In the 2004 parallel, every reference to Macquarrie is absent, and the same words are now used in the context of analyzing the work of Alain Badiou. The textual parallels are entirely hidden to the reader, who likely will assume that the sentences are exclusive and original to the 2004 author of record.

Much like Case 1, the parallels in Case 2 constitute an essential part of the article by the researcher. The portion of parallel text with Carr's article cannot be dismissed as a minor part, since it belongs to the presentation of the history of subjectivity in Badiou's intellectual predecessors and successors. The fact that a few—significant—words have been left out in the textual parallel with the 1995 article is particularly arresting; all references to Macquarrie have been carefully removed. A quotation from Macquarrie's book that explains Heidegger's approach to the Greek word for truth (*alētheia*) is cited in Carr's 1995 article: "when something is presented to us as it really is, without any concealment or distortion" (Macquarrie 1994: 25), but when Macquarrie's words show up in the 2004 article, the reader is

Table 6.3 Text parallels with an earlier article by Thomas K. Carr

[—], "Subjectivity before and after Badiou," *Subject Matters: A Journal of Communications and the Self* (© London Metropolitan University) 1, No. 2 (2004): 67–81, at 76–77.	Thomas K. Carr, "Only A God Can Save Us," *First Things* 55 (1995): 57–62, at 59–60.
'Man' is a German expression for the generic subject - the "they" whose beliefs and behaviours make up the "average everydayness" of human existence. The essence of human nature, proclaims Heidegger in a phrase later exploited by Sartre, is always a product of existence, of being-in-the-world, which is to say the public world of societal norms and rituals. This "publicness," this "being-with-one-another," may be essential to the societal fabric, but it has devastating consequences for human nature as such. For in our "thrownness" into the 'groupthink' of society we come to exist not on our own terms but on those of what Heidegger calls "das Man". As such, most of us thus naturally "fall" into this tempered mode of existence, prefer and are even happy not to think for ourselves, but to follow, instead, the routines and fashions of those around us [...]. Authentic living, according to Heidegger, consists in the fundamental realization that "truth" exists not in the people and institutions among which we are thrown, but is in us as beings who are faithful to our own existence. For Heidegger, famously, truth is *aletheia,* the Greek a-privative signifying the "un-" as in "un-coveredness" or "un-concealment,"; this entails that truth is an "event" which happens "when something is presented to us as it really is, without any concealment or distortion." The accent rides on the "to us," meaning that "the locus of truth is not the proposition", but the *Dasein.* Truth is thus not the correspondence of propositions to what is the case, as it has been traditionally conceived, but is inextricably linked to human existence as a whole. But it is not just any *Dasein* which can, in a facile way, clear space for truth's disclosure. It must be an authentic expression of human being, of human being in its freedom from 'das Man'. In Heidegger's phrasing, "to forego normal choice and to adopt those offered me by the world or other people" is the essence of "in-authenticity." As we said already previously, inauthentic *Dasein* is most at home in the world of rules, rituals, and conventions that disburdens existence of its personal responsibility for choice. In such a world, "everyone is the other and no one himself". How then must we act to recover the authentic self we have lost by living in the world of the other? According to Heidegger the therapy is simple: we must turn from living by the other's rules and habits and project a world of particular significance to ourselves. True Heideggerian selves emerge in their authenticity only on the basis of the continual possibility of recovering themselves from being lost in the world of things that are of no concern, and of creating worlds of meaning around those things that are.	*das Man* (German for the generic subject) —the "they" whose beliefs and behaviors make up the "average everydayness" of human existence [...]. The essence of human nature, proclaims Heidegger in a phrase later exploited by Sartre, is always a product of existence, of being-in-the-world, which is to say the public world of societal norms and rituals. This "publicness," this "being-with-one-another," may be essential to the societal fabric, but it has devastating consequences for human nature as such. For in our "thrownness" into the groupthink of society we come to exist not on our own terms but on those of what Heidegger calls *das Man* [...]. Most of us prefer and thus naturally "fall" into this tempered mode of existence, are happy not to think for ourselves but to follow instead the routines and fashions of those around us [...]. As Macquarrie lays it out, it is straightforward: salvation consists in the fundamental realization that "truth" exists not in the people and institutions among which we are thrown, but is in us as beings who question and think the nature of Being. For Heidegger, famously, truth is *aletheia,* the Greek α-privative signifying the "un-" as in "un-coveredness" or "un-concealment," meaning that truth is an "event" which happens—in Macquarrie's words— "when something is presented to us as it really is, without any concealment or distortion." The accent rides on the "to us," meaning that "the locus of truth is not the proposition, but *Dasein* [...]. Truth is thus not the correspondence of propositions to what is the case, as it has been traditionally conceived, but is inextricably linked to human existence as a whole. "[...]. But not just any *Dasein* proves facile enough to clear space for truth's disclosure. It must be an authentic expression of human being, of human being in its freedom from *das Man.* In Heidegger's phrasing, "to forego normal choice and to adopt those offered me by the world or other people" is the essence of "in-authenticity." Inauthentic *Dasein* is most at home in the world of publicness—the world of rules, rituals, and conventions that disburdens existence of its personal responsibility for choice. In such a world, "everyone is the other and no one himself." How then must we act to recover the authentic self we have lost by living in the world of the other? The therapy is simple: We must turn from living by the other's rules and habits and project a world of particular significance to ourselves [...]. True Heideggerian selves emerge in their authenticity only on the basis of the continual possibility of recovering themselves from being lost in the world of things that are of no concern, and of creating worlds of meaning around those things that are.

left in the dark about their origin. What appears to be an original exposition of Heidegger's philosophy in the 2004 article is really a parallel with Carr's 1995 analysis of Macquarrie's views on Heidegger's theology. There are no quotation marks or footnotes that would tie the later work to the earlier one.

The 2004 article by the researcher offers other examples of textual parallels, as shown in Table 6.4, where this time the parallel is with a 1985 essay by Charles

Table 6.4 Text parallels with an earlier work by Charles Taylor

[—], "Subjectivity before and after Badiou," *Subject Matters: A Journal of Communications and the Self* (© London Metropolitan University) 1, No. 2 (2004): 67–81, at 79–80.	Charles Taylor, "The Person," in *The Category of the Person: Anthropology, Philosophy, History*, ed. Michael Carrithers, Steven Collins, and Steven Lukes (Cambridge University Press, 1985): 257–281, at 263, 261, 264, 265.
The general conception of any person contains a certain reflexive element. A person is an agent who has an understanding of self as an agent, and can make plans for his/her own life; this kind of reflexivity cannot be attributed to animals. But this capacity should not be seen simply as consisting in superior powers of representation: that some animals can only be aware of the objects that surround them immediately, while human animals also have the ability to represent themselves, as well as future states of self and world. This does not seem to be the crucial difference. Rather what seems important about a person's conception of self is that it incorporates a range of significances which have no analogue with non-person agents. For it is not just that we are aware of ourselves as agents that distinguishes us from animals, it is more that we have a sense of certain standards which apply to us as self-aware agents. As such agents, human beings are able to make life-plans, hold values, choose. This doesn't mean that they can represent to themselves different life plans, consider different options and evaluate whether different possibilities are more or less valuable. If human beings are aware of the significance feature, what is striking about being a person is not simply consciousness in the sense of the power to form inner representations. Rather we should say that consciousness goes along with a transformation of the significances subjects live by; and this leads to a view of a man as a self-interpreting animal (cf. Taylor 1989) […]. a significance of a radically different kind than simply any kind of representation of usual things. The self-awareness involved with some event is not just a matter of an independent significance which is now brought within our ken; it is rather a significance bound up with our being self- aware beings, that is, subjects. The same goes for that whole host of emotions which are qualified by a reflective sense of their importance in our lives. In other words, we have to understand the consciousness of subjects not just as a state in consciousness in the sense of the power to form representations of self and world, but much more as the onset of a range of significances which are essentially those of self- aware agents.	The general conception of a person has a certain reflexive element. A person is an agent who has an understanding of self as an agent, and can make plans for his/her own life; this, as against a dog for instance, to whom this kind of reflexivity can't be attributed. But this shouldn't be seen simply as consisting in superior powers of representation: that dogs can only be aware of the objects that surround them immediately, while persons can also have themselves for object, as well as future states of self and world […]. But this doesn't seem to be the crucial difference. Rather what seems important about a person's conception of self is that it incorporates a range of significances which have no analogue with non-person agents. For it is not just that we are aware of ourselves as agents that distinguishes us from dogs, say, It is more that we have a sense of certain standards which apply to us as self-aware agents […]. A person, as we saw above, we conceive of as a special kind of agent, an agent-plus, who can also make life-plans, hold values, choose […]. For the first view, which ignores the significance feature, the crucial difference is consciousness, a power to form inner representations. This means that I can represent to myself different life plans, hence consider different options, and envisage the different possibilities as more or less valuable. But if we are aware of the significance feature, what is striking about being a person is not simply consciousness in this sense. Rather we should say that consciousness goes along with a transformation of the significances we live by; and this leads to a view of man as a self-interpreting animal […]. shame involves a significance of a radically different kind […]. By contrast, the self-awareness involved with shame is not just a matter of some independent significance which is now in our ken; it is rather a significance bound up with our being self-aware beings, that is, persons. The same goes […] for that whole host of emotions which are qualified by a reflective sense of their importance in our lives […]. In other words, we have to understand the step from sub- personal agent to person not just as an increase in consciousness in the sense of the power to form representations of self and world, but much more as of a range of significances which are essentially those of self- aware agents.

Taylor, the distinguished philosopher and recipient of the 2008 Kyoto Prize, the 2006 Kluge Prize, and the 2016 Berggruen Prize.

There are a few cosmetic differences in the later text of 2004: Taylor's 1985 example of "dog" has been inclusively changed to "animals"; Taylor's "persons" have morphed into "human animals"; and all the contractions found in Taylor ("can't," "shouldn't," and "doesn't,") have been spelled out. The sentences at times appear in the later text as re-arranged, inflated, or reduced. There is a single parenthetical reference in the 2004 article that directs the reader to a different work by Taylor, his 1989 classic book *Sources of the Self*, but the textual parallels are not with that work, but with Taylor's completely unreferenced 1985 essay. There are no quotation marks, offsetting, or other normal conventions that would let the reader know of verbatim and near-verbatim parallels with the work by Taylor published almost two decades earlier.

The article printed after this one in the same issue of the journal features a round-table discussion by four European professors, with remarks by the researcher (Baldwin et al. 2004), and the four professors expound upon the first 2004 article at length, apparently unaware that they are discussing a piece of writing that has substantive parallels with various unreferenced texts written by others. The round-table discussion involves at one point the portion of the article that is parallel with Taylor's work, as the researcher responds to a query by presenting an unusual example also found in Taylor, involving a scenario where a rock impedes a man travelling by car. No acknowledgement is given that the example is also found in Taylor. It is difficult not to draw the conclusion that the four professors are entirely unaware that they are engaging with the earlier texts of Carr and Taylor through the proxy of the researcher.

Several difficulties make a retraction request for this article in *Subject Matters* somewhat challenging but not insurmountable. First, the researcher is listed on the editorial board of the journal. As noted in Chap. 5.2.1, many experts now argue that whistleblowers should not contact the author of record for publications for which there is evidence of suspected scientific misconduct, and in this case contacting the journal could involve interaction with the author of record for the article. Minimally, it would be best for the whistleblower to direct any requests for a correction of the scholarly record to other editors at the journal. Second, although the printed version of the journal is held by libraries around the world, the publication of new issues of *Subject Matters* ceased in 2007, 3 years after the issue containing the article for which the researcher is the author of record and the subsequent round-table article. Nevertheless, the first page of every article in the journal lists London Metropolitan University as the copyright holder, and an archive of all the issues of the journal is found on the website of the university (*Subject Matters* 2004–2007). Were the editors of the journal or the copyright-holding university to be persuaded of the need for a correction of the scholarly record in this case, on the evidential basis of the textual parallels, the traditional means of printing a correction on a numbered, citable page of a future issue of the journal would be impossible. An online correction could be published on the webpage for the article on the journal's website, however. If the editors and the university were to issue a correction in such a way, they would be following the best practices for correcting the scholarly record in the digital age.

In May 2017 the four former editors of *Subject Matters* responded to a request for retraction for the article and for the round-table article. The request was denied on the grounds that, in their view, there did not seem to be any improper presentation of ideas. The decision was appealed to the legal department of London Metropolitan University, since the university is the current copyright holder of the journal, but at the time of this writing no final judgment has been issued in the matter. The two articles in the 2004 issue of *Subject Matters* remain firmly within the scholarly record. The first of the two is currently listed on the reading list for a doctoral seminar at a European university and has been cited in later literature.

6.4 Case 3: John Benjamins Publishing Company

The next case for consideration is, like the first two, a work on a philosophical topic. Appearing under the name of the researcher, and listing a university affiliation, Case 3 deals with the notion of subjectivity in ancient Greek philosophical thought, and it was published in an edited collection of essays in 2005 by the commercial Dutch academic press, John Benjamins Publishing Company. Titled "Being in Accordance with Oneself: Moral Self-Controversy in Plato and Aristotle," the chapter presents an interpretation of the views of self-harmony in the thought of two major figures of ancient philosophy, Plato and Aristotle. Portions of the text exhibit parallels with previously published scholarly work by Julia Annas, Anthony Price, and Nancy Sherman, who are established commentators on ancient Greek philosophical texts. Table 6.5 offers a excerpt of textual parallels.

There are considerable parallels between the text of this 2005 chapter and texts found in earlier works by Annas, Price, and Sherman. No quotations marks, extracting, or references are present to indicate to the reader that textual parallels exist with the previously published works by the three scholars. That is, the texts are not presented as quotations. The works by Annas, Price, or Sherman are not listed in the bibliography of sources appended at the end of the chapter. In addition to the verbatim and near-verbatim parallels of text, there are also many close substitutions: "person" becomes "individual" in the 2005 chapter; "one" becomes "single"; "fix on" becomes "focus on"; "reason" and "virtue" become "reasoning" and "virtuous"; "involuntary" becomes "non-chosen", and the like.

Since Case 3 consists entirely of an exegesis of ancient Greek philosophical texts, the chapter does not easily admit of a division into core and non-core parts, a structure that is proposed by some theorists for scientific articles that report empirical research findings. Exegesis is the critical part of examining the history of philosophy; it is not merely supplementary work but constitutes the activity itself. Hence, exegesis cannot be considered a non-core element of a scholarly publication that aims to present an original thesis about Greek philosophy. One cannot avoid the conclusion that the excerpt of textual parallels in Table 6.5 involves the crux of the chapter.

Table 6.5 exhibited textual parallels with previously published works that are nowhere acknowledged in the 2005 chapter; the names and works of Annas, Price, and Sherman do not appear in either citations or in the bibliography. In contrast, Table 6.6 exhibits textual parallels with a volume that is identified in the bibliography and referenced with an in-text citation: Michael Pakaluk's 1998 commentary on Aristotle's *Nicomachean Ethics* in Oxford University Press's "Clarendon Aristotle Series." There appear to be both verbatim and near-verbatim textual parallels between Pakaluk's 1998 volume and the 2005 chapter by the researcher.

There is a named source mentioned in the excerpt from the 2005 article, shown in Table 6.6, as Pakaluk's commentary receives an in-text citation, but the reader has no way of knowing how much exactly is being attributed to Pakaluk. The citation gives a reference to page 170, yet it appears to be merely a bracketed reference for

Table 6.5 Text parallels with earlier secondary literature on ancient Greek philosophy

[—], "Being in Accordance with Oneself: Moral Self-Controversy in Plato and Aristotle," in *Controversies and Subjectivity*, ed. Pierluigi Barrotta and Marcelo Dascal (Amsterdam: John Benjamins Publishing Company 2005): 75–90, at 77, 84.	Julia Annas, *An Introduction to Plato's 'Republic'* (Oxford: Clarendon Press, 1981); Anthony Price, "Friendship (VIII und IX)" in *Die Nikomachische Ethik*, ed. Otfried Höffe (Berlin: Akademie Verlag, 1995), 229–251; Nancy Sherman, *The Fabric of Character: Aristotle's Theory of Virtue* (Oxford: Clarendon Press, 1989).	
Plato [...] uses noncommittal prepositional phrases that seem clumsy in English, like "that by which we desire". However, what Socrates proceeds to outline is the notion that complexity exists in the person, although he does not say *how* that complexity might be realized. In other words, the various kinds of complexity to be found in human behaviour cannot be adequately accounted for unless it is admitted that the person himself is complex. We will not understand a person's actions unless we acknowledge that there is more than one motivational source [...]. Plato stresses the fact that desire is characteristically limited to its object and does not necessarily involve any further considerations. Desire is thought of as manifold and often chaotic, because it can focus on practically any kind of object; the only thing that unifies all cases of desiring is the fact that some particular thing is being sought after Plato's long discussion of desire, which concentrates on basic, easily recognizable examples of desire – especially bodily desires for food, drink, sleep and sex – is not the issue at stake here. It should be mentioned, however, that the emphasis on the basic and forceful nature of desires makes it look as though no cognitive effort at all should be ascribed to this part of the soul. The other part of the soul, usually called "the reasoning part" (*to logistikon*; 439d), has two main functions; the first is to search for the truth and increase one's knowledge in a practical as well as an intellectual sphere; the second is to rule in the soul (441e, 442c). Ruling in the soul implies that this part takes care of the whole soul and not just of itself, whereas the other two parts care only for themselves and not for the whole of which they are parts (441e). Thus, the reasoning part is the source of practical judgement about what is best for the person as a whole. In this area as in many others, Aristotle's ethics is more of an elaboration than a correction of Plato. In his treatments of friendship, in Books VIII and IX of the *Nicomachean Ethics* (EN), as well as in Book VII of the *Eudemian Ethics* (EE), we find him justifying and explaining what Plato had already described metaphorically as psychic health [...]. Although Aristotle has theoretical doubts about any partition of the soul (cf. *De Anima* III, 9), he often follows Plato's division of desires and beliefs as ruled by reason, anger, and appetite [...].	Plato often uses more noncommittal prepositional phrases which are clumsy in English, like 'that by which we desire'.) Plato keeps his vocabulary here perhaps deliberately vague; he is insisting that there is complexity in a single person without saying too much about how that complexity might be realized [...]. The basic point is that the various kinds of complexity to be found in human behaviour cannot be adequately accounted for without allowing that the person is himself or herself complex. We will not understand an individual's actions unless we see that actions do not come from a single motivational source [...]. Plato [...] the fact that desire is characteristically limited to its object without necessarily involving any further considerations. Desire is thought of as manifold and often chaotic because desire can fix on objects of just about any kind; there is nothing that unifies all cases of desiring except that some particular thing is sought for [...]. The long discussion of desire in Book 4 concentrates on basic, easily recognizable examples of desire, especially bodily desires (those one has merely by virtue of having a body, desires for food, drink, sleep, and sex) [...]. All the emphasis on the basic and forceful nature of desires makes it look as though no cognitive performance at all should be ascribed to this part [...]. It is usually called 'the reasoning part' (*to logistikon*, 439d) [...]. It has two main functions. One is that of searching for the truth and increasing one's knowledge [...] in the practical as well as the intellectual sphere [...]. Its other function is to rule in the soul (441e, 442c) [...]. One is that it is the only part that cares for the interests of the whole soul and not just itself, whereas the other two parts care only for themselves and not for the whole of which they are parts (441e) [...]. Reason, then, is the source of practical judgement about what is best for the person as a whole. In this area as in many, Aristotle's ethics is rather an elaboration than a correction of Platonism. In his treatments of friendship, in Books VIII and IX of the *EN*, as with Book VII of the *Eudemian Ethics* (*EE*), we find him justifying and explaining the possibilities of a generous eudaimonism within the framework of his own post-Platonic moral psychology. [...] though he has theoretical doubts about any partition of the soul (*De Anima* III.9), he often follows Plato's division of desires and beliefs between reason, anger, and appetite [...].	Annas 1981: 124. Annas 1981: 129. Annas 1981: 125. Annas 1981: 126 Price 1995: 231, 236
Friendship (*philia*), Aristotle stipulates, is the mutually acknowledged and reciprocal exchange of goodwill and affection among individuals who share an interest in each other on the basis of virtue, pleasure, or utility (EN VIII, 2). Within the sphere of friendship Aristotle also includes the involuntary relations of affection and care that exist among family members and fellow citizens. Although the term is used quite broadly, Aristotle's primary focus is on the paradigmatic case of virtuous friendship [...].	Friendship (*philia*), Aristotle stipulates, is the mutually acknowledged and reciprocal exchange of goodwill and affection that exists among individuals who share an interest in each other on the basis of virtue, pleasure, or utility (NE VIII. 2). In addition to voluntary associations of this sort, Aristotle also includes among friendships the non-chosen relations of affection and care that exist among family members and fellow citizens (cf. NE VIII. 9, VIII 12, IX. 6). The term is thus used quite broadly, though Aristotle's primary focus, and mine as well, is on the paradigmatic case of virtue friendship [...].	Sherman 1989: 124–125.

Table 6.6 Text parallels with an earlier work by Michael Pakaluk

[—], "Being in Accordance with Oneself: Moral Self-Controversy in Plato and Aristotle," in *Controversies and Subjectivity*, ed. Pierluigi Barrotta and Marcelo Dascal (Amsterdam: John Benjamins Publishing Company, 2005): 75–90, at 85–86.	Michael Pakaluk, *Aristotle: 'Nicomachean Ethics' Books VIII and IX. Translated with a Commentary*. Clarendon Aristotle Series (Oxford: Oxford University Press, 1998), at 163, 165, 167, 169, 170.
the relationship between friends is isomorphic to that of a good person with himself. He establishes this thesis by distinguishing four characteristics of friendship (*philika*; 1166a2–10) and then argues that they apply to a good person in relation to himself. The four characteristics of friendship are: (1) wishing and doing good, or things that appear to be good, for another's sake; (2) wishing that another may exist and live, for his sake; (3) spending time with another and choosing the same things; (4) sharing another's joy and sorrow. Applying these characteristics to the intrapersonal relation of the virtuous man, it turns out that he (1) "agrees with himself and desires the same things with his entire soul", and (2) "acts for the sake of himself (since he does so for the sake of his thinking part, which in fact each person seems to be)" (1166a13–14; a16–17). It seems obvious that each of these claims refers to the harmony between the thinking part of the soul and itself or to the disharmony between the thinking part and the non-rational part of the soul. In fact, a few chapters later it becomes clear that Aristotle identifies the non-rational part with the appetites and feelings (1168b20). The Aristotelian distinction between the "thinking part" – sometimes he uses other equivalent expressions like "the part that can think" (to dianoêtikon), "that with which he reasons" (*hôi phronei*), "that which thinks" (*to nooun*; cf. Pakaluk 1998: 170) – and the "non-rational part" can be already found in the first book of EN. A distinction is drawn between the "non-rational" (*alogon*) part of the soul, which includes the "nutritive" (*threptikon*) and "appetitive" (*epithumêtikon kai holôs orektikon*) functions, and the rational part (to logon echon), which can understand and give reasons for action (see I 13). Here, as in Book 9, we find a description of the relationship between the two parts, according to which in a virtuous soul everything is "in agreement" with reason (1166a13; cf. 1102b28): the non-rational part "shares" in reason, the rational part rules over it (1168b35; 1169a1; cf. 1102b30) and the non-rational part should be "obedient" to reason (1168b31; cf. 1102b33).	the relationship between friends is isomorphic to that of a good person in relation to himself [...], Aristotle first distinguishes four 'characteristics of friendship' (*philika*, at 1166^{a}2–10), and then argues that they apply to a good person in relation to himself [...]. He distinguishes four characteristics of friendship: (*Φ*1) wishing and doing good, or things that appear to be good, for another's sake; (*Φ*2) wishing that another exist and live, for his sake; (*Φ*3) spending time with another and choosing the same things; (*Φ*4) sharing in another's joys and sorrows [...]. [W]e should perhaps interpret the first half of each of those claims as referring to the harmony or disharmony between 'the thinking part' of the soul and itself, presumably at different times, and the second as referring to that between the thinking part and the other, non-rational part of the soul, the 'appetites and feelings' mentioned at 1168^{b}20. Understood in this way, each claim would serve to minimize the four distinct points which follow it. How should we understand Aristotle's references to 'the thinking part' for (*Φ*1) and (*Φ*2)? (We should no doubt regard as equivalent, 'the part which can think', *to dianoētikon,* 'that with which he reasons', *hōi phronei,* 'that which thinks', *to nooun,* and 'the understanding', *hē dianoia.*) Is this an unwarranted intrusion of an unexplained philosophy of mind? The simplest answer is to take it as an application of the distinction drawn in I. 13, between the 'non-rational' (*alogon*) part of the soul, which includes the 'nutritive' (*threptikon*) and 'appetitive' (*epithumētikon kai holōs oretlikon*) functions, and the rational part (*to logon echon*), which can understand and give reasons for action. Aristotle's language here, and later in a similar passage in IX. 8, [...]. Moreover, he describes the relationship between the two parts similarly: here as before he holds that the non-rational part 'submits' to reason (1169^{a}18, cf. 1102 ^{b}26); that in a virtuous soul everything is 'in agreement' with reason (1166^{a}13, cf. 1102^{b}28); that the non-rational part 'shares' in reason. when the rational part rules over it (1168^{b}35, 1169^{a}1, cf. 1102^{b}30); and that the non-rational part should be 'obedient' to reason (1168^{b}31, cf. 1102^{b}33).

how to understand the Greek term *to nooun*. The textual parallels begin much earlier, however, with the 2005 article apparently mirroring text beginning on page 163 of Pakaluk's 1998 book. In the absence of quotation marks or block quotation, the example is reminiscent of what is described by theorists as "pawn sacrifice," where a citation is given but without an accurate indication of the extent of the textual overlap.[3] The appropriated text in this case extends well before, and well after, the parenthetical reference. Furthermore, the parallel includes Pakaluk's detailed analysis of Aristotelian terminology in Greek, an English translation, and detailed references to the texts of Aristotle. These references are given—both in the 2005 chapter by the researcher and in Pakaluk's 1998 book—with Bekker numbers, using the

[3] Debora Weber-Wulff describes pawn sacrifice as: "there is a proper attribution of a sentence, but then the text copy continues on, copying the source for additional sentences or even paragraphs without making clear that this is the author of the source speaking and not the purported author" (2014: 10).

precise line indications from the nineteenth-century Berlin edition of Aristotle issued by the Royal Prussian Academy.

Other portions of the 2005 chapter contain textual parallels with another work of Sir Anthony Kenny. That work by Kenny's is indeed listed in the bibliography, but again no quotation marks in the 2005 chapter are used to disclose to the reader that what is presented are Kenny's actual words either verbatim or in close paraphrase. Those parallels with that work of Kenny were pointed out by an anonymous commentator on the post-publication review website *PubPeer* (Peer 1: 2016).

Seventeen months after receiving a request for retraction featuring detailed evidence of parallels in the 2005 chapter with previously published texts, John Benjamins Publishing issued an online correction that can be designated as a corrigendum. On the webpage for the book on the publisher's website is a statement that begins, "It was brought to our attention that the paper [...] has certain severe shortcomings in the references" (2017b). This correction of the scholarly record by John Benjamins Publishing offers a general warning to potential readers that the chapter contains unspecified deficiencies in crediting the work of others. The correction is maximal in availability, as it appears unpaywalled online, but minimal in credit, since none of the authors whose works are unreferenced or under-referenced are identified. The publisher's statement must be judged as medial in determination, since a reader of the correction is left unclear about the extent of the "shortcomings in the references." Nevertheless, the corrigendum succeeds in providing a non-negligible measure of correction to the scholarly record, and one might judge that it compares favorably to the uncorrected status of Case 2.

This corrigendum further states that the publisher "endorses" the view that the researcher "has never appropriated ideas or original research results, passing them as his own. In particular, he has never copied the entirety or essential parts of other people's publications," and it credits the view to the home institution of the researcher. This approach is curious, since apart from the exegesis of the texts of Plato and Aristotle, there are no other scholarly findings presented in the chapter.

6.5 Case 4: The Catholic University of America Press

Theorist Debora Weber-Wulff has argued eloquently that the word *plagiarism* stands for a wide variety of practices that are in violation of academic norms. She describes plagiarism as "a continuous spectrum of text manipulations and not just one particular method of using other people's words" (Weber-Wulff 2014: 14). There are many ways to be deficient in one's use of the words of others, and these deficiencies vary according to gravity. In short, the violations of academic standards for giving credit to the work of others can range from very serious and substantial to quite minor and insubstantial.

Case 4 exhibits textual parallels that arguably are quite minor in comparison with other cases. Nevertheless, it deserves mention since it was the index case for all subsequent consideration of the published writings by the author of record. Case 4

Table 6.7 Text parallels in German and in English with an earlier work by John Monfasani

[—], "Die Kontroverse zwischen Platonikern und Aristotelikern im 15. Jahrhundert: Gemisthos Plethon - Georgios Trapezuntios - Kardinal Bessarion," in *Philosophen der Renaissance*, ed. Paul Richard Blum (Darmstadt: Wissenschaftliche Buchgesellschaft, 1999), 22–33, at 30.	[—], "George Gemistos Plethon (ca. 1360–1454), George of Trebizond (1396–1472), and Cardinal Bessarion (1403–1472): The Controversy between Platonists and Aristotelians in the Fifteenth Century," in *Philosophers of the Renaissance*, ed. Paul Richard Blum and translated by Brian McNeil (Washington, D.C.: The Catholic University of America Press, 2010), 23–32 at 31–32.	John Monfasani, *George of Trebizond: A Biography and a Study of His Rhetoric and Logic* (Leiden: E. J. Brill, 1976), at 228–229.
Mit dem Tode Bessarions in Jahre 1472 endet die Kontroverse, in deren Verlauf eine Vielzahl von platonische, neuplatonische und patristischen Texten veröffentlicht wurde und die nicht nur Anlass zu bewegten Diskussionen unter den Humanisten über philosophische Fragen geführt hatte, sondern durch die Gegenüberstellung von Platon und Aristoteles auch ernst zu nehmende Studien hervorgerufen hatte. Was nach dem Tode des Kardinals etwa in der Gestalt des Florentiner Platonismus unter Marsilio Ficino einsetzt, hat mit dieser im Wesentlichen durch die Byzantiner bestimmten Kontroverse freilich allenfalls am Rande noch etwas zu tun. [Note 21: Einen Nachklang dieser Kontroverse findet man gegen Ende des Jahrhunderts in Pico della Mirandolas Schrift »De ente et uno«*. Vgl. dazu Kristeller 1972, 1–55.	The controversy ended with Bessarion's death in 1472. In the course of the scholarly debates, a large number of Platonic, Neoplatonic, and patristic texts were published. The controversy not only generated heated discussions of philosophical questions among the humanists, but also prompted serious academic investigations thanks to the comparison which was drawn between Plato and Aristotle. Developments after the cardinals death, for example, in the form of Florentine Platonism under Marsilio Ficino, had, however, at most a marginal contact with this controversy, in which the Byzantine theologian had played such a central role. [Note 21: We find an echo of this controversy toward the end of the century in *Pico della Mirandola's* work *De ente et uno*. On this see Kristeller 1970, 1–55.]	The Plato-Aristotle controversy of the fifteenth century ended in 1472 with the death of Cardinal Bessarion. It had exposed a great number of Platonic, Neoplatonic, and patristic texts. It had provoked philosophic discussions among the humanists. It had engendered a series of interesting and, in some instances, very important works. Lastly, it had reestablished the detailed confrontation of Plato and Aristotle as a serious mode of philosophic investigation [...]. After George died [...] Florentine Platonism under the leadership of Marsilio Ficino took a very different line of development from that of the controversy. There were echoes of the controversy at the end of the century after the appearance of Pico della Mirandola's *De ente et uno* and in the works of Ambrosius Flandinus [Note 166: Cf. Kristeller, "Religious Orders," 40–41.

is a chapter on Byzantine philosophy that first appeared in German in 1999 and then in English translation in 2010. A portion of the concluding paragraph of this 1999/2010 chapter has minor parallels, without attribution, to the concluding paragraph of a chapter in a 1976 book by historian of the Renaissance, John Monfasani. The concluding paragraph of both the researcher's 1999/2010 chapter and Monfasani's 1976 chapter summarize the same research findings and provide the same analysis. Table 6.7 sets forth the parallel.

Especially in comparison with Cases 1–3, Case 4 seems minor, but it must be understood that the textual parallel is in the conclusions of the respective works by the researcher and by Monfasani. This textual parallel was initially identified in a published review of the 2010 English translation, which then went on to state, "Should the anticipated success of this collection warrant that the publisher produce a second edition in the future, additional documentation in [—]'s chapter would ameliorate the present reviewer's concerns" (Dougherty 2012: 476). When contacted by a journalist a few years later who was examining the growing number of corrections the researcher was acquiring, the researcher expressed regret about the deficiencies of the chapter, describing the situation as a "severe mistake" and a "shortcoming" and said that a corrective note would be added, presumably to a future edition of the original German text or in a revised translation (Palus 2016a). At the time of this writing, neither a correction nor a corrected new edition has appeared.

6.6 Cases 5 and 6: *Argumentation*

The first two published corrections of the scholarly record to appear in journals concerning the work of the researcher were a retraction and an erratum, both of which were published in late 2015 in Springer's philosophy and communications journal *Argumentation*. These two cases are especially noteworthy: it is good to be reminded that the process of submitting evidence of research misbehavior to journal editors sometimes results in published corrections of the scholarly record that are issued without delay. Since the analysis of cases presented in this chapter is at times critical of the deficient responses by editors, publishers, research integrity offices, and other members of the academic community, it is important to recognize those cases where the mechanisms for repairing damage to the scholarly record operate successfully. The two published corrections in *Argumentation* were issued for a pair of articles published by the researcher in the August 2008 issue of the journal. The articles constitute, respectively, Case 5 and Case 6. Case 5 is a short article that served as a commentary to a longer article by another researcher that appeared in the same issue of journal. An excerpt from the middle of the Case 5 article is given in Table 6.8.

The source of the 2008 text—a 2006 book by Adrian Vermeule—is mentioned in the concluding paragraph on the last page of the article ("As Vermeule (2006) had recently shown [...]") but it is not clear how this mention covers the earlier parts of the article, including the excerpt shown in Table 6.8. The conditions of "pawn sacrifice" text appropriation appear to be fulfilled in this case. Minor differences can be noted: the North American English words "labor" and "laborers" in Vermeule's 2006 book are rendered as the British English "labour" and "labourers" in the 2008 article. Additionally, references that appear as footnotes 9 and 10 in Vermeule's 2006 book are not present in the researcher's 2008 article.

Argumentation published a retraction for Case 5 that states, "This article has been retracted by the Editor-in-Chief, because it does not appropriately acknowledge publication Adrian Vermeule, *Judging Under Uncertainty: An Institutional Theory of Legal Interpretation* (Cambridge, MA, Harvard University Press, 2006), as the primary source of the comments made in the article" (van Eemeren 2015a: 493). This retraction is maximal in determination, credit, and availability, thus providing a full correction of the scholarly record. The retraction confirms and explains the deficiency of the article, gives credit to the original source appropriated in the article, and is available electronically without a paywall on the journal's website. Furthermore, the retracted article and the retraction statement are electronically tethered, and a diagonal watermark with the words "RETRACTED ARTICLE" has

Table 6.8 Text parallels with an earlier work by Adrian Vermeule

[—], "Comments on 'Strategic Manoeuvring with the Intention of the Legislator in the Justification of Judicial Decisions'," *Argumentation* 22.3 (2008): 355–357, at 356.	Adrian Vermeule, *Judging Under Uncertainty: An Institutional Theory of Legal Interpretation* (Cambridge, MA: Harvard University Press, 2006), 91–92.
the opinion of the Supreme Court had initially conceded that "the [church's act] is within the letter of this section," it proceeds on the premise that the legislator intention trumps text: "While there is great force to this reasoning, we cannot think congress intended to denounce with penalties a transaction like that in the present case. It is a familiar rule, that a thing may be within the letter of the statute and yet not within the statute, because not within its spirit, nor within the intention of its makers." In the following part, the Court gives two alternative reasons for concluding that the case falls outside the intent of the statute's makers. Feteris discusses these two grounds later on in her analysis of the opinion under *The justification of the choice of the argumentation scheme*. The first ground is that the critical phrase, which prohibited immigration under contract to perform ''labor or service of any kind,'' was intended to cover only manual labour rather than professional or intellectual work. The second ground is that Congress could not have intended to prohibit the immigration under contract of a Christian minister. In order to support the first ground, the Court relies on several sources. He argues that the statute should be limited to the scope of the evil that the statute was designed to remedy. It was "common knowledge" that the act's "motive" was to prevent an influx of "cheap, unskilled labor" in the form of ''an ignorant and servile class of foreign labourers." Another source that the Court mentions for determining the intention of the legislator is the internal legislative history. This is particularly remarkable, given that in this respect the opinion breaks from traditional doctrine without any further explanation for doing so. Two documents of legislative history are considered. The first is a report from the Committee on labor of the House of Representatives, in which the bill had originated. This report stated that the act wanted to prohibit the immigration or importation of ''[labourers] from the lowest social stratum, [who] live upon the coarsest food… and are certainly not a desirable acquisition to the body politic.'' The second document, which provided strong support for the Court's limitation of the act to manual ''labor or service of any kind,'' was a report of the Senate Committee on Education and Labor	The opinion initially conceded that "the [church's act] is within the letter of this section," yet proceeded on the premise that intent trumps text: "While there is great force to this reasoning, we cannot think congress intended to denounce with penalties a transaction like that in the present case. It is a familiar rule, that a thing may be within the letter of the statute and yet not within the statue, because not within its spirit, nor within the intention of its makers." [9] The Court, however, gave two alternative grounds for concluding that the transaction before it fell outside the intent of the statute's makers. The first ground was that the critical phrase in section 1, which prohibited immigration under contract to perform "labor or service of any kind," was intended to cover only manual labor rather than professional or intellectual work. The second, narrower ground was that Congress could not have intended to prohibit the immigration under contract of a Christian minister. The Court relied on several sources to support the first of these two grounds […]. The Court also argued that the statute should be limited to the scope of the evil that the statute was designed to remedy, as evidenced by "contemporaneous events." It was "common knowledge" that the act's "motive" was to prevent an influx of "cheap, unskilled labor" in the form of "an ignorant and servile class of foreign laborers." […] The Court's next source for determining congressional intent, however, was internal legislative history, and the opinion gave no explanation from that break from traditional doctrine. The Court relied on two items of legislative history. The first was a report from the Committee on Labor of the House of Representatives, in which the bill had originated. The report stated, in general terms, that the act sought to prohibit the immigration or importation of "[laborers] from the lowest social stratum, [who] live upon the coarsest food...and are certainly not a desirable acquisition to the body politic." [10] The second item, which provided the best support for the Court's limitation of the act to manual "labor or service of any kind," was a report of the Senate Committee on Education and Labor

been added to each page of the electronic version of the article. The retraction was also published in a later issue of the print version of the journal on a numbered, citable page, thereby providing a substantive correction of the scholarly record.

The text of the retraction, credited explicitly to the Editor-in-Chief of the journal *Argumentation*, does not indicate whether the researcher agreed to the retraction. The researcher discussed this retraction with a journalist, however, acknowledging "the one or other sentence where quotation marks are missing (but from the context it is clear that I refer to Vermeule), and this is a mistake" (Palus 2016a). Yet, as can be seen from the excerpt in Table 6.8, the issue appears to be more serious than the absence of quotation marks for one or two sentences, and significant textual parallels are found on each page of the short article. The source text by Vermeule is referenced on the last page of the article, but not in any way that discloses to the reader the extensive overlap of text on each page of the article. The reader has no way of determining from the article itself that it largely consists of undocumented extracts from pages 87–102 of Vermeule's book.

Case 6 was published in 2008, likewise in the journal *Argumentation*, where it occasioned a highly unusual and lengthy erratum. The erratum, which appeared in 2015, contains the entire text of the researcher's earlier published 10-page article, but this time the text is supplemented with newly added references to the source materials that had been appropriated without citation in the earlier version. The new, 11-page version of the article in the erratum is now supplemented with quotation marks, in-text citations, and an expanded bibliography.

The 2015 version with the erratum contains a new paragraph authored by the Editor-in-Chief of *Argumentation* that explains that the 2008 version of the paper contained "certain severe shortcomings in the references" and that the revised text is the result of an invitation to the author of record "to prepare a new version of the paper in which these shortcomings have been remedied" (van Eemeren 2015a: 481). Even though the issuance of corrections of the scholarly record for papers that misappropriate the work of others is fairly standard, the republication of a new version of the entire article in an erratum is highly uncommon in philosophy.[4] Since journal space is a limited and valuable commodity, the additional publication arguably precludes the publication of the work of others, for a second time. Journal editors and publishers in philosophy typically restrict the complete publication of a corrected version to those cases in which substantial errors have been introduced by the publisher during the production process, rather than for lapses committed by the author of record.[5]

An example from Case 6, demonstrating clearly how various kinds of deficiencies in documentation were remedied, is found in Table 6.9. The table provides an excerpt of the 2008 deficient article, alongside the 2006 original source text by Richard L. Street, Jr. and the 2015 corrected version that appears in the erratum.

The 2006 original source by Street began with credit to the work of Howard Waitzkin, and this credit that was absent in the 2008 article by the researcher has been given in the 2015 erratum. Unfortunately, some of the additions or amplifica-

[4] The republication of a philosophy paper within the same journal is not unprecedented, however. See Delaney 2007 and Delaney 2008, both of which appeared in *Philosophical Studies*. In that case, the latter paper appears to incorporate the former within consecutive years of the same journal, without attribution to the first.

[5] For examples of errata in philosophy journals that contain a complete republication of an article in corrected form due to production errors (rather than author-generated errors), see Toh 2015: 333: "Prior to the original publication of this paper, a great number of page references were incorrectly changed at the production stage. The correct page numbers have been inserted in this erratum version of the paper"; King 2014: 3379: "A significant number of unfortunate errors have been identified in the above-mentioned article. The full corrected article is republished on the following pages and should be treated as definitive by the reader, replacing the earlier version"; Anonymous 2000c: 113: "Due to a technical error at the printing stage, certain characters, important to a proper understanding of the text, in the above-mentioned article failed to appear in the printed version. The correct version has been reprinted on the following pages."

Table 6.9 Text parallels with an earlier work by Richard L. Street, Jr., with attribution given in a subsequent erratum

[—] et al., "Arguing 'for' the Patient: Informed Consent and Strategic Maneuvering in Doctor–Patient Interaction," *Argumentation* 22.3 (2008): 423–432, at 423.	Richard L. Street, Jr., "Communication in Medical Encounters: An Ecological Perspective," in *Handbook of Health Communication*, ed. Teresa L. Thompson et al. (Mahwah, New Jersey: Lawrence Erlbaum: 2003): 63–94, at 81, 64.	"Erratum," *Argumentation* 29.4 (2015): 481–489, at 481–482.
These forms of investigation provide insight into the complexities of discourse, the nuances of meaning, the human experience of health and the way in which social, political, and historical contexts impinge upon parts of the medical interaction. One line of inquiry examines how patterns of provider–patient communication are related to the attributes of the patients (e.g. education, age, health status), the provider (e.g. gender and medical speciality) and their relationship (e.g. rapport and trust) (Roter et al. 1988; Roter and Hall 1993; Thompson 1994; Ong et al. 1995; Street 2001). A second line of research examines relationships between providers' and patients' communicative actions and the various outcomes resulting from the consultation (e.g. satisfaction with care, commitment to treatment and health improvement) (Roter and Hall 1993; Stewart et al. 1999; Street 2001). These studies have identified, among other things, important sources of variability in the communicative contents of medical encounters, yet they do little to explore	As Waitzkin (1991) has noted, qualitative analyses provide valuable insight into the complexities of discourse, the nuances of meaning, the human experience of health, and the ways in which social, political, and historical contexts impinge upon parts of the medical interaction. // One line of inquiry examines how patterns of provider-patient communication are related to the attributes of the patient (e.g., education, age, health status), the provider (gender, medical specialty), and their relationship (rapport, trust) (for reviews, sec Ong, De Haes, Hoos, & Lammes, 1995; Roter & Hall, 1993; Roter, Hall, & Katz, 1988; Street, 2001; Thompson, 1994). A second area of research examines relationships between providers' and patients' communicative actions and the various outcomes (mostly patient focused) resulting from the consultation (e. g., satisfaction with care, commitment to treatment, health improvement) (for reviews, see Roter & Hall, 1993; Stewart, 1995; Street, 2001). Although these studies have identified sources of variability in the communicative content, structure, and outcomes of medical encounters, they do little to explain	As Street (2003, p. 81) notes, following Waitzkin (1991), these forms of investigation ''provide valuable insight into the complexities of discourse, the nuances of meaning, the human experience of health and the way in which social, political, and historical contexts impinge upon parts of the medical interaction.'' In the context of interpersonal communication, Street (2003, p. 64) distinguishes between two lines of enquiry: ''One line of inquiry examines how patterns of provider–patient communication are related to the attributes of the patients (e.g. education, age, health status), the provider (e.g. gender and medical speciality) and their relationship (e.g. rapport and trust)'' (see also Roter et al. 1988; Roter and Hall 1993; Thompson 1994; Ong et al. 1995; Street 2001). ''A second line of research,'' Street (2003, p. 64) continues, ''examines relationships between providers' and patients' communicative actions and the various outcomes resulting from the consultation (e.g. satisfaction with care, commitment to treatment and health improvement)'' (see also Roter and Hall 1993; Stewart et al. 1999; Street 2001). According to Street (2003, p. 64) ''these studies have identified sources of variability in the communicative content [...] of medical encounters.'' Yet they do little to explore

tions to Street's original text that are made in the 2008 article are now incorrectly ascribed to Street in the erratum, since the researcher has put quotation marks around the sentences by Street in their modified versions, rather than in their true formulation. For example, Street's expression "second area of research" had become "second line of research", and remains that way in the erratum. Similarly, the researcher's 2008 article rendered Street's original wording so that "specialty" had become "speciality," and the researcher's modification—"speciality"—is now credited to Street in the erratum version. The 2008 article also includes many of the secondary references offered in Street's original literature review, and quotation marks are still not given for this textual parallel. These points may seem small, yet given the cost in time, printing, and journal page space in re-presenting the entire 11-page text of the article 2008 article in the 2015 erratum—arguably displacing the work of some author who otherwise would have been published in the journal's limited publication space—it is puzzling that the correction falls short of accuracy in certain respects.

Insofar as this published erratum states that there are "severe shortcomings" with respect to acknowledging the original sources, identifies the sources that were appropriated without sufficient credit, and is available online without a paywall, it must be judged as maximal in determination, credit, and availability.

6.7 Case 7: *Swiss Medical Weekly*

Cases 1–6 each involved publications within the discipline of philosophy. A portion of the publications for which the researcher is the author of record extends outside philosophy to the research area of health communication. Case 7 is a co-authored article on the topic of organ donation that appeared in the journal *Swiss Medical Weekly* in 2007. The article presents the results of a study that sought to correlate organ donation rates in Switzerland with the prevalence of media reports about organ donation in 15 major regional German-, French-, and Italian-language newspapers published from 1999 to 2003. In addition to reporting the results of the study of organ donation rates, the article provides policy recommendations for how to increase these rates. A sidebar on the first page of the article indicates that the research was funded through a Swiss National Science Foundation (SNSF) grant (NRP 46 Raising Public Awareness 4046–58627). The text of the two opening paragraphs of the article is set forth in Table 6.10.

Most of the text of these two opening paragraphs can be found in the earlier works by Besser et al. and Gold et al. Nowhere in the article are these two sources cited. There are some small changes, to be sure; "Switzerland" has replaced "Germany" (and the number of dying patients has been dramatically reduced to reflect the change to the smaller country). The parallels extend to the literature review in the footnotes.

After examining the evidence that I submitted, the home institution of the researcher ruled that there is no plagiarism to be found in this case and the others, but just copyright infringement. This determination was assisted by an outside party who was asked to advise the institution in the matter. In the opinion of the outside party, which was followed by the home university of the researcher, the issue was characterized as a violation of Gold et al.'s rights as the original author. This shift from an investigation into suspected plagiarism to a legal consideration of copyright is notable; it changes the terms under which the textual parallels are analyzed. If the situation is approached primarily as a potential act of copyright infringement, the fundamental focus is on the original author's loss, rather than on the activity of the appropriating authors of record and on the subsequent effects on the reliability of the scholarly record.

In 2017, *Swiss Medical Weekly* issued an erratum for the article. The text of the erratum, which was composed by the researcher and approved by the editor, states that in the introduction of the article "several phrases were taken [...] without

Table 6.10 Text parallels with earlier organ donation literature

[—] et al., "Raising Awareness of Organ Donation," *Swiss Medical Weekly* 137 (2007) (Supplement 155): 132–137.	Avi Besser et al., "Who Signs an Organ Transplant Donor Card? A Study of Personality and Individual Differences in a Sample of Israeli University Students," *Personality and Individual Differences* 36.7 (2004): 1709–1723; Stefan M. Gold et al., *The Organ Donation Process: Causes of the Organ Shortage and Approaches to the Solution: An Analysis of the Content and Methods of Available Studies* (Cologne: Federal Centre for Health Education, 2001).	
Introduction Organ transplantation is considered a marvel of modern technology. For many illnesses, organ transplantation represents the only life-saving therapy, and leads to a substantial improvement in the quality of life. However, modern transplantation medicine creates not only new hope for many terminally ill patients, but also a number of medical, psychosocial, ethical and economic problems. The primary obstacle to the performance of transplantations usually does not lie in their technical feasibility, nor in surgical mortality or rejection problems, but rather in the limited number of donor organs in relation to the number of patients who could be helped by transplantation. In Switzerland, approximately 55 patients died in 2003, 43 in 2004 while waiting for an organ. Despite major educational campaigns among the general public, the supply of organs for transplanting remains far short of demand. Proposed explanations for the organ shortage fall into two categories, one focusing on the medical system, and the other on donor willingness [2–4 Horton RL, Horton PJ 1991; Radecki CM, Jaccard J.1997; Radecki CM, Jaccard J. 1999}]. The present study addresses willingness to sign donor cards, exploring the information sources of the Swiss population between 1999 and 2003. It should be viewed as part of an investigation into the reasons of the small number of people holding a donor card	1. Introduction Organ transplanting, a medical procedure with lifesaving potential, is considered a marvel of modern technology. For many illnesses, organ transplantation represents the only option for life-saving therapy and a substantial improvement in the quality of life [...]. However, the capabilities of modern transplantation medicine create not only new hope for many terminally ill patients, but also a number of medical, psychosocial, ethical and economic problems. The primary obstacle to the performance of transplantations usually lies not in their technical feasibility, nor in surgical mortality or rejection problems, but rather in the limited number of donor organs in relation to the number of patients who could be helped by transplantation. In Germany, approximately 1,000 patients died in 1999 alone while waiting for an organ (DSO, 2000). Despite major educational campaigns among the general public and medical personnel, the supply of organs for transplanting remains far short of demand [...]. Proposed explanations of the organ shortage fall into two categories, one focusing on the medical system and the other on donor willingness. (For an extensive review see Horton & Horton, 1991; Radecki & Jaccard, 1997, 1999.) [...]. The present study addresses willingness to sign donor cards, exploring the effects of psychological aspects and personality variables on awareness and willingness alike. It should be viewed as part of an investigation into the reasons that so few people hold donor cards.	Besser et al. 2004: 1709 Gold et al. 2001: 9. Besser et al. 2004: 1710.

offering proper reference." An apology was offered for "erroneous citations," which were characterized as errors that "concerned phrasing rather than research findings" (Anonymous 2017a). The erratum contained full bibliographical references to the original articles by Besser et al. and Gold et al.

As noted earlier, some institutions in the field of natural sciences distinguish between the core of a research article, namely, its report of original findings, and any non-core items, such as a literature review, footnotes, or introductory material. With this distinction, some institutions restrict their determinations of plagiarism to cases only involving an appropriation of the core of someone else's work. As mentioned in Chap. 3.1.2, this position is found in the guidelines for plagiarism offered by the SNSF, the grant agency that sponsored the Case 7 article. Those guidelines indicate that a "trivial case" of plagiarism can be deemed to have occurred when any of the following conditions are met: (1) "only a few citations are missing," (2) "only a small amount of text is uncited compared to the entire text," or (3) "the content of the uncited text is of a general nature or concerns the state of research" (Swiss National Science Foundation 2016: 6). With regard to Case 7, all three conditions appear to have been fulfilled. The home institution of the researcher, its outside party, and the final editorial position of *Swiss Medical Weekly* supported a determination of copyright infringement rather than plagiarism.

A reader of the 2007 *Swiss Medical Weekly* article would likely assume that the authors of record, rather than the hidden original authors, have done the research of tracking down earlier articles on the topic of organ donation and thereby are the originators of the literature review that appears in the introduction of the article. That is, from the article itself, there is no way of determining that the authors of record are not the authors of origin for the introduction of the article. The writing of a reliable literature review requires an assessment of previously-published scholarship, and this assessment presupposes expertise and critical judgment, for which credit is due. The published erratum therefore assists in correcting the scholarly record, even if it appears to characterize the absence of quotation marks and footnotes as a case of "erroneous citations" for "several phrases." A full correction of the scholarly record would involve a clearer statement of the issues involved. Also, an electronic tether between the 2007 article and the 2016 erratum could assist future readers of the article.

6.8 Cases 8 and 9: Brill Publishing

In 2014, the large Dutch academic publishing house Brill acquired Editions Rodopi, a smaller publisher specializing in humanities disciplines. With the acquisition, Brill inherited the in-print backlist of around 3000 books of Editions Rodopi (Brill 2014). The backlist included volumes in a series called "Probing the Boundaries," two of which were edited collections that each contained one contribution by the researcher. These two contributions, Case 8 and Case 9, were both retracted by Brill in early 2016. Case 8 was actually published twice: first in the collection by Editions Rodopi (Case 8), and then a second time in a four-volume anthology of articles edited by the researcher and published by SAGE Publications (Case 8*). In this way, the same article was added to the scholarly record twice, yet it was corrected only once; even though Brill issued a retraction for Case 8, SAGE left Case 8* unretracted. When notified of the Brill retraction, SAGE responded by placing the 4-volume anthology out of print. For SAGE, a complicating factor appeared to be that the researcher was not only the author of record for the deficient article but also the editor of the anthology.

Table 6.11 provides a short excerpt from Case 8 and 8*; here the original source text is an unreferenced 1985 article by K. Danner Clouser. The differences between the 1985 text by Clouser and the 2006 and 2010 version by the researcher are subtle. For example, the pronoun "he" has been expanded to the more inclusive expression "he or she," and the word "physician" has been changed to "doctor" throughout the

Table 6.11 Text parallels with an earlier work by K. Danner Clouser

[—], "The Communication of Diagnostic Information by Doctors to Patients in the Consultation," in *Bordering Biomedicine*, ed. Vera Kalitzkus and Peter L. Twohig (Rodopi/Brill, 2006): 103–116, at 105–106. Republished in: [—], "The Communication of Diagnostic Information by Doctors to Patients in the Consultation" in *Communication Theory*, vol. 2. ed. [—], SAGE Benchmarks in Communication (Los Angeles: SAGE, 2010), 71–83 at 73–74.	K. Danner Clouser, "Approaching the Logic of Diagnosis," in *Logic of Discovery and Diagnosis in Medicine*, ed. Kenneth F. Schaffner (Berkeley, CA: University of California Press, 1985), 35–55, at 38, 39, 40, 41, 47.
In other words, in working up a patient, doctors are blending at least two activities: one is the gathering of information and the other is the attempt to see a pattern or a constellation in those data. 7[Stoeckle and Barsky] It is the pattern-seeing that might more technically be called the diagnosis. The doctor wants to put the patient in a classification because the doctor's knowledge of causes, treatments, and prognoses is organized that way. In trying to arrive at a classification – that is, a disease or illness label – the process of data gathering and diagnosis are in no set sequence. The heart of any diagnostic reasoning is the relationship between signs, symptoms, laboratory tests, and so on, and the disease of which they are manifestations. What can the presence of the one tell the doctor about the possible presence of the other? That is, given the one, what is the probability of the other's being there? Given the presence of a disease, what is the probability of having a specific symptom? What the doctor usually wants to know in diagnosis is the probability of the patient's having this or that disease, given the symptoms he or she is manifesting. This is just the converse of what we might find out from records or the literature, namely, given the disease, what is the incidence of the symptom. In the diagnostic context, the physician is given the symptom, and is trying to figure out the probability of this or that disease in his patient, given the symptom. Impressionistically, then, if there is a high prior probability (a high incidence of the disease in the population) and if most people with the disease have the symptom in question, then the probability of this particular patient's having the disease seems quite high.	In working up a patient, physicians are blending at least two activities: one is the gathering of information, and the other is the attempt to see a pattern or constellation in those data. It is the pattern-seeing that might more technically be called the diagnosis. The physician is wanting to put the patient in a classification because the physician's knowledge of the causes, treatment, and prognoses is organized that way. In trying to arrive at a classification, that is, as disease or illness label, the process of data gathering and diagnosis are in no set sequence [...]. We now turn to the heart of any diagnostic reasoning: the relationship between signs, symptoms, lab tests, and so on, and the disease of which they are manifestations. What can the presence of the one tell us about the possible presence of the other, statistically? That is, given the one, what is the probability of the other's being there? Given the disease, what is the probability of having such and such as symptom? [...] What we usually want to know in diagnosis is the probability of the patient's having this or that disease, given the symptoms he is manifesting. This is just the converse of what we might find out from records or the literature, namely, given the disease, what is the incidence of the symptom. In the diagnostic context, the physician is given the symptom, and is trying to figure out the probability of this or that disease in the patient, given the symptom. Impressionistically, then, if there is a high prior probability (i. e., a high incidence of the disease in the population) and if most people with the disease have the symptom in question, then the probability of this particular patient's having the disease seems quite high.

excerpt, and the term "lab" has been expanded to "laboratory." A footnote has been created for the first sentence that directs the reader to the work John Stoeckle and Arthur Barsky, but this reference appears to be irrelevant, since Clouser, who remains unreferenced in the researcher's 2006 text, is the author of the original 1985 text.

In the introductory blurb to the four-volume SAGE collection *Communication Theory* that republishes Case 8 as Case 8*, the publisher states that volumes in the series "will bring together the best of the best from across the disciplines" and in the process will be "setting out the debates and defining the subfields of research." In the introductory essay to the collection, the researcher apparently continues the publisher's rhetorical line by stating that the point of the collection is:

> To provide readers—students and academics—with access to a free-standing body of theoretical work which is applicable to a range of different topics within communication sciences, media, and cultural studies ([—] 2010a: lxvii).

In fact, a different four-volume collection, published by Routledge 4 years earlier and similarly titled as *Communication Theories*, makes an almost identical claim, saying that it

> provides academics and students with access to a free-standing body of theoretical work which is applicable to a range of different topics within communications, media and cultural studies. (Cobley 2006)

The overlap in the description of the 2010 and the 2006 anthologies is unmistakable.

The example of Case 8 and 8* highlights certain difficulties in correcting the scholarly record. Given the retracted status of the first and the unretracted status of the second, it is arguable that the integrity of the scholarly record has not been fully restored. Content that is now-retracted is presented in an anthology that purports to offer to students and researchers the best literature in the field.

Case 9—the other chapter appearing in another volume of the "Probing the Boundaries" series—had similar the textual parallels to the ones found in Case 7 from *Swiss Medical Weekly*. Texts found in Besser et al. and Gold et al. constituted portions of the beginning of the article, and again quotation marks and proper citation of these articles were lacking. Like Case 7, this article was also published with credit to the SNSF for the National Research Programme grant, NRP 46 "Implants and Transplants."

The wording of the respective retractions published by Brill for Cases 8 and 9 was nearly identical. Published in January and February of 2016, the retractions state that "direct reference and citation of the works of other scholars is often inconsistent and in some cases totally lacking," and that the two articles do "not meet standards currently expected of an academic publication" (Anonymous 2016a, b). The two retractions appearing on the Brill website are maximal in determination and availability, but minimal in credit, as the original source texts (e.g., Clouser or others) are not identified in the retractions.

6.9 Cases 10 and 11: *Studies in Communication Sciences*

The journal *Studies in Communication Sciences* issued one retraction and one corrigendum for a pair of articles published by the researcher (Case 10 and Case 11). The corrections appeared more than a year after one of the editors of the journal stated to a journalist, "We at *SComS* take allegations of plagiarism very seriously. The issue is under investigation" (Palus 2016b). The situation was complicated by two factors: the journal is co-sponsored by the department of the researcher's home institution, and the researcher had served in various editorial capacities for the journal. Thus, for the issue containing the now-retracted article, the researcher was listed on masthead of the journal as one of the five members of the Editorial Board, and for the issue containing the article now subject to the corrigendum, the researcher was listed as a member of the Advisory Board of the journal.

Table 6.12 Text parallels in a literature review

[—], "Maximizing Health Outcomes through Optimal Communication," *Studies in Communication Sciences* 6.2 (2006): 215–232, at 217.	Scott C. Ratzan, J. Gregory Payne, and Carol Bishop, "The Status and Scope of Health Communication," *Journal of Health Communication* 1.1 (1996): 25–41, at 28.
two main levels of communication have received the most attention—interpersonal and mass communication. Generally, the study of interpersonal communication focuses on face-to-face communication between two or more people. Much of the literature in health communication addresses this level, examining the patient-provider relationship (Arnston et al. 1978; Ballard-Reisch 1990; Beisecker & Beisecker 1993 ; Brenders 1989; Evans et al. 1992; Garko 1992; Korsch et al. 1968; President's Commission 1982; Quill 1983; Sharf 1984; Street & Wiemann 1987; Wyatt 1991). Defining the field from an interpersonal perspective, Costello (1977) asserted that relationships in health communication should be the central component of a definition: "Communication research may usefully begin to focus on the dynamics by which the health professional and patient establish, maintain, exploit and/or alter their relationships" (p. 565) On this view, the success of the physician-patient encounter depends in part upon how information about such factors is discussed and, ultimately, how it shapes the behavior of patients. The other popular level of analysis of health communication is *mass communication*. This level focuses on effective message dissemination for health promotion, disease prevention, and health-related messages transmitted through mediated channels, including health marketing and policy-making (Signorelli 1993; Cumberbatch & Negin 1992; Backer et al. 1992; Jackson 1992; Atkin & Wallack 1990; Payne et al. 1989). Among those whose definition of health communication is representative of the mass communication emphasis are Maibach et al. (1993), who have asserted that "health communication is a crucial element in disease prevention and health promotion campaigns because the provision of relevant and persuasive health information is the primary social process that can empower individuals to take charge of their own health" (p. 15).	two levels of communication have received the most attention: interpersonal and mass communication [...]. Generally, the study of interpersonal communication focuses on face-to-face communication between two or more people. Much of the literature in health communication addresses this level, examining the patient-provider relationship (Arnston et al., 1978; Ballard-Reisch, 1990; Beisecker & Beisecker, 1993; Brenders, 1989; Evans et al., 1992; Garko, 1992; Korsch et al., 1968; President' s Commission, 1982; Quill, 1983; Sharf, 1984; Street & Wiemann, 1987; Wyatt, 1991). Defining the field from an interpersonal perspective, Costello (1977) asserted that *relationships* in health communication should be the central component of a definition: "Communication research may usefully begin to focus on the dynamics by which the health professional and patient establish, maintain, exploit and/or alter their relationships" (p. 565) [...]. The other popular level of analysis of health communication is *mass communication*. This level focuses on effective message dissemination for health promotion, disease prevention, and health-related messages transmitted through mediated channels, including health marketing and policy-making (Jackson, 1992; Payne et al., 1989). Among those whose definition of health communication is representative of the mass communication emphasis are Maibach et al. (1993), who have asserted that "health communication is a crucial element in disease prevention and health promotion campaigns because the provision of relevant and persuasive health information is the primary social process that can empower individuals to take charge of their own health" (p. 15).

The retracted article, Case 10, which was published in 2006, does not present any research findings, but states that its twofold goal is to "examine briefly the field of Health Communication" and to identify research projects for "maximizing health outcomes through optimal communication" ([—] 2006b: 215). Table 6.12 provides an excerpt from the 2006 article with textual parallels to a 1996 article by Scott C. Ratzan et al.

In this excerpt, the only significant text portion not parallel with Ratzan et al. is a single sentence. Most of the review of secondary literature is identical, although there are added a few references in the 2006 text. The work of Ratzan et al. is not mentioned in the bibliography or elsewhere in the 2006 text, and there are no quotation marks to suggest that the 2006 text is in any way dependent upon the 1996 text.

The retraction of the 2006 article appeared in *Studies in Communication Sciences* in August 2017, in the journal's newly launched online archive of past issues. On the landing page for the article is the statement, "This article has been retracted." In place of the full article is now to be found just the first page that presents the abstract, and the word "RETRACTION" has been added in red letters to the four corners of the page (Anon 2017c). No additional information is provided, but a news article discussing the case identified the reason for retraction as the "misappropriations of the expressions of other scholars" (Stern 2018a, b). The removal of the rest of the retracted article beyond the first page is surprising, since retraction and removal are distinct acts, and editors generally restrict the latter to exceptional circumstances (see Hames 2007: 194). The retraction must be judged as minimal in determination and credit, since it neither indicates to a reader the reason for the retraction nor identifies the authors of the appropriated texts in the now-retracted article. The retraction is maximal in availability, however, given its free online distribution.

This retraction of the 2006 article generated, at least for a short time, what might be called a secondary or "collateral" retraction. The next article in the same issue of *Studies in Communication Sciences* was a short two-page commentary on the now-retracted article. Written by another researcher, the text is titled "Comments on [—]" and opens with the declaration that the researcher "has produced a well written and thought provoking article" (Mackay 2006: 143). The other researcher then provides critical comments, but also praises the article, saying,

> I find his vision intellectually attractive and I believe that the concepts he has started to develop may well have important consequences for the whole field of health communication (143).

In mid-August 2017, this short commentary article was unavailable on the website of *Studies in Communication Sciences*. In place of the abstract appeared the brief notice, "Please note that the referred article has been retracted. [SComS Editorial Office]." I assumed that the reason for the retraction statement and the accompanying removal of the two-page article was its commendation of the Case 10 article to readers. When I wrote to the editors of *Studies in Communications Sciences* to ask for a contextualization of this apparently collateral retraction, the short commentary article was restored to the website within a day, and the aforementioned retraction notice that had stated "Please note that the referred article has been retracted. [SComS Editorial Office]" had been removed from the website. Apparently this secondary or collateral retraction had been retracted.

Unlike Case 10, Case 11 in the same journal is a co-authored article that presents new research findings. The discovered textual parallels generating the corrigendum for Case 11 concern exclusively the method statement and the literature review, both of which appear in the introductory portion of the article and precede the presentation of the data and the analysis of the research findings. The study reported in Case 11 was conducted by approaching women in grocery stores and on trains in Switzerland and Amsterdam to inquire about their knowledge of breast health. The article reports that the investigators "studied 480 women in all, aged between 15 and 84," and that "trained female research assistants" approached the women in grocery stores and on trains with a questionnaire featuring queries about breast health and the availability of check-ups, asking them questions on such topics as the "frequency of breast self-examination." For some of the participants, "a few oral questions were asked." The authors of record reported a response rate of "more than 90%" for the women approached in supermarkets and trains in Switzerland to talk about their breast health practices, and similarly "up to 95%" for the Netherlands" ([—] et al. 2009: 251, 252).

The corrigendum for Case 11 issued by *Studies in Communication Sciences* was published in August 2017. Signed by the author of record and a co-author, the corrigendum appears on a new page added electronically before the first page of the article. It presents two bibliographical references to works in *The Lancet* and *The European Journal of Epidemiology* that were not referenced in the article, and it offers the conditional sentence, "If we violated the authors' rights, we regret this and apologize for the missing citations" (Anonymous 2017d: 248a). The corrigendum appears to invoke the core/non-core distinction in characterizing the defects of the article, stating that "these errors concerned phrasings rather than research findings" and that the defects do not "affect the empirical data of the study, nor the interpretation of these data nor the conclusions that we draw." As this corrigendum gives full bibliographical references to the previously unreferenced source material and is freely available online, it can be judged as maximal in credit and maximal in availability.

6.10 Case 12: Elsevier's *Patient Education and Counseling*

Much like Case 11, Case 12 also involves an article that presents empirical research findings. Published in 2006 in the Elsevier journal *Patient Education and Counseling*, there are textual parallels with the earlier published works of others, and again these parallels pertain to what some might define as the non-core elements of the article. The parallels involve the introductory literature review and the statement of methodology. Table 6.13 provides excerpts from the literature review and from the methodology description.

The textual parallels for the literature review include the footnotes, 24 of which are common to the 2006 article and the previously published works of others. Some of the textual parallels in the footnotes involve works by authors already mentioned above, such as Besser et al. and Radecki et al. The article by Radecki et al. is footnoted on another page of the article, but there it is cited in support of another, totally unrelated point. Surprisingly, one of the exhibited parallels involves an article that appeared in *Patient Education and Counseling* a year before Case 12, namely, Sander et al. 2005.

In the very detailed method statement, the 2006 text also specifies the languages used in the study. However, whereas the earlier 2005 article by Siegel et al. refers to Spanish as the language in which the interviews were conducted, in the 2006 article this has been changed to "German, French, or Italian."

The journal *Patient Education and Counseling* had listed the author of record as a "Deputy Editor" on its masthead. This 2006 article that constitutes Case 12 was, in addition to Cases 7 and 9, the third by the researcher published with acknowledgment of funding from the SNSF grant. According to the Swiss newspaper *Neue Zürcher Zeitung*, the resultant controversy over articles involving SNSF funding led

Table 6.13 Text parallels in a literature review and in a method statement

[—], "More than Nation and Knowledge: Cultural Micro-Diversity and Organ Donation in Switzerland," *Patient Education and Counseling* 64 (2006): 294–302, at 295 and at 296.	Avi Besser et al., "Who Signs an Organ Transplant Donor Card? A Study of Personality and Individual Differences in a Sample of Israeli University Students," *Personality and Individual Differences* 36.7 (2004): 1709–1723; Carmen M. Radecki and James Jaccard, "Psychological Aspects of Organ Donation: A Critical Review and Synthesis of Individual and Next-of-Kin Donation Decisions," *Health Psychology* 16.2 (1997): 183–195; Shannon L. Sander and Barbara Kopp Miller, "Public Knowledge and Attitudes regarding Organ and Tissue Donation: An Analysis of the Northwest Ohio Community," *Patient Education and Counseling* 58.2 (2005): 154–163.	
Several recent studies of organ donation have emphasized the importance of cultural context in understanding attitudes and willingness towards organ donation [5–12 Alden and Cheung, 2000; Martinez et al., 2001; Shih et al., 2001; Yeung, Kong, & Lee, 2000]. Cultural factors are those attitudes relevant to organ donation that derive from the broader culture inherent in one's ethnic group membership. Several studies have addressed the role of such attitudes for Native Americans [13 Blagg, Helgerson, and Warren, 1992], Asian [14–16], Hispanic [17,18] and African American cultures [19].	An increase in cultural studies of organ donation (Alden & Cheung, 2000; Martinez et al., 2001; Shih et al., 2001; Yeung, Kong, & Lee, 2000) recently emerged, emphasizing the importance of the cultural context in understanding attitudes and willingness towards organ donation. Cultural beliefs are those beliefs relevant to organ donation that derive from the broader culture inherent in one's ethnic group membership. Although a few studies have addressed the role of such beliefs for Native Americans, (Blagg, Helgerson, & Warren, 1992; Sills, 1992 Popkin, & Najarian, [...]), most studies have focused on African American, Asian, and Hispanic cultures.	Besser et al. 2004: 1710. Radecki et al. 1997: 184.
Researchers have reported considerable variation in the willingness and commitment to donate among various subgroups of the U.S. population. The two largest minority cultures in the United States – the African American and Hispanic communities – are the focus of a number of studies of organ and tissue donation. A study of attitudes, beliefs, and behaviors toward donation among African Americans found that nearly 87% of respondents support donation [20 Park 1998]; yet, significantly lower African American families as compared to Caucasians [21,22 Evers, Lewis, and Schaeffer, 1999]. Hispanics are more likely than African Americans to express their support and willingness to donate organs as well as to grant consent for the donation of a deceased family member's organs [23 Gallup Organization, 1993]. Studies also indicate that younger people, those high in socioeconomic status, and Whites are more likely to sign an organ donor card [24,25 DeJong et al., 1998; Sanner 1994]. African Americans are less likely to sign an organ donor card because of distrust of the medical system [7, 26 Davidson and Devney 1991, 27 Lange 1992].	Researchers have reported considerable variation in the willingness and commitment to donate among various subgroups of the population [...]. The two largest minority cultures in the United States, the African American and Hispanic communities, are the focus of a number of studies in the literature on organ and tissue donation. A study of attitudes, beliefs, and behaviors toward donation among African Americans found that nearly 87% of respondents support donation [16 Park 1998], yet significantly lower consent rates for organ donation among African American families as compared to Caucasians have been reported [8 Evers, Lewis, and Schaeffer, 1999] [...]. Hispanics are more likely than African Americans to express their support and willingness to donate organs as well as to grant consent for the donation of a deceased family member's organs [6 Gallup Organization, 1993].Findings indicate that younger people, those high in socioeconomic status, and Whites are more likely to sign an organ donor card (DeJong et al., 1998; Sanner, 1994), and African Americans are less likely to sign an organ donor card because of distrust of the medical system (Davidson & Devney, 1991; Lange, 1992).	Sander et al. 2005: 155.
All telephone interviews were conducted using computer-assisted telephone interviewing (CATI). Trained interviewers used the ''most recent birthday technique'' to assure random selection within each household. To be considered a valid contact, the telephone number dialed had to belong to a private residence. Once contact with the eligible respondent was made – in German, French or Italian – the trained interviewer introduced himself or herself and invited the person to take part in a telephone survey. If the designated respondent, the person over 18 with the most recent birthday, was not home or if the call had occurred at an inconvenient time, the interview was rescheduled. Furthermore, to obtain the highest possible response rates on selection of the eligible respondent, a minimum of five callback attempts were made. Callbacks were routinely made at different times and on different days. Eligible respondents were assured participation would be voluntary and confidential.	All telephone interviews were conducted by using computer-assisted telephone interviewing (CATI).[//] Trained bilingual interviewers used the "most recent birthday technique" to assure random selection within each household (O'Rourke & Blair, 1983). To be considered a valid contact, the telephone number dialed had to belong to a private residence [...]. Once contact with the eligible respondent was made —in Spanish—the trained bilingual interviewer introduced himself or herself and invited the person to take part in a brief telephone survey in the language of his or her choosing. If the designated respondent, the person over 18 with the most recent birthday, was not home or if the call had occurred at an inconvenient time, the interview was rescheduled. Furthermore, to obtain the highest possible response rates on selection of the eligible respondent, a minimum of five callback attempts were made. Callbacks were routinely made at different times and on different days. Eligible respondents were assured participation would be voluntary and confidential;	Siegel et al. 2005: 245.

to the researcher no longer being listed as a research councilor (*Forschungsrat*) of the SNSF, and the newspaper cited competing accounts over whether the departure was due to a resignation or a sanction (Hafner 2017). In July 2018, the journal published a corrigendum (Anonymous 2018c). The author of record's position as Deputy Editor of the journal had been brought to an end earlier in the year after new revelations of plagiarism.

6.11 Cases 13 and 14: Two Additional Corrections

Two additional cases have earned corrections. Case 13 is a co-authored article that appeared in a 2012 collection of essays published by Springer, and it discusses the notion of health literacy by presenting a study of antibiotic use in Switzerland. Some of the literature review on antibiotic misuse that precedes the discussion of the article's empirical findings overlaps with previously-published research ([—] et al. 2012). During an investigation of this article and of two other works, the research integrity officer at the home institution of the co-author of record made an interesting discovery. According to his final report, the officer determined not only that "substantial plagiarism has occurred" but that, regarding the Case 13 article, the co-author "did not recognize it, and said he was not even aware that it existed" (Virginia Tech Research Integrity Office 2016: 1, 2). To support this claim, the co-author had provided to the officer "his entire Curriculum Vitae, which does not include this reference. He suspects his name was added as 'a courtesy', but that he had never seen it" (2). According to the research integrity officer, the article appears to suffer from more than one authorship violation: plagiarism and gift authorship. In June 2018, Springer republished a corrected version with the book containing a revised table of contents, a revised chapter (that omitted the name of the second author of record), and a new addendum of sources. An accompanying explanatory erratum—online and unpaywalled—states that the second author of record "should not be named as author to this chapter, as he did not fulfill all authorship criteria" (Anonymous 2018a). The erratum revealed further that the home institution of the first author of record held that in the chapter an "extract was not cited appropriately, but that any other overlap in this chapter was not significant" (Anonymous 2018a: E1).

Case 14 is a 2015 co-authored article on breast cancer screening programs that appeared in *Journal of Communication in Healthcare*, a quarterly published by Taylor and Francis ([—] et al. 2015). This article was one of the three works investigated by the research integrity office of one of the co-authors of record, and the university ruled that the researcher should provide a correction of the article and should do so in consultation with the first author of record. In April 2018, the journal issued a corrigendum that supplied quotation marks and bibliographical references to two sources, with the qualification, "Although we did not violate the authors' rights, we regret the missing citations" (Anonymous 2018b: 81). The short one-page corrigendum is behind a paywall, and non-subscribers are informed that they can purchase 24 h of access to the corrigendum for $54.00.

6.12 Twelve Corrections of the Scholarly Record

Cases 1–14 generated 12 corrections of the scholarly record by publishers and editors: five retractions (Cases 1, 5, 8, 9, 10), three errata (Cases 6, 7, 13), and four corrigenda (Cases 3, 11, 12, 14). These corrections of the scholarly record were issued by major international publishers as well as smaller, specialized academic presses. Beyond the affected editors and publishers, other involved parties included the research integrity offices of three institutions. In August 2016, the research integrity office of the researcher's home institution ruled (as noted above) that there was no plagiarism to be found but only "minor violations of copyright [violazioni minori del diritto d'autore]" and "incorrect methods of referring to sources" which however were considered to be "of minor importance" [modalità scorrette di citazione delle fonti di minor gravità] (USI 2016b, see also Illarietti 2018). Whether that research integrity office supported or did not support any of the published corrections of the scholarly record is not public knowledge. The research integrity officer at the American university, which is the home institution of one of the co-authors of record for three articles, determined in April 2016 that "substantial plagiarism" had been committed. As noted in Chap. 5.5, that finding was:

> After careful review of the evidence, I agree with the complainant that substantial plagiarism has occurred that extends beyond reasonable repetition of common statements, that there is misrepresentation of who has done the research stated in these articles, there is a lack of proper citations, and there is extensive use of others' statements verbatim. (Virginia Tech Research Integrity Office 2016: 1).

The American research integrity officer forwarded this judgment to the research integrity office of the researcher's home university and to the respective editors and publishers of the three co-authored articles. In April 2017, the research integrity office at the Dutch home institution of another co-author of record completed its investigation, ruling that principles of scientific research had indeed been violated, but not by the co-author of record himself. The research integrity office declared that the complaint "is found to be justified" [gegrond wordt verklaard] and directed the co-author to pursue two published corrections, and, if possible, to do so in consultation with the first author of record (College van Bestuur van de Universiteit van Amsterdam 2017: 2). The research integrity office was unable to come to a conclusion regarding the third article it had examined. In that instance, the suspected victim—a former graduate student of the co-author of record—declined to participate in the investigation.

6.13 Using the Correct Tools of Correction

Cases 1–6 involved articles in the field of philosophy, and some of the editors, publishers, and investigating research integrity offices appeared to approach the articles using a distinction between core elements (i.e., presentation of empirical research findings) and non-core elements (i.e., introduction, literature review, or method statement). This distinction applies to many articles in the empirical sciences but does not seamlessly transfer to most articles in humanities disciplines, including philosophy. To ask whether the undocumented text parallels in an article in philosophy concern the presentation of original research findings, or to try to distinguish between core and non-core elements of a philosophy article, is to commit a category mistake. Articles that are exercises of exegesis of classic philosophical texts or articles that trace the development of ideas through the history of philosophy cannot be approached in the same manner as one would approach a published report based on empirical lab experiments. In short, articles in the humanities do not divide into the "core" and "non-core" elements in the same way a publication of the results of empirical research might. To fail to see that research articles in humanities are not like the presentation of published conclusions in the sciences leads investigating bodies to approach questions of research misconduct with the wrong tools. In philosophy, for example, an article that traces the history of subjectivity among Continental philosophers of the twentieth century should not be approached as a "literature review." Such an article could constitute a genuine, original contribution to the history of philosophy. Research integrity offices and international publishers who use the same analytical tools for examining articles in the empirical sciences and in the humanities, failing to attend to the differences between the fields, their methodology, and their research ends, invariably follow the pernicious Law of the Hammer: *If your only tool is a hammer then every problem appears as a nail.*

6.14 Responsibility for Problems with the Scholarly Record

Who is responsible when the scholarly record is marred by faulty articles, books, and book chapters? In the view of some theorists, the blame extends to two parties. Most obviously, the authors of record who publish deficient papers and who neglect to support published corrections are culpable. Theorist Jaime Teixeira da Silva explains:

> Authors who publish papers that carry errors and who take laurels for such publications, but who then fail to correct reported errors at a post-publication stage should be classified as unethical, and their act or irresponsibility as misconduct. (Teixeira da Silva 2016b: 220)

However, the fault extends beyond the authors of record who pollute the scholarly record with deficient articles and do not support their correction; editors and publishers who withhold corrections in the presence of evidence of research misconduct are also culpable. Teixeira da Silva adds:

> Similarly, editors and publishers who fail to revise erroneous literature when claims are factually correct, independent of the volume of claims, are also in violation of their codes of conduct and professional responsibilities towards the scientific community and society. (2016b: 220)

In other words, the valuable good of a reliable scholarly record is threatened by uncorrected faulty research. Philosopher Roberto Casati has proposed an auditory metaphor to characterize the disruption created by published works that are products of plagiarism. Each plagiarized work is described as "extremely noisy as it confounds all possible assessments of the author of the original paper" and it "introduces noise [...] in the body of knowledge" (2010: 197, 196). The task of editors and publishers in such cases should be "to block the production of noise" by issuing published corrections to restore the scholarly record (198).

Readers of academic journals and books do not expect to find a pastiche of the undocumented work of others in the pages they read. A host of conventions exist for authors of record to signal to others when texts originate elsewhere, and these conventions include quotation marks, block quoting, footnotes, and in-text citations. When authors of record forgo the use of these conventions, the scholarly record is damaged. For Cases 1–14, the 12 published corrections of the scholarly record have improved the body of published literature.

References

[—]. 1999. Die Kontroverse zwischen Platonikern und Aristotelikern im 15. Jahrhundsssert: Gemisthos Plethon – Georgios Trapezuntios – Kardinal Bessarion. In *Philosophen der Renaissance*, ed. Paul Richard Blum, 22–33. Darmstadt: Wissenschaftliche Buchgesellschaft. (Case 4.)

———. 2001. Subjectivity from a semiotic point of view. In *Ecosemiotics: Studies in environmental semiosis, semiotics of the biocybernetic bodies, human/too human/post human*, Nordic-Baltic Summer Institute for Semiotic and Structural Studies, Part IV, ed. Eero Tarasti, Richard Littlefield, Lotta Rossi, and Maija Rossi, 149–159. Imatra: International Semiotics Institute. [Retracted in: Martinelli 2018a and 2018b.] (Case 1.)

———. 2004. Subjectivity before and after Badiou. *Subject Matters: A Journal of Communications and the Self* 1 (2): 67–81. (Case 2.)

———. 2005. Being in accordance with oneself: Moral self-controversy in Plato and Aristotle. In *Controversies and subjectivity*, ed. Pierluigi Barrotta and Marcelo Dascal, 75–90. Amsterdam: John Benjamins Publishing Company. [Corrigendum in: Anonymous 2017b.] (Case 3.)

———. 2006a. The communication of diagnostic information by doctors to patients in the consultation. In *Bordering biomedicine*, ed. Vera Kalitzkus and Peter L. Twohig, 103–116. Leiden: Rodopi/Brill. [Retracted in: Anonymous 2016a.] (Case 8.)

———. 2006b. Maximizing health outcomes through optimal communication. *Studies in Communication Sciences* 6 (2): 215–232. [Retracted in: Anonymous 2017c.] (Case 10.)

———. 2006c. More than nation and knowledge: Cultural micro-diversity and organ donation in Switzerland. *Patient Education and Counseling* 64: 294–302. [Corrigendum in: Anonyous 2018c.] (Case 12.)

[—], et al. 2007. Raising awareness of organ donation. *Swiss Medical Weekly* 137 (Supplement 155): 132–137. [Erratum in: Anonymous 2017a.] (Case 7.).

[—]. 2008a. Media treatment of organ donation: A case study in Switzerland. In *Social studies of health, illness and disease: Perspectives from the social sciences and the humanities*, ed. Peter Twohig and Vera Kalitzkus, 229–243. Leiden: Brill/Rodopi. [Retracted in: Anonymous 2016b.] (Case 9.)

———. 2008b. Comments on 'Strategic manoeuvring with the intention of the legislator in the justification of judicial decisions'. *Argumentation* 22.3: 355–357. [Retracted in: van Eemeren 2015a.] (Case 5.)

[—], et al. 2008c. Arguing 'for' the patient: Informed consent and strategic maneuvering in doctor–patient interaction. *Argumentation* 22 (3): 423–432. [Erratum in: van Eemeren 2015b.] (Case 6.).

———. 2009. Knowledge, information sources and awareness regarding breast cancer screening. *Studies in Communication Sciences* 9 (1): 249–264. [Corrigendum in: Anonymous 2017d.] (Case 11.).

[—]. 2010a. Editor's introduction. In *Communication theory*, 1, ed. [—], xxi–lxvi. SAGE Benchmarks in Communication. Los Angeles: SAGE.

———. 2010b. The communication of diagnostic information by doctors to patients in the consultation. In *Communication theory*, 2, ed. [—], 71–83. SAGE Benchmarks in Communication (Los Angeles: SAGE. (Case 8*.)

———. 2010c. George Gemistos Plethon (ca. 1360–1454), George of Trebizond (1396–1472), and Cardinal Bessarion (1403–1472): The controversy between Platonists and Aristotelians in the fifteenth century. In *Philosophers of the Renaissance*, ed. Paul Richard Blum, trans. Brian McNeil, 23–32. Washington, DC: The Catholic University of America Press. (Case 4.)

[—], et al. 2012. The concept of health literacy. In *Science|environment|health: Towards a renewed pedagogy for science education*, ed. Albert Zeyer and Regula Kyburz-Graber, 69–84. Dordrecht: Springer. [Erratum in Anonymous 2018a.] (Case 13.).

———. 2015. Differential appraisal of age thresholds for mammographic screening in Holland and Switzerland. *Journal of Communication in Healthcare* 8 (1): 32–44. [Erratum in Anonymous 2018b.] (Case 14.).

Annas, Julia. 1981. *An introduction to Plato's 'Republic'*. Oxford: Clarendon Press.

Anonymous. 1953. Erratum. *The Philosophical Quarterly* 3 (10): 96.

———. 1968. Erratum. *Analysis* 28 (5): 158.

———. 1984. Erratum. *Mind* 93 (369): 154.

———. 1989. Errata in the general index, volumes 1 to 50. *The Thomist* 53 (1): 171–172.

———. 1990. Corrigendum. *Phronesis* 35 (2): 223.

———. 1993a. Erratum. *The British Journal for the Philosophy of Science* 44 (2): 377.

———. 1993b. Erratum. *The Journal of Medicine and Philosophy* 18 (3): 346.

———. 1998. Erratum. *Business Ethics Quarterly* 8 (3): 613–615.

———. 1999a. Erratum. *The British Journal for the Philosophy of Science* 50 (2): 324.

———. 1999b. Erratum. *The British Journal for the Philosophy of Science* 50 (4): 791.

———. 2000a. Errata. *Journal of the History of Ideas* 61 (1): 158.

———. 2000b. Erratum. *Mind* 109 (433): 127.

———. 2000c. Erratum. *Philosophical Studies* 98 (3): 113.

———. 2003. Errata. *Idealistic Studies* 33.2 (3): 237.

———. 2009a. Errata. *Theoria: Revista de Teoría, Historia y Fundamentos de la Ciencia* 24 (1): 107–111.

———. 2009b. Erratum. *The Monist* 92 (2): i.

———. 2012a. Erratum. *The Monist* 95 (1): 2.

———. 2012b. Erratum. *The Journal of Medicine and Philosophy* 37 (5): 505.

———. 2013. Erratum. *The Philosophical Quarterly* 63 (253): 856.

———. 2014. Erratum. *The Monist* 97 (4): ii.

———. 2016a. *It has been brought to our attention [...]*. Leiden: Brill. https://brill.com/abstract/title/27457. Accessed 6 July 2018.

———. 2016b. *It has been brought to our attention [...]*. Leiden: Brill. https://brill.com/view/title/27478. Accessed 6 July 2018.

———. 2016c. Corrigendum. *The Journal of Medicine and Philosophy* 41 (6): 699.

———. 2016d. Erratum/corrigendum: Expression of concern. *The Journal of Physiology* 594 (17): 5031–5032.

———. 2017a. Erratum. *Swiss Medical Weekly* 147 (14406): 1. https://smw.ch/en/resource/jf/journal/file/download/article/smw/en/smw.2017.14406/SMW14406.pdf. Accessed 6 July 2018.

———. 2017b. *It was brought to our attention [...]*. Amsterdam: John Benjamins Publishing Company. https://doi.org/10.1075/cvs.1.05sch Accessed 6 July 2018.

———. 2017c. Retraction. *Studies in Communication Sciences* 6 (2): 215. http://www.hope.uzh.ch/scoms/article/view/1603. Accessed 6 July 2018.

———. 2017d. Corrigendum/*Scoms* correction note. *Studies in Communication Sciences* 9 (1): 248A. http://www.hope.uzh.ch/scoms/article/view/1616. Accessed 6 July 2018.

———. 2017e. Addenda. *Journal of the American Philosophical Association* 3 (1): 127.

———. 2017f. Erratum. *Ergonomics* 2: 175. https://doi.org/10.3998/ergo.12405314.0002.008 Accessed 6 July 2018.

———. 2017g. Erratum. *Journal of Business Ethics* 140 (3): 607.

———. 2017h. Erratum. *Mind* 126 (501): 319.

———. 2017i. Erratum. *Journal of Business Ethics* 141 (2): 433.

———. 2017j. Erratum. *Journal of Philosophical Logic* 46 (5): 539.

———. 2017k. Editorial apology. *The Heythrop Journal*. https://doi.org/10.1111/heyj.12834 Accessed 6 July 2018.

———. 2018a. *Erratum*. Cham: Springer. https://doi.org/10.1007/978-90-481-3949-1_11 Accessed 6 July 2018.

———. 2018b. Corrigendum. *Journal of Communication in Healthcare* 11 (1): 81.

———. 2018c. *Patient Education and Counseling*. https://doi.org/10.1016/j.pec.2018.07.014. Accessed 30 Aug 2018.

Bailin, Sidney C. 2011. Retraction of 'A normalization theorem for set theory'. *The Journal of Symbolic Logic* 76 (3): 1096.

Baldwin, Jon, Paul Cobley, Nick Haeffner, and Jenny Harding. 2004. Response. *Subject Matters: A Journal of Communications and the Self* 1 (2): 83–89.

Barbour, Virginia, Theodora Bloom, Jennifer Lin, and Elizabeth Moylan. 2017. Amending published articles: Time to rethink retractions and corrections? *bioRxiv archive*. https://doi.org/10.1101/118356 Accessed 6 July 2018.

Besser, Avi, Marianne Amir, and Sharon Barkan. 2004. Who signs an organ transplant donor card? A study of personality and individual differences in a sample of Israeli university students. *Personality and Individual Differences* 36 (7): 1709–1723.

Blair, Ann. 2007. Errata lists and the reader as corrector. In *Agent of change: Print culture studies after Elizabeth L. Eisenstein*, ed. Sabrina Alcorn Baron, Eric N. Lindquist, and Eleanor F. Shevlin, 21–41. Amherst: University of Massachusetts Press.

Boyle, Leonard E. 1976. Optimist and recensionist: 'Common errors' or 'common variations'? In *Latin script and letters A. D. 400–900*, ed. John J. O'Meara and Bernd Naumann, 264–274. Leiden: E. J. Brill.

Brill. 2014. Brill completes acquisition of editions Rodopi. Brill, June 12. https://brill.com/fileasset/downloads_static/static_investorrelations_brill_press_release_rodopi_12_june_2014_eng_final.pdf. Accessed 6 July 2018.

Buckner, Edward, and Jack Zupko (trans.). 2014. *Duns Scotus on time and existence: The questions on Aristotle's 'De interpretatione'*. Washington, DC: The Catholic University of America Press.

Carr, Thomas K. 1995. Only a God can save us. *First Things* 55: 57–62.

Casati, Roberto. 2010. On publishing. *Social Epistemology* 24 (3): 191–200.

Clouser, K. Danner. 1985. Approaching the logic of diagnosis. In *Logic of discovery and diagnosis in medicine*, ed. Kenneth F. Schaffner, 35–55. Berkeley: University of California Press.

Cobley, Paul, ed. 2006. *Communication theories*. Vol. 4. London: Routledge.

Cohen, L. Jonathan. 1967. Corrigendum. *British Journal for the Philosophy of Science* 17 (4): 352.

College van Bestuur van de Universiteit van Amsterdam, 2017. Besluit: Definitief oordeel op klacht. April 20. N. 2017cb0099. https://retractionwatch.com/wp-content/uploads/2018/01/PDF_2_University_of_Amsterdam.pdf. Accessed 6 July 2018.

Delaney, Neil Francis. 2007. A note on intention and the doctrine of double effect. *Philosophical Studies* 134 (2): 103–110.

———. 2008. Two cheers for 'closeness': Terror, targeting and double effect. *Philosophical Studies* 137 (3): 335–367.

René Descartes. 1984–1991. *The philosophical writings of Descartes*. Trans. John Cottingham, Robert Stoothhoff, Dugald Murdoch, and Anthony Kenny), 3 vols. Cambridge: Cambridge University Press.

Deutsch, Harry. 2014. Corrigendum: Resolution of some paradoxes of propositions. *Analysis* 74 (4): 629.

Dougherty, M.V. 2012. Review of: *Philosophers of the Renaissance*. *The Thomist* 76 (3): 473–477.

Dutton, Denis. 2001. Erratum. *Philosophy and Literature* 25 (1): iv. https://doi.org/10.1353/phl.2001.0019 Accessed 6 July 2018.

van Eemeren, Frans H. 2015a. Retraction note. *Argumentation* 29 (4): 493.

———. 2015b. Erratum. *Argumentation* 29 (4): 481–491.

Elsevier. 2016. Policy and best practice: Errata & corrigenda. https://www.elsevier.com/editors/perk/policy-and-best-practice-errata-And-corrigenda. Accessed 6 July 2018.

Etkin, Adam, and Ivan Oransky. 2016/2017. Dopey dupe retractions: How publisher error hurts researchers. *International Society of Managing and Technical Editors* 9 (11): 12.

Fanelli, Daniele, John P.A. Ioannidis, and Steven Goodman. 2018. Improving the integrity of published science: An expanded taxonomy of retractions and corrections. *European Journal of Clinical Investigation* 48 (4): e12898. https://doi.org/10.1111/eci.12898 Accessed 6 July 2018.

Gaifman, Haim. 2011. Erratum. *Synthese* 179 (3): 510–502.

Gold, Stefan M., Karl-Heinz Schulz, and Uwe Koch. 2001. *The organ donation process: Causes of the organ shortage and approaches to the solution: An analysis of the content and methods of available studies*. Cologne: Federal Centre for Health Education. http://www.bzga.de/botmed_60809070.html. Accessed 6 July 2018.

Hafner, Urs. 2017. Hexenjagd im Reich der Vernunft. *Neue Zürcher Zeitung*, January 21. https://www.nzz.ch/feuilleton/plagiat-hexenjagd-im-reich-der-vernunft-ld.142092. Accessed 6 July 2018.

Hames, Irene. 2007. *Peer review and manuscript management in scientific journals: Guidelines for good practice*. Malden: Blackwell Publishing.

Hu, Guangwei. 2017. Authorship of retraction notices: 'If names are not rectified, then language will not be in accord with truth'. *Publications* 5 (2): 1–3. https://doi.org/10.3390/publications5020010 Accessed 6 July 2018.

Illarietti, Davide. 2016. Plagio all'Usi: sospeso il prof. *Ticino Online*, August 29. http://www.tio.ch/News/Ticino/Attualita/1103037. Accessed 6 July 2018.

———. 2018. Il docente dell'Usi ha copiato anche il Papa. *Ticino Online*, January 18. https://www.tio.ch/ticino/attualita/1236159/il-docente-dell-usi-ha-copiato-anche-il-papa. Accessed 6 July 2018.

Kenny, Anthony. 1988. *The self. The Aquinas lecture, 1988*. Milwaukee: Marquette University Press.

King, Nathan L. 2014. Erratum to: Perseverance as an intellectual virtue. *Synthese* 191 (15): 3779–3801.

Lee, Patrick, and Robert P. George. 2009. Erratum.In *Body-self dualism in contemporary ethics and politics*, x. Cambridge, UK: Cambridge University Press.

Mackay, James. 2006. Comments on [—]. *Studies in Communication Sciences* 6 (2): 143–144. http://www.hope.uzh.ch/scoms/article/view/1620/463. Accessed 6 July 2018.

Macquarrie, John. 1994. *Heidegger and Christianity: The Hensley Henson lectures*, 1993–1994. London: SCM Press.

Marcus, Adam, and Ivan Oransky. 2015. We gave bad advice: Better not contact authors first. *Lab Times*, November 22. http://www.labtimes.org/labtimes/ranking/dont/2015_06.lasso. Accessed 6 July 2018.

———. 2017. Is there a retraction problem? And, if so, what can we do about it? In *The Oxford handbook of the science of science communication*, ed. Kathleen Hall Jamieson, Dan M. Kahan, and Dietram A. Scheufele, 119–126. New York: Oxford University Press.

Martinelli, Dario. 2018a. Statement of retraction. International Semiotics Institute, January 8. https://www.facebook.com/isisemiotics/posts/880483692121348. Accessed 6 July 2018.

———. 2018b. Concerning our statement of retraction. International Semiotics Institute, January 18. https://www.facebook.com/isisemiotics/posts/883179261851791. Accessed 6 July 2018.

McCook, Alison. 2016. Retract – and replace? JAMA may expand use of this tool. *Retraction Watch*, June 20. http://retractionwatch.com/2016/06/20/retract-and-replace-jama-may-expand-use-of-this-tool. Accessed 6 July 2018.

Michalos, Alex C. 2010. Erratum. *Journal of Business Ethics* 93 (3): 495.

Monfasani, John. 1976. *George of Trebizond: A biography and a study of his rhetoric and logic*. Leiden: E. J. Brill.

Morgan, Susan E., and Jenny K. Miller. 2001. Beyond the organ donor card: The effect of knowledge, attitudes, and values on willingness to communicate about organ donation to family members. *Health Communication* 14 (1): 121–134.

Morton, Adam. 2016. Corrigendum. *Analysis* 76 (4): 445.

Nature. n.d. Correction and retraction policy. *Nature*. http://www.nature.com/authors/policies/corrections.html. Accessed 6 July 2018.

Oggiscienza. 2018. In filosofia, caccia alle bufale = caccia alle streghe. *Oggiscienza: La ricerca e i suoi protagonisti*, March 9. https://oggiscienza.it/2018/03/09/filosofia-caccia-bufale-caccia-streghe. Accessed 6 July 2018.

Oxford English Dictionary. 1971. *Vol. IX: S–Soldo*. Oxford: Oxford University Press.

Pakaluk, Michael. 1998. *Aristotle: 'Nicomachean Ethics' books VIII and IX. Translated with a commentary*. Oxford: Oxford University Press.

Palus, Shannon. 2016a. Communications researcher regrets 'severe shortcomings' in three publications. *Retraction Watch*, January 18. http://retractionwatch.com/2016/01/18/communications-researcher-regrets-severe-shortcomings-in-three-publications. Accessed 6 July 2018.

———. 2016b. Communications researcher loses two book chapters, Investigated for plagiarism. *Retraction Watch*, April 18. http://retractionwatch.com/2016/04/18/communications-researcher-loses-two-book-chapters-investigated-for-plagiarism. Accessed 6 July 2018.

Peer 1. 2016. Being in accord with oneself. *PubPeer*, January 24. https://pubpeer.com/publications/A171003E647551A13745F649EAE523. Accessed 6 July 2018.

Price, Anthony. 1995. Friendship (VIII und IX). In *Die Nikomachische Ethik*, ed. Otfried Höffe, 229–251. Berlin: Akademie Verlag.

Radecki, Carmen M., and James Jaccard. 1997. Psychological aspects of organ donation: A critical review and synthesis of individual and next-of-kin donation decisions. *Health Psychology* 16 (2): 183–195.

Ratzan, Scott C., J. Gregory Payne, and Carol Bishop. 1996. The status and scope of health communication. *Journal of Health Communication* 1 (1): 25–41.

Rydzyński, Konrad. 2006. Message from the Editor-in-Chief: Expression of concern due to plagiarism. *International Journal of Occupational Medicine and Environmental Health* 19 (2): 149–150.

Ryle, Gilbert. 1932. Erratum. *Mind* 41 (161): 138.

Sander, Shannon L., and Barbara Kopp Miller. 2005. Public knowledge and attitudes regarding organ and tissue donation: An analysis of the northwest Ohio community. *Patient Education and Counseling* 58 (2): 154–163.

Sherman, Nancy. 1989. *The fabric of character: Aristotle's theory of virtue*. Oxford: Clarendon Press.

Siegel, Jason T., Eusebio M. Alvaro, and Sara Pace Jones. 2005. Organ donor registration preferences among Hispanic populations: Which modes of registration have the greatest promise? *Health Education & Behavior* 32 (2): 242–252.

Souder, Lawrence. 2010. A rhetorical analysis of apologies for scientific misconduct: Do they really mean it? *Science and Engineering Ethics* 16 (1): 175–184.

South, James B. 2018. Martinelli letter, January 2. *Retraction Watch*. https://retractionwatch.com/wp-content/uploads/2018/01/Martinelli-Letter.pdf. Accessed 6 July 2018.

Stern, Victoria. 2018a. A cardinal sin? Communications researcher accused of plagiarizing former Pope. *Retraction Watch*, January 12. http://retractionwatch.com/2018/01/12/cardinal-sin-communications-researcher-accused-plagiarizing-former-pope. Accessed 6 July 2018.

———. 2018b. University defends researcher accused of plagiarizing former Pope. *Retraction Watch*, January 31. https://retractionwatch.com/2018/01/31/university-defends-researcher-accused-plagiarizing-former-pope. Accessed 6 July 2018.

Street, Richard L., Jr. 2003. Communication in medical encounters: An ecological perspective. In *Handbook of health communication*, ed. Teresa L. Thompson et al., 63–94. Mahwah: Lawrence Erlbaum.

Subject Matters: A Journal of Communications and the Self. 2004–2007. http://archive.londonmet.ac.uk/jcamd/research/groups/communicationsand-subjectivity/subject-matters/index.html. Accessed 6 July 2018.

Swiss National Science Foundation. 2016. *Regulations on scientific misconduct (Research integrity regulations, RI regulations)*. http://www.snf.ch/sitecollectiondocuments/ueb_org_fehlverh_gesuchstellende_e.pdf. Accessed 6 July 2018.

Taylor, Charles. 1985. The Person. In *The category of the person: Anthropology, philosophy, history*, ed. Michael Carrithers, Steven Collins, and Steven Lukes, 257–281. New York: Cambridge University Press.

———. 1989. *Sources of the self: The making of modern identity*. Cambridge, MA: Harvard University Press.

Teixeira da Silva Jaime, A. 2016a. Silent or stealth retractions, the dangerous voices of the unknown, deleted literature. *Publishing Research Quarterly* 32 (1): 44–53.

———. 2016b. An error is an error… is an erratum: The ethics of not correcting errors in the science literature. *Publishing Research Quarterly* 32 (3): 220–226.

Toh, Kevin. 2015. Erratum to: Four neglected prescriptions of Hartian legal philosophy. *Law and Philosophy* 34 (3): 333–368.

USI. 2016a. Un'inchiesta per sospetto plagio. http://www.usi.ch/en/comunicati-stampa/1971. Accessed 6 July 2018.

———. 2016b. Violazioni del diritto d'autore ma nessun dolo. Conclusa l'inchiesta per comportamento scientifico scorretto. http://www.usi.ch/en/comunicati-stampa/1894. Accessed 6 July 2018.

Vermeule, Adrian. 2006. *Judging under uncertainty: An institutional theory of legal interpretation*. Cambridge, MA: Harvard University Press.

Virginia Tech Research Integrity Office. 2016. Response to complaint regarding potential plagiarism. http://retractionwatch.com/wp-content/uploads/2018/01/PDF_1_Virginia_Tech-1.pdf. Accessed 6 July 2018.

Weber-Wulff, Debora. 2014. *False feathers: A perspective on academic plagiarism*. Heidelberg: Springer.

Weinberg, Justin. 2018a. Plagiarizes again—and is caught by philosophy prof.'s class (updated). *Daily Nous: News for and About the Philosophy Profession*, January 15, http://dailynous.com/2018/01/15. Accessed 6 July 2018.

———. 2018b. Plagiarist's university issues criticism …of the whistleblower. *Daily Nous: News for and About the Philosophy Profession*, February 1. http://dailynous.com/2018/02/01/plagiarists-university-issues-criticism-whistleblower. Accessed 6 July 2018.

Wojtyła, Cardinal Carol [Pope John Paul II]. 1978. Subjectivity and the irreducible in man. In *The human being in action: The irreducible element in man, Part II: Investigations at the intersection of philosophy and psychiatry*, ed. Anna-Teresa Tymieniecka, 107–114. Dordrecht: D. Reidel Publishing.

Wojtyła, Cardinal Karol [Pope John Paul II]. 1979. *The acting person*. Trans. Andrzei Potocki, ed. Anna-Teresa Tymieniecka. Dordrecht: Springer.

Wojtyła, Karol [Pope John Paul II]. 1993. Subjectivity and the irreducible in the human being. In *Person and community: Selected essays*, (trans: Theresa, Sandok), 209–217. New York: Peter Lang.

Xu, Shaoxiong (Brian), and Guangwei Hu. 2018. Retraction notices: Who authored them? *Publications* 6 (1): 1–18. https://doi.org/10.3390/publications6010002. Accessed 6 July 2018.

Chapter 7
Contested Authorship, Self-Plagiarism, and the Scholarly Record

Abstract This chapter considers recent cases of contested authorship in two articles involving an American philosophy professor. The examination of these unusual publishing circumstances, which include allegations of self-plagiarism, places in high relief the requirements of genuine authorship, the complexity of some self-plagiarism claims, and the challenges of requesting corrections of the scholarly record. Differing standards of evidence are often proposed in considering whether to retract a published article for authorship violations, and the chapter evaluates these standards in light of potential benefits to the larger research community. Researchers, editors, and publishers sometimes conflate standards from legal, moral, and scientific domains, and a lack of clarity about the applicable standard can significantly delay a proper correction of the scholarly record.

Keywords Authorship · Facilitated communication · Disability studies · Philosophy · Retractions

On June 9, 2017, philosophy professor Anna Stubblefield had her conviction on two counts of first-degree aggravated sexual assault overturned and was granted the right to a new trial. The original sentencing had been announced by the prosecutor's office in a press release titled, "Former Chairwoman of Rutgers University Philosophy Department Sentenced to 12 Years for Raping Disabled Man" (Essex County Prosecutors Office 2016: 1). The disabled man in this case, John Doe, is nonverbal and suffers from cerebral palsy.[1] Throughout his life, he has been recognized by medical experts as suffering from severe intellectual as well as physical disabilities, and two of his family members are his legal guardians (for background, see Engber 2015, 2016). During the criminal trial, the state had experts testify that John Doe "did not have the capacity to independently make meaningful medical, legal, residential or vocational decisions" and "did not appear capable of giving consent to sexual activities" (Superior Court of New Jersey 2017: 6). One expert

[1] Although John Doe's real name has appeared in newspaper articles, online discussions, and published literature concerning the Stubblefield case, for the sake of privacy it will not appear in this chapter.

M. V. Dougherty, *Correcting the Scholarly Record for Research Integrity*, Research Ethics Forum 6, https://doi.org/10.1007/978-3-319-99435-2_7

testified that he was unable to elicit meaningful communication from John Doe (Wichert 2015b). Stubblefield maintained, however, that she was uniquely able to communicate with John Doe and that she was in love with him, despite the long-standing insistence by his family that he is incapable of language.

Stubblefield claimed to communicate with him through the technique of facilitated communication, a method by which a facilitator supports the hand of a disabled person in the use of a portable keyboard, letter board, or other communicative device. In overturning the sentence, the presiding judges for the appeal ruled that Stubblefield was unnecessarily limited during her trial in contextualizing her "unorthodox perception" of John Doe's capabilities (Superior Court of New Jersey 2017: 20). The appeals ruling did not affect Stubblefield's civil case, where she had been ordered to pay compensatory and punitive damages of four million dollars to John Doe's family (Moriarty 2016).

On March 19, 2018, Anna Stubblefield avoided a retrial on criminal charges when she pleaded guilty to third-degree aggravated criminal sexual contact (Moriarity 2018). In accepting the plea deal, Stubblefield acknowledged that she should have known that John Doe was legally incapable of consent (Burke 2018; Flaherty 2018). Third-degree aggravated criminal sexual contact is specified as an act of sexual contact with a vulnerable victim, including situations in which

> The victim is one whom the actor knew or should have known was physically helpless or incapacitated, intellectually or mentally incapacitated, or had a mental disease or defect which rendered the victim temporarily or permanently incapable of understanding the nature of his conduct, including, but not limited to, being incapable of providing consent. (New Jersey Revised Statutes 2014; NJ Rev Stat § 2C:14-2 2014)

Despite the wishes from John Doe's family that the plea deal be vacated, the overseeing judge accepted it, and Stubblefield was sentenced to prison time already served—under 2 years—with the requirement that she register as a sex offender (Napoliello 2018; Engber 2018).

The widespread newspaper coverage of Stubblefield's criminal trial and civil case, the overturning of her conviction, and her plea deal, has increased the visibility of Stubblefield's published research, and her articles and book chapters in the area of disability studies have received attention far outside that field of scholarship. Two publications, one in *Disability Studies Quarterly*, and another in an edited collection published by Cambridge University Press, raise questions of authorship. They provide striking case studies for examining the respective roles that researchers, editors, and publishers exercise in the process of issuing retractions for suspected authorship violations.

7.1 A Controversial Technique

According to a mainstream view, there is no scientific validity to facilitated communication; rather than facilitating any communication, the facilitator unknowingly guides the hand of the disabled person in order to give the appearance of communication by way of an ideomotor or "ouija board" effect (Sherry 2016: 974–975). A meta-analysis of existing literature published in 2014 concluded that there is "unequivocal evidence for facilitator control: messages generated through FC are authored by the facilitators rather than the individuals with disabilities" (Schlosser et al. 2014: 359; see also Mostert 2010, 2012). More succinctly, the study concludes: "FC is a technique that has no validity" (359). The list of medical, professional, educational, and governmental organizations that repudiate its use is quite lengthy (see *Behavior Analysis Association of Michigan*, n.d.; see also Sherry 2016: 974). Despite this criticism, the technique has had passionate adherents since its origins in the mid-to-late 1970s. It has been called "the single most scientifically discredited and professionally repudiated intervention" for intellectual disabilities, and yet it garners new generations of devotees (Todd 2016: 372). According to some critics, the persistence of interest in facilitated communication provides a distinctive example of the great difficulty in expunging pseudo-science from educational systems and institutional settings. The harm generated by the continued use of the technique includes the waste of valuable resources as well as lost opportunities for individuals who would have benefited from the use of evidence-based interventions (Travers 2017; Travers et al. 2014). In extreme cases, facilitated communication has been transformed by some religious sects into a "mystical device through which autistic children disclose otherworldly messages" (Bilu and Goodman 1997: 375), and it has been used in wrongly prosecuting innocent parents for child abuse (Brasier and Wisely 2014; Campbell 2016; Burke 2016; Valle and Gurney 2018). Well before the Stubblefield case became well known, one critic of facilitated communication warned that a facilitator "might author, then carry out, expressions of intimacy with his or her own facilitatee" (Todd 2012: 42).

Even though the presiding judge in Stubblefield's criminal trial did not allow the presentation of some forms of support for facilitated communication, that exclusion was not the basis for overturning the conviction. The appellate judges ruled that Stubblefield had been unfairly precluded from allowing one of founders of the facilitated communication movement, Rosemary Crossley, to testify about an evaluation she had conducted with John Doe. The appellate judges did not find fault with the court's position that facilitated communication did not meet the court's standard for scientific evidence (Wichert 2015a; Engber 2017). Indeed, there was much discussion of facilitated communication by the defense during the criminal trial. Eastern Michigan University psychology professor James Todd was brought in as a rebuttal witness by the state, and he testified to the unscientific status of facilitated communication before the case went to the jury.

One of Stubblefield's research areas is the philosophy of disability, and Stubblefield has published widely on the topic. The field of disability studies is interdisciplinary and draws scholars from many academic disciplines. The field overlaps not only with philosophy, but with history, cultural studies, sociology, communications, and literature, among other disciplines. Scholars working in disability studies have provided strong support for Stubblefield throughout her criminal trial and civil case, and that support has come under criticism. Thus, disability studies scholar Mark Sherry has sharply rebuked others in his field for "failing to critically explore whether support for Stubblefield means they have taken on the role of *de facto* rape apologists" (Sherry 2016: 978). According to Sherry, "a lack of critical reflection from disability studies scholars" has misled them to focus their attention on Stubblefield, when they should have instead been assisting victim John Doe (978). After the criminal conviction was overturned, Sherry re-iterated his point to a journalist, stating that Stubblefield's "rescue fantasy and need for power resulted in the rape of this intellectually disabled man, justified through the sham technique of facilitated communication" (Flaherty 2017).

The two previously mentioned articles involving Stubblefield claim to present content obtained through facilitated communication. They not only provide distinctive case studies for identifying the requirements of academic authorship, but they exemplify as well the difficulty in securing corrections of the scholarly record.

7.2 *Disability Studies Quarterly*

In 2011, the peer-reviewed online journal *Disability Studies Quarterly* published a guest-edited issue with the theme "Mediated Communication." The issue featured essays that in various ways promote facilitated communication as an authentic disability intervention. One article in the issue, titled "Sound and Fury: When Opposition to Facilitated Communication Functions as Hate Speech," was authored by Anna Stubblefield, and she argues that some forms of opposition to facilitated communication are sufficiently oppressive to constitute detraction and hate speech. The article likens some researchers who oppose facilitated communication to pornographers who abuse women and Ku Klux Clan members who burn crosses on the lawns of minorities (Stubblefield 2011). In Stubblefield's view, the rhetoric of opposition to the technique by some critics is a violation of the rights of "FC users." Some of the researchers who were named targets in Stubblefield's article attempted to publish a response in *Disability Studies Quarterly*, but this request was denied by the editor on the grounds that the journal does not publish rebuttals (Lilienfeld et al. 2014: 80). One of the targeted scholars, James Todd, later remarked that the journal "is so infused with advocacy, fabrication, and fantasy that the legitimate items are tainted by association" (Leiter 2017b). Stubblefield's article on facilitated communication received criticism not only in academic venues, but also in more popular ones, including a *Slate* piece titled "Facilitated Communication Is a Cult That Won't Die" (Auerbach 2015).

Much more substantive controversy, however, has attended another article in the same issue of *Disability Studies Quarterly*. Published under John Doe's real name, followed by the annotation "c/o Anna Stubblefield," the article is titled, "The Role of Communication in Thought" (Doe 2011b). This piece appears to be a scholarly work authored by John Doe through the use of facilitated communication, with Stubblefield acting as his facilitator. The article discusses the works of Jean Piaget and Steven Pinker, among others, and it attempts an academic defense of facilitated communication. Perhaps more significant than the article's content is its existence: its presence in the journal functions as a demonstration of the success of facilitated communication. The article itself is ostensibly meant to display that the method actually works, not only in everyday communication, but in the production of scholarship. The article thus claims to be a scholarly argument as well as a performative exhibition of the reliability of the method. The promotion of John Doe as the article's author of record is enhanced by a biographical blurb that appears in the list of contributors to the special issue, and states that he is "31 years old and has been using facilitated communication for two years" (Disability Studies Quarterly 2011). The blurb concludes by attributing past conference presentations to him.

7.2.1 The First Request for a Statement of Retraction

On October 24, 2015, several weeks after Stubblefield's criminal conviction, the editors of *Disability Studies Quarterly* received a letter from me requesting that the journal publish a statement of retraction of the article for which John Doe is the author of record. The letter, later released online by the philosophy news site *Daily Nous* and also discussed in part on the post-publication review website PubPeer, set forth two distinct arguments for retraction. The first argument stated that the presumption of John Doe's authorship "requires that one uphold as trustworthy the unique word of his convicted rapist" (Weinberg 2015; Peer 1: 2016). The letter explained that no court expert, legal guardian, or family member vouches for any ability of John Doe to express himself propositionally, let alone write articles for academic journals. The letter states in part:

> The continued availability of the abovementioned article would imply that the Editors of *Disability Studies Quarterly* and the Directors of the Society for Disability Studies find the unique word of John Doe's rapist to be sufficiently trustworthy for maintaining the article as part of the repository of peer-reviewed published research. Such trust [...] would not appear to satisfy the minimum conditions of scientific integrity of an article for a peer-reviewed academic journal. The article in question continues to be indexed in a variety of academic databases for various disciplines, including those in philosophy.

The argument centered primarily on the integrity of the scholarly record. The claim of John Doe's authorship hinges exclusively on the mediation of Stubblefield, as those who know John Doe best (e.g., his family and guardians) do not consider John Doe to be linguistically communicative. The apparent trust that the editors of *Disability Studies Quarterly* have in the unique word of Stubblefield does not come close to the standard of evidence required for scientific, scholarly, or academic endeavors. Furthermore, the scholarly record is corrupted by the fact that the article in question is indexed in—and thereby promoted by—a variety of academic databases.

The letter continued with an additional argument for retraction, one that was independent of the first that concerned the integrity of the scholarly record:

> Additionally, the ongoing unqualified availability of the article by *Disability Studies Quarterly* is arguably a further injustice to the rape victim in this case, who is a member of the disability community.

For the journal to continue to promote publicly John Doe's real name as the author of record is arguably a further victimization of John Doe. The article's availability continues to present to readers the appearance that Stubblefield is a unique agent who can unlock his thoughts. As *Disability Studies Quarterly* is an open-access journal, the ascription of authorship to John Doe with the article's continued unqualified online presence could be viewed as an enduring injustice to the victim, in addition to failing to meet plausible academic requirements of authorship. The letter concluded by requesting "without delay" a published retraction of the online article by the editors.

I later learned that I was not the only one seeking a retraction on these grounds. A pseudonymous commentator ("Noetica") on the website Feminist Philosophers recounts writing to the journal at least by early October about the article attributed to John Doe. The online post states:

> I've read the paper that was published under his name at DSQ, and in fact, I wrote to them the other day to ask if they were considering retracting it because if it's true that Stubblefield was (I'll grant, very possibly unconsciously) manipulating the communications, I think it would be an insult to both justice and academic integrity to let an article in which the very practice used to victimize him is defended remain under his name. (Feminist Philosophers 2015a)

This comment argues for retraction on the basis of the ongoing injustice to John Doe as well as for reasons of academic integrity. The pseudonymous commentator then added to the comment in late October indicating that no reply has been received for the retraction inquiry (Feminist Philosophers 2015b).

The editors of *Disability Studies Quarterly* and the Society for Disability Studies took up the matter of the retraction requests at a board meeting on October 27, 2015. According to the minutes of the meeting, two agenda items were discussed. The first concerned a request for an amicus brief for the Stubblefield criminal trial. The board considered the possibility of polling the society's members on the matter, but the board members could not decide on a course of action. The second was the request "from multiple people who are asking that *DSQ* retract the article(s) written by Stubblefield and [John Doe]" (Society for Disability Studies 2015a: 1). Apparently the journal had received additional requests for retraction, beyond my October 24, 2015 letter and the query from the pseudonymous online commentator, both for Stubblefield's hate speech article and for the article appearing under the authorship of "[John Doe] c/o Anna Stubblefield." In a newspaper article that appeared the following month, board chairwoman Brenda Brueggemann was quoted as saying, "we have had a number of queries asking us if we would be reevaluating the articles (either one or both) and considering either 'retraction' or what the publishing industry calls an 'expression of concern'" (Wichert 2015c). Brueggemann added, "In a few cases we have had not just queries but demands for retraction" (ibid.).

The request for a correction of the article for which John Doe is the author of record gained some supporters. Disabilities scholar Mark Sherry was the first to argue for some action by the journal in print. In a footnote to his 2016 article in *Disability & Society* he wrote, "Some critics have demanded the removal of this paper from *Disability Studies Quarterly*, given the court findings that he was incapable of writing it. This is a position I would support" (Sherry 2016: 980, n. 2).

The board of the Society for Disability Studies considered the status of Stubblefield's hate speech article together with the article for which John Doe is author of record. This joint consideration was unfortunate, since they had been criticized for different reasons. Stubblefield's hate speech article was being criticized for the way it seemed to label unfairly those who maintain that facilitated communication is not an evidence-based disability intervention. The article attributed to John Doe was being criticized because it appeared to be unscientific and unscholarly; critics suggested that it appeared to be either fraudulent or the result of incompetence, in addition to being unjust to John Doe. Psychologist James Todd later summarized the problem as: "*DSQ* is a journal that publishes what any competent editor should assume are forgeries" (Leiter 2017b). The initial conflation by the Society of Disability Studies created impediments to a clear analysis of the relevant issues of authorship, research integrity, and the reliability of the scholarly record. The only links between the two articles were the common involvement of Stubblefield and the common theme of facilitated communication, but the reasons put forward for corrections for each were unrelated. To put the matter another way: any general argument for retraction based solely on a claim of the unscientific status of facilitated communication would have applied to the entire special issue of *Disability Studies Quarterly*. The specific calls for retraction for independent and particular reasons concerned only the two articles in question.

The minutes of the October 27, 2015 board meeting of the Society for Disability Studies also include a rather curious remark about the board's understanding of the requests for retraction: "The board discusses what this would mean and whether or not it's possible to fully retract an article, even if the website was changed" (Society for Disability Studies 2015a: 1). This remark reveals an apparent confusion among members of the board about the difference between a published statement of retraction and the removal of an online article. As was discussed in Chap. 4.3, the two actions can be performed separately: a journal can publish a statement of retraction for an article without ever removing it, and a journal can remove an article without ever publishing a statement of retraction. Retraction corrects the scholarly record by changing the status of the version of record of a published work; removal is merely the termination of accessibility. At least initially, the board did not distinguish between issuing a statement of retraction for an article and removing it completely. This misperception is somewhat understandable; in a handbook of publishing guidelines Irene Hames observes that the term *retraction* "can be a confusing term because it does not in the great majority of cases mean that an article is physically withdrawn" (Hames 2007: 194). Perhaps it can be argued that in the special case of the paper in *Disability Studies Quarterly* for which John Doe is named as the author of record, removal (in addition to retraction) would be warranted. A request for removal, however, was not made as part of my October 24, 2015 request for retraction. A year later, on October 27, 2016, *Disability Studies Quarterly* announced a new retraction policy that clearly disambiguates retraction from removal: "A retraction is defined as a public disavowal, not an erasure or removal" (Disability Studies Quarterly 2016). In 2015, however, these two courses of action had not yet been distinguished by the board. The minutes of the October 27, 2015 board meeting concluded with the unanimous approval of a new published statement in response to the requests for retraction.

On November 3, 2015, that statement appeared on the website of *Disability Studies Quarterly*. It says:

> The Society for Disability Studies (SDS) Board of Directors, as the final oversight and decision-making body of *Disability Studies Quarterly*, is aware of the many questions and debate regarding several articles published in the 2011 (31.4) issue. As an intellectual community, centered on scholarship, research, and learning, we are paying significant attention to the issues raised. We have not yet come to a decision. The case itself, regarding the authors, is not yet concluded. (Disability Studies Quarterly 2015)

The statement appears to satisfy the conditions of an expression of concern. In it, the editors and publisher acknowledge publicly—albeit obliquely—a receipt of claims made against some of the contents of the 2011 special issue on facilitated communication. Furthermore, it clearly asserts a position of neutrality, postulating that there are "many questions and debate," and it implies a future resolution by noting that the matter has not yet been decided.

Several parties immediately noticed that there were problems with this expression of concern. The first was the ambiguity of the expression "several articles." In response to a query by *Retraction Watch*, board chairwoman Brueggemann clarified that the published statement referred to the two articles for which Stubblefield was involved, namely, the hate speech article by Stubblefield and the article attributed to John Doe; the other 20 articles appearing in the special issue of the journal were not intended to fall under the statement. Brueggemann offered a justification for the ambiguity by saying, "No specific articles are mentioned with the statement because the case surrounding the authors is still in process and privacy is being respected as well" (Palus 2015). From the expression of concern itself, however, the range of articles falling under the notice appeared to be unspecified, making the expression of concern less than ideal.

A potential problem with the expression of concern was the use of the plural "authors," which was repeated in the clarification by Brueggemann to *Retraction Watch*. The very reason for the requests for retraction of the article attributed to John Doe was the claim that there could be only one author, namely Stubblefield. The use of the plural "authors" in the published expression of concern seems implicitly to ascribe authorship to John Doe, unless the restricted sense of "author" as *author of record* was to be understood, instead of "author" in the plain sense of *genuine* or *originating author*.

Another problem with the published expression of concern was the lack of an electronic tether between it and the articles falling under it. Someone could read both the hate speech article and the article attributed to John Doe on the *Disability Studies Quarterly* website without ever knowing that they were subject to the expression of concern.

Table 7.1 Some textual parallels between Anna Stubblefield and John Doe

Anna Stubblefield, "Sound and Fury: When Opposition to Facilitated Communication Functions as Hate Speech," *Disability Studies Quarterly* 31.4 (2011) (http://dsq-sds.org/article/view/1729/1777).	[John Doe], "The Role of Communication in Thought," *Disability Studies Quarterly* 31.4 (2011) (http://dsq-sds.org/article/view/1717/1765).
people who cannot speak and have trouble regulating their body movements by presuming that they are profoundly intellectually impaired	people who cannot speak and have trouble controlling their body movements [...] are labeled profoundly intellectually impaired
Detractors have contended that the facilitator [...] rather than the FC user [...] is the author	detractors claim that the facilitator rather than the FC user authors

The journal's expression of concern also neglected to specify what kind of additional information could lead to a resolution of the issue to the satisfaction of the editors of the journal. *Online* discussions had noted that there were certain verbal formulations in the two articles that, although minor, still seemed noteworthy and could be used to argue for there being one author (Weinberg 2015). Table 7.1 exhibits some textual parallels.

These text parallels were certainly not dispositive for the determination of authorship. They could, however, be seen as an indication of Stubblefield's authorship, since these particular verbal formulations are not used in any of the other articles on the common theme of facilitated communication in the special issue of *Disability Studies Quarterly*.

In January 2016, the expression of concern quietly disappeared from the website of *Disability Studies Quarterly*. Since it had included the assertion that the board members "have not yet come to a decision," its disappearance after 2.5 months suggested that the board had come to a decision. The minutes of the December 15, 2015 meeting of the Society for Disability Studies reveal that the removal had been broached by board chairwoman Brueggemann:

> Brenda [Brueggemann] opens a discussion about taking down the statement about the Stubblefield case from our website. (Society for Disability Studies 2015b: 1)

The minutes of the January 13, 2016 meeting of the Society for Disability Studies further reveal that the expression of concern had been removed by the two new editors of the journal, who were at work on a new upcoming issue of *Disability Studies Quarterly* and "are taking down the statement about the Stubblefield pieces" (Society for Disability Studies 2016: 1). I wrote again to the journal on January 18, 2016, asking for a decision on the pending request for a published statement of retraction and noting the disappearance of the expression of concern. I did not receive any answer, but the statement of retraction silently re-appeared on the website just as before.

On October 27, 2016, just over 1 year after *Disability Studies Quarterly* first received my request for retraction, the journal editors posted the previously mentioned retraction policy on its home page. They thereby explicitly committed to the guidelines offered by the *Committee on Publication Ethics* in resolving any issues, and they invited readers to contact the journal with any concerns about published articles. Two conditions were specified, each of which could trigger a retraction of an article from the journal:

> Retractions will occur if the editors and editorial board find that the main conclusion of the work is undermined or if subsequent information about the work comes to light of which the authors or the editors were not aware at the time of publication. (Disability Studies Quarterly 2016)

The first condition stipulates that retractions would be published if the "main conclusion of the work is undermined." It is worth considering whether this condition is met with the article attributed to John Doe. The article was obviously published with the knowledge that it was the product of facilitated communication. The themed issue in which it appears is quite favorable in general to the disability intervention, so the editorial board of the journal and its editors would likely not consider the endorsement of the method in the article attributed to John Doe to be an undermining factor. Nevertheless, Stubblefield's criminal trial and the extensive news reporting about it had placed into the public domain particular details about Stubblefield's use of the method that were likely unknown at the time of publication. These details include the remarks of John Doe's family that the views that Stubblefield attributed to John Doe through facilitated communication are contradicted by his known preferences, particularly with regard to such things as music, alcohol consumption, and relationships with caretakers (Engber 2015). Even if one believed that facilitated communication is an evidence-based disability intervention, its failure in the case of John Doe to reveal reliable information with regard to normal everyday matters would seem to cast substantive suspicion on its ability to generate a scholarly article. It can be argued that in this particular case, the failure of facilitated communication to confirm indisputably known facts about John Doe's preferences creates doubt regarding the article attributed to him, and hence one could argue that the "main conclusion of the work is undermined."

The second condition that the journal identified as capable of triggering a retraction is exceedingly general, as it simply references "subsequent information" that is discovered after publication. This condition leaves the required gravity of such newly revealed information unspecified, however.

7.2.2 The Second Request for a Statement of Retraction

Almost a year-and-a-half had passed since my request for retraction had been sent to *Disability Studies Quarterly*, and still there was still no explicit judgment by the journal editors. A new retraction policy had appeared online, but my inquiry requesting an update on the retraction request remained unanswered and the expression of concern still appeared on the website of the journal. I crafted a new, independent request for retraction. Observing that *Disability Studies Quarterly* had a policy that prohibits duplicate or redundant publication (i.e. self-plagiarism), there seemed to be grounds for another, entirely new argument for retraction. In its guidelines for prospective authors, the editors state that the journal "does not consider manuscripts under review elsewhere or that have been previously published" (Disability Studies Quarterly n.d.). In the interim, I had discovered that the article attributed to John Doe had appeared in print with the exact title in another venue, almost verbatim, months earlier (in February 2011), while the article in *Disability Studies Quarterly* had appeared afterward (in October 2011). The earlier appearance of the article had occurred in the newsletter of the Autism National Committee, a non-profit advocacy group friendly to facilitated communication. This original publication of the article in *The Communicator: A Publication of the Autism National Committee* was accompanied with a photo featuring Stubblefield holding John Doe's wrist in front of an electronic communicative device (Doe 2011a). The publication in *Disability Studies Quarterly*, therefore, was the second issuance of the same material, and its presence there seemed to be in conflict with the journal's stated prohibition of duplicate or redundant publication.

On March 13, 2017, I sent this argument along with a second retraction request to the editors of *Disability Studies Quarterly*, this time on the basis of duplicate publication. In it I referenced the published "Editorial Policies" section on the *Disability Studies Quarterly* website. The journal policies state that the journal follows the guidelines of *Committee on Publication Ethics* in the matter of retractions, and a link is offered to the organization's website (Disability Studies Quarterly 2016). The *Committee on Publication Ethics* guidelines direct editors to contact the author of record for a response when the editors of a journal receive a documented request for retraction by a third party that contains evidence of an ethical lapse, duplicate publication, or some other potential violation of scholarly integrity. For example, the steps "Contact corresponding author in writing" and "Contact author to explain your concerns" are each recommended in the respective guidelines for editors, "Suspected Redundant Publication in a Published Manuscript" and "Suspected Fabricated Data in a Published Manuscript" (Committee on Publication Ethics 2006–). I wondered how this requirement of contacting the author of record in this case—John Doe—could be fulfilled by the editors of *Disability Studies Quarterly*.

This new retraction request would seem to place the editors in a dilemma: they would either have to contact the author of record or not contact the author of record. For the editors to contact the author of record would require the assumption that the author of record is communicative and that facilitated communication is a successful disability intervention for him. The continued presence of the article on the journal's website appeared to be an implicit concession of these assumptions. Yet, it would seem that any contact by the editors would not only be ineffectual but excessively burdensome to his guardians. On the other hand, if the editors of *Disability Studies Quarterly* did not contact the author of record, on the basis of John Doe's inability to provide a response, the editors would be implicitly conceding the premises that formed the basis for the first request for retraction, as that first request had sought a retraction on the grounds that the author of record cannot be the author of a scholarly article.

The Society for Disability Studies took up the second request for retraction at a board meeting held on April 6, 2017. The board decided that it would be inappropriate to act on this new request for retraction, and the board remanded responsibility for the decision to the two editors of the journal. In a chance occurrence, 2 days later—on April 8, 2017—the Stubblefield case was again ushered in the journalistic spotlight, as philosophers Jeff McMahon and Peter Singer published an essay in *The New York Times* with the provocative title, "Who Is the Victim in the Anna Stubblefield Case?" In the article, the philosophers suggested that the decision by the judge in Stubblefield's criminal trial to limit consideration of facilitated communication had precluded Stubblefield from providing an adequate defense. In particular, the philosophers criticized the judge's decision to disallow Rosemary Crossley to testify as a defense expert. In this respect, the two philosophers presciently anticipated the grounds for the overturning of Stubblefield's initial conviction by the appeals judges that occurred 2 months later. Immediate responses to the essay by McMahon and Singer were varied, and in some instances they were quite critical. David Auerbach, who had authored the previously mentioned 2015 critique of facilitated communication in *Slate*, noted that:

> Crossley is the inventor of Facilitated Communication, something the authors [McMahon and Singer] seem to be unaware of. Complaining that her evidence of competence was not admitted is like complaining that Uri Geller's evidence of someone's psychic abilities was not admitted. (Leiter 2017a)

The most intense opposition to the essay by McMahon and Singer was directed to a portion in which the two philosophers claimed that "if Stubblefield wronged or harmed him [John Doe], it must have been in a way that he is incapable of understanding and that affected his experience only pleasurably" (McMahon and Singer 2017). To some commentators, this assertion seemed tantamount to justifying the sexual abuse of those with disabilities, among other vulnerable populations (see Feminist Philosophers 2017). Philosopher Eric Schliesser noted:

> I have to admit that I was genuinely shocked by the last paragraph of the passage that I quoted above. It seems clear to me that the line of reasoning by McMahon and Singer is the kind of thing that may well be used to justify many forms of sexual abuse of minors and cognitively impaired. (2017)

James Todd, the final witness in Stubblefield's criminal trial, also responded to the essay, stating:

> Singer and McMahon also leave out many other important details necessary to understand the crime and verdict--such as Stubblefield's many instances of deception and prevarication in committing the rape, why Crossley's ill-conceived forensic examination was excluded, that there is really no reasonable medical doubt about how severely cognitively disabled the victim is, and that Howard Shane, a world leader in adaptive communication technology, also evaluated the victim, confirming that he is essentially infantile in his capabilities. Even if FC worked perfectly, it would not work for the victim. (Todd 2017)

Kevin Mintz described the view of McMahon and Singer as "appalling and dangerous" in a journal article in *Disability & Society* (2017: 1669). He explained:

> Their logic is flawed because it supposes that for someone to be harmed, they have to actually perceive the harm being done to them. This would also imply that sex crimes against anyone who is incapable of perceiving harm are not explicitly harmful. What would that mean for cases involving children who might not understand when harm is being done to them, the unconscious, or the intoxicated? (ibid.)

A later interview with McMahon in which he expounded his views on the Stubblefield case does little to assuage the specific concern expressed by Schliesser, Mintz, and others (Goldhill 2017).

The appearance of the essay by McMahon and Singer drew increased attention to the ongoing unresolved issue of the article attributed to John Doe in *Disability Studies Quarterly*. The expression of concern still appeared on the journal's website preserving a limbo-status for the article. Auerbach commented, "Despite saying they were reviewing the matter, *Disability Studies Quarterly* never added any sort of disclaimer to the article purportedly authored by [John Doe] but written 'through' Stubblefield via Facilitated Communication" (Leiter 2017b). This unexpected renewal of attention to the Stubblefield case, brought about by the essay by McMahon and Singer, may have contributed to a partial resolution of the matter of correcting the scholarly record.

7.2.3 The Retraction

On April 14, 2017, *Disability Studies Quarterly* notified me that a decision had been made regarding my request, which was characterized then by the editors as "a possible copyright/plagiarism" issue, and I was told that the journal would be removing the paper. The use of the term *plagiarism* seemed to be shorthand for self-plagiarism; i.e., the unacknowledged reuse of one's previously published work. On April 18, 2017, I received another notification, this time stating that the decision has been made to retract the article, rather than remove it, and a statement of retraction appeared later that day on the article's webpage. Titled "Retraction Notice," it explains that the "decision to retract is due to major overlap with a previously published work" and that the "retracted article will remain online to maintain the scholarly record, and will be marked as retracted" (Editors-in-Chief of Disability Studies Quarterly 2017). The wording indicates that the reason for retraction is entirely based on duplicate or redundant publication—self-plagiarism—which had been the basis for the second, more recent request for retraction rather than the original request. As explained above, the original request had sought a correction because of (1) an apparently substantive defect in scholarly reliability, and (2) the possible ongoing injustice of authorial mis-attribution to John Doe.

The retraction statement itself can be read as a rejection of the original request for retraction insofar as it doubles down on the claim of John Doe's authorship. The retraction uses John Doe's real name and provides a web link to the original newsletter article in *The Communicator* that features a photo of Stubblefield and John Doe together. The first retraction request had been sent to *minimize* attribution of the article to John Doe; in the statement of retraction, the editors of *Disability Studies Quarterly* offer a re-affirmation of John Doe's authorship of the article and link to the document containing the photo. A subsequent *Retraction Watch* article that covered the case noted that the retracted article was "supposedly penned by a man with severe disabilities" and reported that the editor-in-chief of the journal declined to provide any additional information about the change in status of the article (Stern 2017).

Whether the editors of *Disability Studies Quarterly* complied with the journal's stated commitment to follow the *Committee on Publication Ethics* guidelines in determining whether to issue the retraction is not publicly known. Did the editors contact John Doe, the author of record, in fulfillment of the guidelines? To date, the expression of concern from November 2015 is still maintained online by the journal. Since it was meant to refer both to the now-retracted article for which John Doe is the author of record, as well as Stubblefield's hate speech article, the expression of concern's presence may be offered by the journal only with regard to the latter. If so, the expression of concern leaves the ultimate status of the hate speech article still unresolved.

Despite these difficulties, the retraction does provide some measure of correction to the published literature. One might argue that the text's first publication in *The Communicator* did not place it firmly within the scholarly record, since the newsletter version does not appear to fulfill the basic conditions that are met by undisputed items of the scholarly record, such as the *Publication*, *Library*, and *Database* conditions discussed in Chap. 2.1. The publication of the article in *Disability Studies Quarterly* is now attended by a statement of retraction, and the change in status of the article provides at least some measure of correction by making future unqualified citations to it in the downstream literature to be less likely.

Newspaper coverage of Stubblefield's conviction reversal, however, continued to cast doubt on John Doe's authorship of the *Disability Studies Quarterly* article. An *Inside Higher Ed* report on the story, for example, stated that "[John Doe] supposedly authored an academic paper, with Stubblefield's help" in discussing the granted request for a retrial, and it noted the retraction in *Disability Studies Quarterly* (Flaherty 2017).

7.3 Cambridge University Press

In 2014, Cambridge University Press published a collection of essays, *Disability and the Good Human Life*, based in part on papers presented at a 2010 conference in Basel, Switzerland (Bickenbach et al. 2014). The volume features essays from many disciplines on topics of disability. One chapter, titled "Living a Good Life … in Adult-Sized Diapers," was authored by Anna Stubblefield (Stubblefield 2014). It provides a defense of facilitated communication that largely consists in anecdotal accounts of those with disabilities. The chapter includes several references to Stubblefield's earlier article in *Disability Studies Quarterly* that characterizes some forms of opposition to facilitated communication as hate speech and it quotes the words of others obtained through facilitated communication.

The chapter is written in an easygoing style, mixing personal accounts and autobiography with references to scholarly literature, and it can be characterized as a work in autoethnography. In light of the title of the chapter, it is unsurprising that the chapter ends with a discussion of an individual who requires assistance in changing his adult incontinence undergarments. This person is unnamed, but Stubblefield references him as "a dear friend" (Stubblefield 2014: 239). She describes him as one who "was labeled as severely intellectually impaired well into adulthood, when he finally acquired access to a means of communication" and as one who "finds it best to manage bodily waste elimination by wearing diapers." She also adds, "had he not had his impairments, it is unlikely that we would have met" (240). These facts overlap to some degree with known elements of Stubblefield's interaction with John Doe. During Stubblefield's criminal trial, the press (at times uncharitably and with sensationalizing effects) emphasized John Doe's use of incontinence undergarments. For instance, the British tabloid *Daily Mail* published one article about the trial with the headline, "Rutgers Professor Accused of Raping Disabled, Nonverbal, Diaper-Wearing Man with Cerebral Palsy Says He Wanted Sex and They Were in Love" (Boyle 2015). Mark Sherry has argued that John Doe has been subject to additional abuse during the trial, since "to parade a disabled person as an exhibit is eerily reminiscent of the 'freak shows' of yesteryear" (Sherry 2016: 977). Kevin Mintz concurs, noting that John Doe "should not have been infantilized, especially in light of how briefly he was paraded to the jury" (Mintz 2017: 1668). While the evidence that the last portion of the chapter is referencing John Doe is only circumstantial, Stubblefield explicitly mentions John Doe in the early and middle parts of the chapter. In light of the explicit as well as possibly implied references to John Doe in this chapter, one can consider whether there is a case for a correction of the scholarly record.

Stubblefield's chapter begins with an epigraph attributed to John Doe, whose real name is used in print. The epigraph says:

> The right to communication is the right to hope.... I am jumping for joy knowing I can talk, but don't minimize how humiliating it can be to know people jump to the conclusion I am mentally disabled. If people understand the punishment of perceiving other people as inhuman, then things will get better. – [John Doe] (219).

In a lengthy footnote to this epigram, Stubblefield characterizes her own work of facilitation with John Doe by stating that he had been diagnosed with severe mental retardation prior to "gaining—at age twenty-eight—access to a form of alternative communication" (219 n. 1). She continues:

> In the spring of 2011, [John Doe]'s access to his means of communication was taken from him, and he is once again treated as severely intellectually impaired by those who have control over his life.

This line regarding the removal of Doe's means of communication is an apparent reference to the decision of John Doe's family to shield him from Stubblefield. The decision was made after Stubblefield had revealed to the family her sexual acts upon him. The family contacted authorities shortly after Stubblefield had attempted to arrange a secret meeting with John Doe at his afternoon day program, after they had tried to remove him from her influence (Engber 2015). The epigram footnote then ends with a dedication of the chapter to John Doe, "in the hope that he will one day regain his voice and his freedom." This epigraph is mentioned briefly again later in the chapter (233).

7.3.1 The Request for Retraction

On March 13, 2017, the editors of *Disability and the Good Human Life* received a request from me for a published statement of retraction. The request was similar to the one received by *Disability Studies Quarterly*, arguing that the unique word of Stubblefield "is not sufficiently trustworthy for maintaining the article as part of the repository of scholarly published research." The retraction request continued by explaining that trust in Stubblefield's use of the method

> would not appear to satisfy the minimum conditions of scientific integrity of an article for a scholarly volume. The chapter in question continues to be indexed in a variety of academic databases for various disciplines. Additionally, the ongoing unqualified electronic availability of the chapter on the Cambridge University website is arguably a further injustice to the rape victim in this case, who is a member of the disability community.

Patterned after the request made to *Disability Studies Quarterly*, this letter identifies the integrity of the scholarly record and the injustice to John Doe as reasons to justify the issuance of a statement of retraction. The request also notes that discussions about the need to correct the scholarly record have already appeared in print, since Mark Sherry had raised in general the issue of the victimization of John Doe in Stubblefield's scholarship in his article in *Disability & Society* (2016: 980, n. 2).

As discussed in Chaps. 1 and 4, humanities disciplines—including philosophy—often fall short of the best practices found in scientific disciplines for correcting the scholarly record. As to why philosophy falls short, perhaps one might consider that the ill effects of deficient publications in philosophy do not always manifest themselves as quickly as they might in other disciplines. The need to retract a plagiarized article on ancient Greek political theory or on medieval metaphysics, for example, appears less urgent than to correct deficient articles in fields like structural engineering, pharmacology, or oncology. In the natural sciences, the discovery of a manipulated image in a published journal article is typically sufficient to elicit a published correction by that journal. Such corrections occur not only in cases of outright fraud, but even when the manipulations are deemed to have been inadvertent or unintentional on the part of authors of record (Bik et al. 2016; Oransky 2012). Published corrections for manipulated images are needed to maintain the reliability of the content of scientific journals for the community of scholars served by them. For example, a molecular biology journal will typically issue a correction of an article when a western blot figure has been discovered to have been manipulated. An editor of a microbiology journal who would forego issuing a published correction after learning of such a manipulation would likely be considered guilty of editorial misconduct by the journal's readership. One might consider the words attributed to John Doe in Stubblefield's Cambridge University Press chapter to be similar to manipulated images that appear in a molecular biology journal article. Both types of defective elements—manipulated images and unreliable quotations—should be considered as "forgeries" (to use James Todd's expression again) that vitiate the reliability of the scholarly record. Corrections for both would warn readers that the defective elements should not be considered to be trustworthy from a scientific or academic standpoint.

A correction of Stubblefield's chapter has not appeared from Cambridge University Press; its status within the scholarly record remains unchanged. The chapter has been cited in the subsequent neurodiversity literature, and it is currently available for download on the publisher's website. In an errant entry in Google Scholar, John Doe himself (under his real name)—rather than Stubblefield—is inexplicably misidentified as the author of record for the chapter.

7.4 An Analogy

An analogy can elucidate the current situation regarding the publications appearing in *Disability Studies Quarterly* and in the Cambridge University Press volume that each ascribe words to John Doe. A short story by F. Scott Fitzgerald, "The I. O. U.," is told from the perspective of a successful publisher of books whose recent best-seller is a psychic researcher's account of his posthumous communications with his deceased nephew, who has been killed in the First World War (Fitzgerald 2017). The book is a great success until the nephew is revealed to have survived the war, and his appearance provides strong evidence that the purported spiritual communications alleged by the uncle in the book are not veridical. Throughout the story, the publisher is depicted in somewhat unscrupulous terms; he is more concerned with avoiding financial and reputational harm than with informing the book-buying public that the central claim of the book has been refuted. The mercenary publisher considers briefly whether changing the advertised genre of the book from non-fiction to fiction might save the volume, but ultimately he is unsuccessful in keeping the secret of the nephew's survival from appearing in the mainstream press.

Fitzgerald's short story is fiction, but the question it occasions is real: what should editors and publishers do when serious evidence comes to light that calls into question the attribution of words to others? In scholarly publishing, the goods are not primarily financial, but academic; journals and books contribute to the body of knowledge, and a reliable scholarly record preserves these contributions for students and researchers. The case for John Doe's authorship of the article in *Disability Studies Quarterly* not only hinges on accepting the reliability of facilitated communication in general; one must also accept that it worked in the particular case of John Doe. Even if one is sympathetic to the use of facilitated communication, the use of it in this case is extremely suspect, given the assessment of his abilities by the medical community, and furthermore, by John Doe's family's observation that the words attributed to him were inconsistent with what they knew about his preferences and dislikes. Even if one assumes that facilitated communication is an evidence-based disability intervention, there remains the problem that any professional-client relationship with John Doe was vitiated by sexual activities in violation of the practices that standardly govern patient-client relationships, and this substantive violation calls into question the reliability and efficacy of the therapy in this case. In such a scenario, a professional-client relationship involved what would have been the most extreme of power disparities, as Stubblefield claimed to be in the unique position of unlocking and manifesting the thoughts of John Doe that have remained hidden to everyone else, including the medical community and his family.[2] For the article in *Disability Studies Quarterly*, the evidence strongly favors a non-identity of the author of record and the author of origin, and such a non-identity is almost universally disallowed in scholarly publishing today.

The Cambridge University Press chapter presents a stronger parallel with Fitzgerald's story in some respects. John Doe is not the author of record in this case, but rather words are attributed to him in the chapter. For the same reasons offered above, any reliability to the words attributed to him falls short of an academic or scholarly standard. The ongoing unqualified ascription of these words to John Doe in the currently available electronic and print versions of the chapter arguably constitutes a further injustice to him. What is more, there is possibly a secondary injustice to readers, whose default position is likely to be one of trust that the attribution of words to John Doe is veridical.

Since the publication of Mark Sherry's 2016 critical article in *Disability & Society*, other publications in the field of disability studies have begun to appear that consider the Stubblefield case, and they largely focus on John Doe. One recently published work asserts that the controversy in the Stubblefield case exists primarily because John Doe "has not been offered platform or access to share his story" (Brown 2017: 175). That is,

> in the entire time discussion of the case spread like wildfire through the disability community and even hit the front page of the *New York Times* [...], no one has attempted to converse with [John Doe] to learn whether he consented to a relationship or separately to have sex (174–175).

A similarly themed view appears also in a recent article in *Disability & Society* that was published in response to Mark Sherry's earlier article in the same journal. In it, Kevin Mintz asserts that "justice will not be served in this case until ableist assumptions are put aside and our legal system enables [John Doe], after far too many years of silence, to find his voice" (2017: 1670). Arguments such as these, that express

[2] Some disabilities theorists do not appear to find sexual activity between professionals and clients in this context to be a problem, however. Kevin Mintz writes in *Society & Disability*, "Romantic relationships develop between disability-related professionals and their clients with relative frequency, and families of those with disabilities sometimes have trouble accepting those relationships" (Mintz 2017 1668–1669).

confidence that John Doe will be able to communicate, appear to place trust in the merits of the method of facilitated communication as a disability intervention.

Philosopher Marcia Baron has recently published a study of the Stubblefield case in which she defends the view that even though Stubblefield acted wrongly for a variety of reasons, "a suggestion that Stubblefield believed on reasonable grounds that [John Doe] consented to sex with her is not preposterous" (Baron 2018: 429). Baron faults Stubblefield for straying beyond her role as facilitator and for "not considering the possibility that [John Doe] might be incapable of consenting to sex" (449). She argues that Stubblefield lacked any ill will but still was culpably indifferent in allowing her ideological commitment to facilitated communication "to go too far" (ibid.). In discussing the John Doe article in *Disability Studies Quarterly*, Baron observes that its publication by the editors constitutes an implicit endorsement of the paper as a product of facilitated communication, and she notes that the paper's retraction was "not for reasons concerning authorship" (439).

7.5 In Sum

The words ascribed to John Doe in *Disability Studies Quarterly* and in the Cambridge University Press chapter appear as both a demonstration of facilitated communication and a disclosure of the thoughts of John Doe. These claims cannot be replicated. In this respect, a research or academic standard is not met. In some disciplines, testimony is an acceptable form of evidence, but in these cases there are sufficient reasons to treat testimony to John Doe's views as falling short of a research or academic standard. Even though the *Disability Studies Quarterly* article has been retracted for a particular kind of authorship violation ("overlap with previously published work" or *self-plagiarism*), I have argued that correction for an entirely different kind of authorship violation is required. In short, it is highly unlikely that John Doe is the genuine author of the article. Additionally, the Cambridge University Press chapter should be corrected because it is highly unlikely that the words attributed to John Doe in it are his. In my view, the correction of both articles is warranted to preserve the integrity of the scholarly record for students and researchers.

References

Auerbach, David. 2015. Facilitated communication is a cult that won't die. *Slate*, November 12. http://www.slate.com/articles/health_and_science/medical_examiner/2015/11/facilitated_communication_pseudoscience_harms_people_with_disabilities.html. Accessed 6 July 2018.

Baron, Marcia. 2018. Sexual consent, reasonable mistakes, and the case of Anna Stubblefield. *Ohio State Journal of Criminal Law* 15 (2): 429–449.

Behavior Analysis Association of Michigan. n.d. Resolutions and statements by scientific, professional, medical, governmental, and support organizations against the use of facilitated communication. http://www.baam.emich.edu/baamsciencewatch/baamfcresolutions.htm. Accessed 6 July 2018.

Bickenbach, Jerome E., Franziska Felder, and Barbara Schmitz, eds. 2014. *Disability and the good human life*. New York: Cambridge University Press.

Bik, Elisabeth M., Arturo Casadevall, and Ferric C. Fang. 2016. The prevalence of inappropriate image duplication in biomedical research publications. *MBio* 7 (3): e00809–e00816. https://doi.org/10.1128/mBio.00809-16. Accessed 6 July 2018.

Bilu, Yopam, and Yehuda C. Goodman. 1997. What does the soul say? Metaphysical uses of facilitated communication. *Ethos* 25 (4): 375–407.

Boyle, Louise. 2015. Rutgers professor accused of raping disabled, nonverbal, diaper-wearing man with cerebral palsy says he wanted sex and they were in love. *Daily Mail*, September 25, updated September 26. http://www.dailymail.co.uk/news/article-3249484/Professor-takes-stand-defense-against-claims-raped-disabled-nonverbal-diaper-wearing-man-cerebral-palsy-says-love.html. Accessed 6 July 2018.

Brasier, L. L., and John Wisely. 2014. Jury awards $3M to Wendrows in wrongful prosecution case. *Detroit Free Press*, Nov. 7. https://www.freep.com/story/news/local/michigan/oakland/2014/11/06/jury-awards-wendrow-millions-oakland/18611789. Accessed 20 Mar 2018.

Brown, Lydia X.Z. 2017. Ableist shame and disruptive bodies: Survivorship at the intersection of queer, trans, and disabled existence. In *Religion, disability, and interpersonal violence*, ed. Andy J. Johnson, J. Ruth Nelson, and Emily M. Lund, 163–178. Cham: Springer.

Burke, Michael. 2016. Double Talk: Syracuse University institute continues to use discredited technique with dangerous effects. *The Daily Orange*, April 11. http://dailyorange.com/2016/04/double-talk-syracuse-university-institute-continues-to-use-discredited-technique-with-dangerous-effects. Accessed 6 July 2018.

———. 2018. Educator trained in discredited communication method at SU pleads guilty to criminal sexual contact. *The Daily Orange*, March 19. http://dailyorange.com/2018/03/educator-trained-discredited-communication-method-su-pleads-guilty-criminal-sexual-contact. Accessed 20 Mar 2018.

Campbell, Kerry. 2016. Controversial technique led to allegations of sexual abuse against father of woman with autism. *Canadian Broadcasting Corporation*, March 16. http://www.cbc.ca/beta/news/canada/prince-edward-island/pei-health-minister-court-1.3494662. Accessed 6 July 2018.

Committee on Publication Ethics. 2006–. Flowcharts. https://publicationethics.org/resources/flowcharts. Accessed 6 July 2018.

Disability Studies Quarterly. 2011. Contributors' notes. *Disability Studies Quarterly* 31.4. http://dsq-sds.org/article/view/1730/1784. Accessed 6 July 2018.

———. 2015. Note About Volume 31, Issue 4 (2011). http://dsq-sds.org/announcement. Accessed 6 July 2018.

———. 2016. Retraction policy. http://dsq-sds.org/about/editorialPolicies#custom-3. Accessed 6 July 2018.

———. n.d. Author guidelines. http://dsq-sds.org/about/submissions#onlineSubmissions. Accessed 6 July 2018.

[Doe, John]. 2011a. The role of communication in thought. *The communicator: A publication of the Autism National Committee* 20 (1): 4–5 http://www.autcom.org/pdf/AutcomNLWinter2011.pdf. Accessed 6 July 2018.

———. 2011b. The role of communication in thought. *Disability Studies Quarterly* 31.4. https://doi.org/10.18061/dsq.v31i4.1717. Accessed 6 July 2018. [Retracted in: Editors-in-Chief of *Disability Studies Quarterly* 2017.]

Editors-in-Chief of Disability Studies Quarterly. 2017. Retraction notice. *Disability Studies Quarterly*. https://doi.org/10.18061/dsq.v31i4.1717. Accessed 6 July 2018.

Engber, Daniel. 2015. The strange case of Anna Stubblefield. *The New York Times Magazine*, October 20. https://www.nytimes.com/2015/10/25/magazine/the-strange-case-of-anna-stubblefield.html. Accessed 6 July 2018.

———. 2016. What Anna Stubblefield believed she was doing. *The New York Times*, February 3. https://www.nytimes.com/2016/02/03/magazine/what-anna-stubblefield-believed-she-was-doing.html. Accessed 6 July 2018.

———. 2017. Talking without talking. *Slate*, April 11. http://www.slate.com/articles/health_and_science/science/2017/04/will_anna_stubblefield_get_a_new_trial_in_her_facilitated_communication.html. Accessed 6 July 2018.

———. 2018. The strange case of Anna Stubblefield, revisited. *The New York Times Magazine*, April 5. https://www.nytimes.com/2018/04/05/magazine/the-strange-case-of-anna-stubblefield-revisited.html. Accessed 10 July 2018.

Essex County Prosecutors Office. 2016. Former chairwoman of Rutgers University philosophy department sentenced to 12 years for raping disabled man. http://www.njecpo.org/?p=3535&pdfex_dl=3535. Accessed 6 July 2018.

Feminist Philosophers. 2015a. Update on the Stubblefield sexual assault case, October 7. https://feministphilosophers.wordpress.com/2015/02/23/update-on-the-stubblefield-sexual-assault-case/#comment-142622. Accessed 6 July 2018.

———. 2015b. Update on the Stubblefield sexual assault case, October 22. https://feministphilosophers.wordpress.com/2015/02/23/update-on-the-stubblefield-sexual-assault-case/#comment-143079. Accessed 6 July 2018.

———. 2017. A reply to McMahan and Singer on the Stubblefield case, April 3. https://feministphilosophers.wordpress.com/2017/04/03/a-reply-to-mcmahan-and-singer-on-the-stubblefield-case. Accessed 6 July 2018.

Fitzgerald, F. Scott. 2017. The I.O.U. *The New Yorker*. http://www.newyorker.com/magazine/2017/03/20/the-i-o-u. Accessed 6 July 2018.

Flaherty, Colleen. 2017. Second chance for fallen philosopher. *Inside Higher Ed*, June 12. https://www.insidehighered.com/news/2017/06/12/professor-accused-raping-disabled-man-sees-her-convictions-overturned. Accessed 6 July 2018.

———. 2018. Former professor admits to assaulting disabled man. *Inside Higher Ed*, March 23. https://www.insidehighered.com/quicktakes/2018/03/23/former-professor-admits-assaulting-disabled-man. Accessed 23 Mar 2018.

Goldhill, Olivia. 2017. An Oxford philosopher's moral crisis can help us learn to question our instincts. *Quartz*. https://qz.com/1102616/an-oxford-philosophers-moral-crisis-can-help-us-learn-to-question-our-instincts. Accessed 6 July 2018.

Hames, Irene. 2007. *Peer review and manuscript management in scientific journals: Guidelines for good practice*. Malden: Blackwell Publishing.

Leiter, Brian. 2017a. Singer and McMahon on the Stubblefield case. *Leiter Reports: A Philosophy Blog*, April 3. http://leiterreports.typepad.com/blog/2017/04/singer-and-mcmahan-on-the-stubblefield-case.html. Accessed 6 July 2018.

———. 2017b. Disability studies appears to be an advocacy field, not a scholarly one: The case of Anna Stubblefield. *Leiter Reports: A Philosophy Blog*, April 6, http://leiterreports.typepad.com/blog/2017/04/disability-studies-appears-to-be-an-advocacy-field-not-a-scholarly-one-the-case-of-anna-stubblefield.html. Accessed 6 July 2018.

Lilienfeld, Scott O., Julia Marshall, James T. Todd, and Howard C. Shane. 2014. The persistence of fad interventions in the face of negative scientific evidence: Facilitated communication for autism as a case example. *Evidence-Based Communication Assessment and Intervention* 8 (2): 62–101.

McMahon, Jeff and Peter Singer. 2017. Who is the victim in the Anna Stubblefield case? *The New York Times*, April 3. https://www.nytimes.com/2017/04/03/opinion/who-is-the-victim-in-the-anna-stubblefield-case.html. Accessed 6 July 2018.

Mintz, Kevin. 2017. Ableism, ambiguity, and the Anna Stubblefield case. *Disability & Society* 32 (10): 1666–1670.

Moriarty, Thomas. 2016. Rutgers prof convicted in sex assault of disabled man ordered to pay $4M. *New Jersey On-Line*, October 25. http://www.nj.com/essex/index.ssf/2016/10/judge_awards_4_million_to_family_of_anna_stubblefi.html. Accessed 6 July 2018.

———. 2018. Ex-Rutgers prof admits it was a crime to have sex with disabled man. *New Jersey On-Line*, March 19. http://www.nj.com/essex/index.ssf/2018/03/ex-rutgers-newark_prof_admits_criminal_sexual_cont.html. Accessed 20 Mar 2018.

Mostert, Mark P. 2010. Facilitated communication and its legitimacy—twenty-first century developments. *Exceptionality: A Special Education Journal* 18 (1): 31–41.

———. 2012. Facilitated communication: The empirical imperative to prevent further professional malpractice. *Evidence-Based Communication Assessment and Intervention* 6 (1): 18–27.

Napoliello, Alex. 2018. No more prison for ex-Rutgers professor who sexually assaulted disabled student. *New Jersey On-Line*, May 11. https://www.nj.com/essex/index.ssf/2018/05/anna_stubblefield_sentenced_for_second_time.html. Accessed 6 July 2018.

New Jersey Revised Statutes. 2014. New Jersey Legislature. http://lis.njleg.state.nj.us/nxt/gateway.dll/statutes%2F1%2F2706%2F2824. Accessed 20 Mar 2018.

Oransky, Ivan. 2012 Can we trust Western blots? *Retraction Watch*, April 3. https://retractionwatch.com/2012/04/03/can-we-trust-western-blots. Accessed 6 July 2018.

Ovalle, David, Kyra Gurney. 2018. How a teacher's 'junk science' landed a dad in jail on charge of raping his autistic son. *Miami Herald*, March 8. http://www.miamiherald.com/news/local/crime/article204034704.html. Accessed 20 Mar 2018.

Palus, Shannon. 2015. Journal reviewing papers by researcher who sexually assaulted disabled author. *Retraction Watch*, November 3. http://retractionwatch.com/2015/11/03/journal-reviewing-papers-by-researcher-who-sexually-assaulted-co-author. Accessed 6 July 2018.

Peer 1. 2016. The role of communication in thought. PubPeer, October 31. https://pubpeer.com/publications/61602FCFECD8E4B449F9B6952D3DF8. Accessed 6 July 2018.

Schliesser, Eric. 2017. On philosophical authority (McMahan & Singer on the Stubblefield case). *Digressions & Impressions*, April 4. http://digressionsnimpressions.typepad.com/digressionsimpressions/2017/04/on-philosophical-authority-mcmahan-singer-on-the-stubblefield-case.html. Accessed 6 July 2018.

Schlosser, Ralf W., Susan Balandin, Bronwyn Hemsley, Teresa Iacono, Paul Probst, and Stephen von Tetzchner. 2014. Facilitated communication and authorship: A systematic review. *Augmentative and Alternative Communication* 30 (4): 359–368.

Sherry, Mark. 2016. Facilitated communication, Anna Stubblefield and disability studies. *Disability & Society* 31 (7): 974–982.

Society for Disability Studies. 2015a. Society for Disability Studies board meeting, October 27. http://disstudies.org/wp-content/uploads/2016/09/2015.10-October-Board-Meeting-Minutes.docx. Accessed 6 July 2018.

———. 2015b. December board meeting, December 15. http://disstudies.org/wp-content/uploads/2016/09/2015.12-December-Board-Meeting-Minutes.docx. Accessed 6 July 2018.

———. 2016. Society for Disability Studies board meeting, Mid-year board meeting, January 13. http://disstudies.org/wp-content/uploads/2016/09/2016.1-January-Board-Meeting-Minutes-1.docx. Accessed 6 July 2018.

Stern, Victoria. 2017. After researcher is convicted of sexual assault, Journal retracts her co-author's paper. *Retraction Watch*, May 4. http://retractionwatch.com/2017/05/04/author-sexually-assaulted-journal-retracts-paper-duplication. Accessed 6 July 2018.

Stubblefield, Anna. 2011. Sound and fury: When opposition to facilitated communication functions as hate speech. *Disability Studies Quarterly* 31(4). https://doi.org/10.18061/dsq.v31i4.1729. Accessed 6 July 2018.

———. 2014. Living a good life . . . in adult-sized diapers. In *Disability and the good human life*, ed. Jerome E. Bickenbach, Franziska Felder, and Barbara Schmitz, 219–242. New York: Cambridge University Press.

Superior Court of New Jersey. 2017. State of New Jersey (Plaintiff-Respondent) v. Marjorie Anna Stubblefield (Defendant-Appellant). *Superior Court of New Jersey Appellate Division*, Docket No. A–2112–15T1, June 9. http://judiciary.state.nj.us/attorneys/assets/opinions/appellate/published/a2112-15.pdf. Accessed 6 July 2018.

Todd, James T. 2012. The moral obligation to be empirical: Comments on Boynton's 'Facilitated communication—what harm it can do: Confessions of a former facilitator'. *Evidence-Based Communication Assessment and Intervention* 6 (1): 36–57.

———. 2016. Rapid prompting, facilitated communication, science, ethics, and the history of magic. In *Controversial therapies for autism and intellectual disabilities: Fad, fashion, and science in professional practice*, ed. Richard M. Foxx and James A. Mulick, 2nd ed., 372–409. New York: Routledge.

Todd, James A. 2017. Comment. *The New York Times*, April 11. https://www.nytimes.com/2017/04/03/opinion/who-is-the-victim-in-the-anna-stubblefield-case.html#permid=22021835. Accessed 6 July 2018.

Travers, Jason C. 2017. Evaluating claims to avoid pseudoscientific and unproven practices in special education. *Intervention in School and Clinic* 52 (4): 195–203.

Travers, Jason C., Matt J. Tincani, and Russell Lang. 2014. Facilitated communication denies people with disabilities their voice. *Research and Practice for Persons with Severe Disabilities* 39 (3): 195–202.

Weinberg, Justin. 2015. Disability studies quarterly is reviewing Stubblefield's articles. *Daily Nous: News for and About the Philosophy Profession*, November 4. http://dailynous.com/2015/11/04/disability-studies-quarterly-is-reviewing-stubblefields-articles. Accessed 6 July 2018.

Wichert, Bill. 2015a. Expert can't prove she can communicate with disabled man who may be victim of sex assault, judge rules. *New Jersey On-Line*, February 20, updated September 09. http://www.nj.com/essex/index.ssf/2015/02/judge_bars_defense_testimony_on_victim_analysis_at.html. Accessed 6 July 2018.

———. 2015b. Disabled man could not consent to sex in professor's abuse case, expert says. *New Jersey On-Line*, September 18. http://www.nj.com/essex/index.ssf/2015/09/disabled_man_could_not_consent_to_sex_in_professor.html. Accessed 6 July 2018.

———. 2015c. Professor looks to overturn conviction for sex assault of disabled man. *New Jersey On-Line*, November 8. http://www.nj.com/essex/index.ssf/2015/11/professor_asks_judge_to_throw_out_conviction_for_s.html. Accessed 6 July 2018.

Chapter 8
Beyond the Published Retraction

Abstract This chapter considers ways in which published corrections of the scholarly record are disseminated throughout the research community. Even when editors and publishers issue corrections, these notifications can still be minimized to the point of irrelevance if they are not reflected in the research practices of other members of the scholarly community. In some humanities disciplines, including philosophy, entries for articles in specialized research databases are not updated to reflect when the status of an article has been changed by a publisher through a retraction, erratum, corrigendum, or expression of concern ("The Database Problem"). At times an article is retracted by a publisher, but then the article is reprinted in another venue without acknowledgment of the retracted status of the original publication ("The Anthology Problem"). Some publishers will correct the electronic version that it hosts in one system but will leave other electronic versions unchanged ("The Platform Problem"). Quite frequently the original uncorrected versions of articles are uploaded on secondary websites outside the control of a publisher. These copies are more accessible to researchers than the corrected versions on the proprietary platforms of publishers, and so researchers download and use the more easily accessible but uncorrected versions ("The Repository Problem"). Furthermore, the authors of record and their institutions at times keep quiet about the changes in status of articles and reference uncorrected versions ("The Pretend-It-Didn't-Happen-Problem"). I propose clear solutions to these problems and note some positive developments.

Keywords Retractions · Humanities · Scholarly record · Research databases · Publishing

Published retractions for academic plagiarism that are both are maximal in determination (by stating unequivocally that an article is plagiarized) and maximal in credit (by identifying the original sources that were misappropriated) can still fail to correct the scholarly record adequately if they are deficient in availability. When statements of retraction are hidden behind paywalls, blocked with registration pages, not linked to the retracted articles, or issued only in leaflet form to be deposited in copies of unsold volumes, they are severely limited in the extent of their availability.

M. V. Dougherty, *Correcting the Scholarly Record for Research Integrity*, Research Ethics Forum 6, https://doi.org/10.1007/978-3-319-99435-2_8

Such defects in availability decrease the likelihood that researchers will come to know that a work has been subject to a change in status. But even when published retractions are fully maximal in availability, the optimal correction of the scholarly record still requires that the status of a work *as retracted* be reflected in the various gateways through which researchers access the published literature.

If these gateways fail to reflect the published corrections of the scholarly literature, the deficient works will continue to enjoy a presumption of trustworthiness on the part of students and researchers who are left unaware that a publisher has changed the status of a published work. This concluding chapter considers the issue of promulgating the status changes of works after a publisher has issued retractions, errata, corrigenda, or expressions of concern. All those engaged in the world of learning—especially researchers themselves—have an indispensable role in promulgating status changes of published works.

8.1 The Database Problem

In some fields, research databases are never modified to reflect when a publisher has changed the status of an article, creating what can be called the *Database Problem*. Ideally, database entries should indicate whenever the version of record of any article has been subject to a retraction or any other change of status. If one considers how the databases accessed by researchers in the respective fields of philosophy and medicine compare in regard to the *Database Problem*, philosophy will be seen to fall substantially short of best practices. A researcher who accesses *MEDLINE*—the curated bibliographical database for biomedical articles maintained by the U. S. National Library of Medicine—will find that entries for articles in the database are subject to ongoing revision to include updates whenever there is a post-publication change to the status of an article indexed in the database. The *MEDLINE* entry will not only be updated to indicate when an article has been subject to a publisher's retraction, erratum, corrigendum, or expression of concern, but also to indicate duplicate publication, revisions in a subsequent publication, republication in another venue, and new conflict-of-interest statements that appear after initial publication (U. S. National Library of Medicine 2017; see Schmidt 2018). In comparison, a philosopher using either of the two standard databases of the field of philosophy, *The Philosopher's Index* or *PhilPapers*, will not find entries for articles updated when there are any post-publication status changes for articles or books indexed in those databases. Since *The Philosopher's Index* and *PhilPapers* are the primary databases for the published literature in philosophy, their failure to reflect the status changes of articles and books means that researchers who would benefit most from an awareness of the status changes may remain ignorant of them.

Founded in 1966, *The Philosopher's Index* is a traditional curated database. Entries for articles are restricted to those appearing in journals that have been vetted by an editorial team, and entries for books are restricted to volumes issued by established publishers in the field. The catalogue entries in *The Philosopher's Index* are for items that belong indisputably to the scholarly record, and they represent items that generally fulfill the *Knowledge*, *Authorship*, *Publication*, *Library*, and *Discipline* conditions that were discussed in Chap. 2. In contrast to *The Philosopher's Index*, *PhilPapers* is an exceptionally inclusive database insofar as it maintains entries for items housed in open-access archives, in addition to entries for traditional books and journal articles. Established in 2009, *PhilPapers* allows individuals to create entries in the database for unpublished manuscripts, and the website will host copies of these manuscripts and make them accessible to users. The complete census of works catalogued in *PhilPapers* therefore includes items that indisputably belong to the scholarly record, those that are marginal to it, as well as those that appear to fall entirely outside it. Occasionally some unusual unpublished manuscripts that flout traditional scholarly norms appear in *PhilPapers*, having been uploaded by their authors, but researchers can avoid them by restricting their database search queries to published materials.

One can exhibit how the discipline of philosophy is impacted by the *Database Problem* by considering whether the published corrections discussed in the previous chapters of this book are represented in *The Philosopher's Index* and *PhilPapers*. For the various pseudonymously published articles in philosophy that were discussed in Chap. 2, involving Amélie Oksenberg Rorty ("Leila Tov-Ruach"), David Lewis ("Bruce Le Catt"), and Neven Sesardić ("Carmen de Macedo"), there is currently no representation of the corrections of authorship in either database. Furthermore, none of the 23 corrections discussed in Chap. 4 are identified in either database, and none of the entries for the articles subject to these 23 corrections have been updated to reflect the changes in status brought about by the various publishers. For the now-retracted article in *Disability Studies Quarterly* attributed to John Doe discussed in Chap. 7, there is an entry in *PhilPapers* for the paper, but no acknowledgment that the status of the article has been changed by the retraction for duplicate publication.

For a closer look at the *Database Problem* in philosophy, one might consider three articles discussed in Chap. 6 above. The author of record has published articles both in the field of philosophy and in the field of health communications, and some of the researcher's articles are catalogued in the standard databases in either philosophy or medicine. This makes possible a comparison of how databases in these respective fields disclose (or fail to disclose) post-publication changes in the status of the articles.

The first of the three articles, designated as Case 7, and published in 2007 in *Swiss Medical Weekly*, was subject to an erratum in 2017. When one accesses the *MEDLINE* database entries for this article through the standard portals (e.g., *PubMed*, *Web of Science*, or *Ebsco*) one finds that each entry has been updated with a note that states "Erratum" under a subheading titled "Comments" or "Comments & Corrections." Full bibliographical references to the published erratum, with links to the respective database entries for the published erratum, were added to each article entry for the 2007 article shortly after the publication of the erratum in 2017 (PMID: 17874518). Furthermore, the database entries for the published erratum have a link back to the database entries to the original article: a two-way tether thus exists between the entries for the article and the entries for the erratum in the three portals to the *MEDLINE* database (PMID: 28102881). Researchers who access the entry for the Case 7 article will therefore be informed that the article has been subjected to a correction issued by the publisher. For the benefit of researchers, the *MEDLINE* database promulgates fully the change in status of the article by the journal, providing a complete disclosure of the journal's correction of the scholarly record.

In contrast to *MEDLINE*, databases in philosophy fall short. One can take, for example, Cases 5 and 6 of the same researcher's published works that both appeared in the philosophy journal *Argumentation*. The first of these, Case 5, received a retraction, and the second, Case 6, received an erratum. In *The Philosopher's Index*, there are database entries for each of the two articles, but no database entries for the published retraction and the published erratum, even though these two corrections were published in *Argumentation* on numbered, citable pages of the journal 7 years after the original articles appeared. Given the way *The Philosopher's Index* represents Cases 5 and 6, database users remain uninformed that the publisher has retracted one article and issued an erratum for the other. As was seen in Chap. 6.6, the erratum for Case 6 contains the complete revised text of the article. A user of *The Philosopher's Index* is not alerted to the new, fully augmented version of the article contained within the erratum.

From the perspective of promulgating the corrections of the scholarly record, the representation of the two articles in *Argumentation* fares only slightly better in the other major philosophy database, *PhilPapers*. There are database entries for the two articles, and also database entries for the retraction and the correction, but there are no two-way links to tether the entries for corrections with the corresponding article entries. Furthermore, the text of the two database entries for the articles have not been updated to inform readers that the two articles in *Argumentation* have been corrected by the publisher.[1]

Databases form the primary pathways for students and researchers to find items of the published literature. When an entry for a published item is not updated to show that a publisher has changed the status of an item, the databases thereby commend research that has authoritatively been established as deficient. There is no reason why the discipline of philosophy should fall short of the standards found in the biomedical sciences regarding *The Database Problem*. In addition to updating entries to reflect all corrections issued by a publisher, a further improvement to *The Philosopher's Index* and to *PhilPapers* would be to follow the model of *MEDLINE* for updating entries to reflect when an article has been re-issued in another form, such as in an anthology or in revised form as a contribution to a monograph.

---

[1] For every database entry in *PhilPapers*, there appears at the bottom of each page a section titled "Similar Books and Articles" listing 15 other works in the database that happen to have similar titles. In the case of the entries for the two articles in *Argumentation*, the relevant retraction or correction does show up in that section, but the connection appears to be accidental and merely due to similarity in title. No updating of the entries for the articles themselves has been made to reflect the change in status of the articles.

8.2 The Anthology Problem

Should deficient articles be retracted a second time when they have been re-added to the scholarly record through republication? Occasionally a journal article will be published a second time when it is anthologized in an edited collection, either by the same publisher or a different one. When the original article is retracted, some editors and publishers apparently hold that the statement of retraction for the first published version is sufficient to cover the second published version as well. In my view, a second statement of retraction is needed to cover the second, anthologized version to provide a complete and reliable correction of the scholarly record, since a researcher could use the more recent anthologized version without knowing that the first version has been retracted. Unfortunately, the practice of leaving the status of the second published version unchanged is relatively common. This phenomenon was seen in Chap. 6.8 with Cases 8 and 8*, when the publisher of the later anthologized version did not issue a retraction but instead stopped sales of the anthology containing Case 8*, after being informed that the original publisher had issued a retraction for Case 8. Ceasing to distribute a work is not a correction of the scholarly record, however. Without a public declaration to the research community of a reason for the action, quietly halting sales is indistinguishable to an observer from the normal lifecycle of a book going out of print. A statement of retraction by one publisher should not be viewed as "covering" the second publication of the same material by a different publisher, in a different venue, at a later time.

A variant of the *Anthology Problem* occurs when the second published version of an item is subject to a change in status by the publisher, but the first version remains uncorrected. In Chap. 4, I noted two such cases. In the first, a portion of an article published in a journal from Fordham University Press in 2006 (Case 28) was republished in part in an edited collection by Leuven University Press in 2009 (Case 39), and only the latter received a change in status, when Leuven University Press issued an expression of concern in 2010. Similarly, a portion of a chapter in a volume published by Harrassowitz Verlag 2007 (Case 33) was republished in a volume by Felix Meiner Verlag in 2011 (Case 43), and only Felix Meiner Verlag issued an expression of concern. In both cases, one and the same material was added twice to the scholarly record by different publishers, and the change in status was only issued by the second publisher.

Philosophy is not the only humanities discipline struggling with the *Anthology Problem*. In 2014, *Journal of Markets & Morality* (*JMM*) issued a retraction for plagiarism for a 2005 article on the topic of early modern scholastic economic theory (Anonymous 2014). The electronic version of the article was also removed by the publisher, and a red strikethrough line was added to the title and table-of-contents pages of the issue to indicate further the status change (*Journal of Markets & Morality* 2005: 161, 163). In an editorial titled "Plagiarism in a Digital Age," the editor explained that during the process of preparing to republish the article as part of a planned book, the editors had discovered "unattributed dependence" in "a number of direct, substantial, and nearly verbatim sections" (Ballor 2014: 350, 351). This discovery not only led to the retraction of the *JMM* article for plagiarism but also prevented the article's reappearance in the planned volume, which was eventually published—minus the plagiarized article—a year later (Luis de Molina 2015). Even though the detection of the substantial defects of the 2005 article was made in time to prevent the further dissemination in the 2015 anthology, it turns out that the article had already been anthologized in an earlier 2007 collection issued by Lexington Books, an imprint of Rowman & Littlefield (Grabill 2007: 111–135). When a retraction was sought for the republication of the plagiarized material in the 2007 anthology, an editor at Lexington denied the request to issue a retraction for the chapter, but told a journalist, "Yes, we did stop selling this book due to plagiarism" (Palus 2015; see also Dougherty 2015).

In this scenario, the correction of the scholarly record was only partially successful. On the positive side, the first version of the plagiarized article was retracted, and its potential re-appearance in the 2015 anthology was forestalled. On the negative side, the republished version of the plagiarized material in the 2007 anthology remains unretracted. Furthermore, although the 2015 anthology avoided the planned republication of the plagiarized chapter, the anthology nevertheless managed to commend to its readers other plagiarized material. The introductory essay in the anthology references and quotes another publication by the same author of record of the retracted *JMM* article, namely a 1998 chapter that was later retracted by another publisher because "direct reference and citation of the works of other scholars is often inconsistent and in some cases totally lacking" (Anonymous 2015). The new anthology was not entirely successful in directing readers away from plagiarized content.

The *Anthology Problem* can be solved by issuing discrete statements of retraction each time the same plagiarized item has been added to the scholarly record by a publisher. The presence of distinct corrections for each instance is more likely to succeed in informing readers that a publication is deficient or untrustworthy. The lax practice in many humanities disciplines—including philosophy—to allow the single retraction of the first or last published version of an item to extend actively or retroactively to other published versions of the same item is not sufficient.

8.3 The Platform Problem

Similar to the *Anthology Problem* is the *Platform Problem*, which occurs when a publisher's correction does not follow the corrected work through each of a publisher's venues for disseminating that work. The *Anthology Problem* can be conceived as a vertical problem: the correction of a work does not extend to the downstream republication of that work in subsequent years after the initial publication. The *Platform Problem* can be conceived as a horizontal problem: corrections do not extend through each of the ways in which a publisher diffuses the work that it publishes. Ideally, each portal through which a reader can access a work will reflect the change that a publisher makes to the status of that work, but at times a work will remain without an update on one or more portals. Such portals are many: an article may exist in the print pages of a journal as well as on the electronic proprietary platform of the publisher (e.g., SpringerLink, Taylor & Francis Online, or ScienceDirect) or on a contracted multi-publisher platform (e.g., JSTOR, Questia, or Scribd). A book may exist in print and in a variety of ebook forms, and also issued through the proprietary electronic platform of a publisher (e.g., Oxford Scholarship Online, or Cambridge Core) or on a multi-publisher platform (e.g., ACLS Humanities E-Book). Furthermore, previews on secondary sites like Google Books or Amazon.com also present published material to students and researchers. These venues are not always updated to reflect when a publisher changes the status of a work, or sometimes the change is made with a significant delay, thereby increasing the chances that readers will encounter a deficient work without knowing that its status has changed. A satisfactory correction of the scholarly record requires that every change that a publisher makes to the status of a work should be registered in each of publisher's proprietary or contracted venues.

Solving the *Platform Problem* requires that retractions and other corrections be reflected with minimal delay across all portals through which readers access published works. This problem is solvable in principle given a publisher's control of a work in various platforms, either through ownership of the platforms themselves or through contracts with distributing parties. Publishers who are committed to the integrity of the scholarly record can insure that corrections follow a work in its various presentations. Chapter 4 gave examples of corrections by publishers that followed works in their appearance in JSTOR (Cases 10 and 29), Google Books (Case 34), and Amazon.com (Case 26). Not all secondary venues are equally successful in showing a publisher's change to the status of a work, however. Thus, for example, a red watermark stating "THIS ARTICLE HAS BEEN RETRACTED" is found on each page of the PDF version of the Case 35 article that is hosted on the publisher's website, but no such watermark appears on the version of the same article on JSTOR. The Case 4 article was entirely removed from the publisher's website and substituted with a retraction statement, but the article is present on JSTOR without any acknowledgment of its change of status. In these two cases, readers who access the article on JSTOR are therefore not informed that the publisher has changed the status of the article. Furthermore, the text of the retraction by Brill for Case 9, discussed in Chap. 6.8, is found on the webpage for the volume on Google Books under an electronic fold titled "More »" (Google Books 2016), but the retraction issued by Brill for Case 8 is not found on the Google Books webpage for that volume. Even though publishers have achieved some success in solving the *Platform Problem*, published corrections are not uniformly reflected in all platforms, and to that extent the promulgation of corrections of the scholarly record is limited.

8.4 The Repository Problem

The proliferation of new venues for researchers to access published works independently of publishers is arguably the most significant threat to keeping researchers up to date about any change in the status of a work. These venues, wholly outside governance of publishers, separate a work from the ongoing stewardship by its publisher, thereby to a great extent limiting the efficacy of publishers' corrections.

Consider, for example, a scenario in which a reader wishes to access a copy of an article that—unbeknownst to the reader—has been subject to a retraction issued by a publisher for plagiarism. On the traditional route, the reader would log in to the publisher's website, often with an institutional or a personal subscription, and then download a PDF of the article. In doing so, the reader would optimally discover the publisher's statement of retraction tethered to the article. In this scenario, the publisher's activity of promulgating the correction of the scholarly record would suffice to inform the researcher about the change in status of the article. Nevertheless, accessing articles in this traditional way may no longer be the most typical manner for present-day researchers to access published work.

There are many ways by which a reader could access the article independently of the publisher's proprietary platform. A reader might download the article from Sci-Hub, the widely used illegal repository containing around 62 million scholarly articles, which represents over 85% of all paywalled research articles (Himmelstein et al. 2017; Greshake 2017). Some researchers use Sci-Hub to gain access to articles in journals for which they lack institutional subscriptions, yet even some subscribers still prefer Sci-Hub simply for the ease of bringing up articles by means of DOIs alone, without having to go through the steps of inserting log-in information and passwords. In short, Sci-Hub offers a lower "click burden" (Faust 2016: 15a; see also Heathers 2016). One publishing consultant has remarked on "the interesting possibility that some of Sci-Hub's enormous traffic is generated by researchers who are indeed authorized to get access to the material they seek," and who only use Sci-Hub "because of its more congenial user experience" (Esposito 2016). Another study has noted that "Sci-Hub users in the United States seem to congregate near universities and likely have institutional access to the articles they request" (Bohannon 2016: 510). By accessing the article through Sci-Hub, a reader would encounter the article separated from its retraction, with no indication that its status has been authoritatively changed by the publisher.

Researchers who have concerns about using the Sci-Hub or other popular academic piracy websites can find copies of retracted articles freely available online elsewhere in their final formatted versions, with no indication of the change in publication status. Authors of record commonly upload copies of the final versions of their published papers to online servers (e.g., PhilArchive, Social Science Research Network, or HAL Archives Ouvertes), academic social networks and scholarly collaboration networks (e.g., Academia.edu, Mendeley, or ResearchGate), or their personal homepages and open-access institutional repositories. This proliferation of electronic versions of articles is at times in violation of signed author agreements with publishers that seek to limit modes of distribution of published research. Frequently a PDF of the published version of record of an article will appear on a large swath of easily accessible secondary sites. A recent study concluded that 51.3% of non-open access articles on ResearchGate "infringed the copyright and were noncompliant with publishers' policy" (Jamali 2017: 241). If a reader accesses a published item outside of a publisher's platform, whether it be an early manuscript version, proof, or the final version of record, it will not, in general, exhibit any changes in status that have been imposed by the publisher since the article's publication.

Some researchers have characterized the increased influence of academic social networks in hosting publications as a positive "disruption" in academic publishing (Laakso et al. 2017: 125), but without acknowledging that this new form of distribution severely impacts a publisher's ability to promulgate any of the corrections it has issued. The founder of the academic social network site Academia.edu, philosopher Richard Price, has stated that "the mission of Academia.edu is to put every academic paper that has ever been written on the internet, available for free" by having researchers upload "non-final versions of their papers" to the site (Price 2017). One might speculate that, at the present time, most published articles are more easily available outside of a publisher's control than under it. Researchers who put out general online calls asking for a PDF of a particular article, using the hashtag #ICanHazPDF, or who request a particular article in *The Philosophical Underclass* group on Facebook, can never be certain that the results will produce the corrected version of an article rather than its deficient unaltered predecessor. Furthermore, some readers might access pirated unmodified copies of published items on the darknet using peer-to-peer networks. Readers also might access electronic versions of articles that are simply photocopied scans of the original print version and do not reflect a publisher's notification of a change in status; such is the case when users access scanned journals in the Hathitrust online database, in the text repository archive.org, or in Google Books.

Following the principles of the Digital Millennium Copyright Act, publishers sometimes will send takedown requests when copies of an article's version of record are hosted on independent online repositories, such as academic social media sites, institutional repositories, or the personal homepages of authors. These requests are seen by many researchers as controversial. In defense, publishers have invoked what has been termed here as the *Repository Problem*. After a public backlash following a series of takedown requests in 2013, the publisher Elsevier responded that their actions were necessary not only to insure accurate usage metrics, but "to protect the quality and integrity of the scientific record" (Reller 2013).

More recently, the American Psychological Association defended a round of takedown notices in the summer of 2017 by describing the action as expressive of its commitment to "preserving the scientific integrity of the research we publish, including linking the authoritative document with any ensuing corrections or retractions so that readers have the most updated information" (American Psychological Association n.d.). After some pushback from authors and as a reaction to the attending negative publicity, the association "refocused" its program to avoid targeting individual authors who have uploaded their own papers (American Psychological Association 2017; see also McCook 2017). The *Repository Problem* exists because of the increasing availability of copies of articles that are removed from the correcting reach of publishers. The online world of freely available PDFs, severed from the control of publishers, has been viewed by many researchers as liberating, but not always with the recognition that outside the traditional channels maintained by publishers the various published corrections can easily be missed.

8.5 The Pretend-It-Didn't-Happen Problem

Not all authors of record like to advertise either that their publications are deficient or that these have been retracted by publishers. Some critics have proposed that a "failure by authors to correct their mistakes should be classified as scientific misconduct" (Kamoun and Zipfel 2016: 173). The lack of significant incentives for authors of record to report errors or to acknowledge subsequent changes in the status of published articles causes many deficient works to appear reliable. One often encounters a *curriculum vitae* that lists published articles but fails to indicate that one or more of them have been retracted or corrected in some way by a publisher. Some institutions are similarly at fault, hosting online bibliographies that present the research output of their various faculties that are never updated to indicate which articles have been subject to a change in status issued by a publisher. Some authors of record will even continue to cite the original version of their articles in their later publications without referencing the issued corrections for them. One study of retractions in biomedical disciplines revealed that "18% of authors self-cite retracted work post retraction with only 10% of those authors also citing the retraction notice" (Madlock-Brown and Eichmann: 2015: 127). Dubious practices such as these lessen the influence and practical effectiveness of corrections of the scholarly record.

8.6 A Partial Answer

A partial solution has been the adoption by some publishers of the CrossMark service offered by the non-profit association CrossRef. Publishers that participate in the service can register DOI-based metadata for each article or book chapter they publish, and then attach a CrossMark logo to the electronic version of record for each article or book chapter. When online readers click on the logo, they will be directed to a page that indicates whether the document is current or has been subject to some change, such as a retraction, expression of concern, corrigendum, erratum, or new conflict-of-interest statement. This service allows publishers to re-establish the most authoritative version of a text and allows readers to know of all changes (see Meyer 2011).

The CrossMark service can be a helpful way of identifying modifications that publishers make to the version of record of a published item. Some limitations, however, include a dependency upon publishers to participate in the paid service, and for those that do, to update the metadata of an article whenever a published item has been modified. The CrossMark service originally applied to the version of record and did not apply to versions not attached to a publisher, such as non-final versions (e.g., proofs or edited manuscripts) that may appear on academic social networks or pre-print servers, but now the service allows for the registration of some manuscripts appearing on preprints servers. Presently the CrossMark service does not apply to articles without a DOI or to print versions of articles. For the articles designated in Chap. 6 above as Cases 5 and 6 that appeared in the journal *Argumentation*, the CrossMark logo appears only on the respective erratum and retraction statements, but not on the articles they are correcting. In many humanities disciplines, articles published many years or even decades earlier are still considered relevant—and therefore citable—literature, and so it may be a while before the CrossMark service covers a significant part of the most relevant literature in some fields.

Increased litigation may also have a significant impact on the *Repository Problem*. Ongoing legal challenges to Sci-Hub, ResearchGate, and other venues may result in extensive limitations on the way independent platforms distribute the proprietary content of publishers (Matthews 2017; Kwon 2017b; Chawla 2017). In 2017, the publishers American Chemical Society, Brill, Elsevier, Wiley, and Wolters Kluwer formed a coalition that sought "to take formal steps to remedy the illicit dissemination of millions of published articles on the ResearchGate site," explaining that the practice of issuing millions of take-down notices is "not a viable long-term solution" (Coalition for Responsible Sharing 2017). Notably, one of the coalition's stated objections to ReseachGate is the claim that "where corrections or retractions are issued, it fails to update articles accordingly on its site, undermining research integrity" (ibid.). The final outcome of these legal challenges will have implications for the ability of publishers to maintain control over the distribution not only of articles but also of corrections.

8.7 Informal Corrections of the Scholarly Record

As noted in Chap. 4 above, the gold standard for a correction of the scholarly record for a journal article is an unpaywalled statement issued by a publisher on a numbered, citable page that declares fully and unambiguously the manner in which a published item falls short of acceptable research standards. Such published corrections are optimally tethered electronically to the item that has been corrected. In the previous chapters of this book, I have offered numerous examples of published corrections of the scholarly record that fall short in determination, credit, or availability. The fact that publishers have an indispensable role in issuing formal corrections of the scholarly record does not mean that there are no other useful methods for declaring that an article is deficient. Certain corrections that are issued by academic editors—but without the institutional backing of the original publisher—can be designated as "informal." In Chap. 4, I examined four examples of this practice for Cases 37, 39, 41, and 42. Informal corrections do provide some measure of rectification, but they do not constitute an authoritative alteration of the version of record. Informal corrections have a less than optimal chance of informing readers of the deficiencies of a published article, but in cases where a publisher does not act, they are sometimes cited by researchers to protect the downstream literature.

Various types of informal corrections of the scholarly record exist. Since 2011, the philosophy annual *Heidegger Studies*, for example, has published in each issue a list titled “Errata and Omissions” for problems in the editions and translations of the works of Martin Heidegger that are found in volumes issued by other publishers (*Heidegger Studies* 2011–). These published errata do not modify the versions of record of the respective editions and translations. Yet, since they appear in a venue for a specialized audience—the community of Heidegger researchers—there is an increased chance that the relevant researchers will be exposed to the corrections. Perhaps the approach to such informal corrections by *Heidegger Studies* is inspired by the longstanding practice in some disciplines for book reviewers to offer a list of misspellings, omissions, or other errors.

Informal corrections can be valuable in situations where a genuine author has accidentally published in a pseudo-academic, predatory, or vanity publisher. Since some of these outlets will either charge authors exorbitant retraction fees or refuse to allow their work to be withdrawn, one proposed solution is for authors themselves to “upload the retraction letters attached to their papers on institutional or personal archiving services” (Balehegn 2017: 99). Sometimes the name of a well-established journal is “hijacked” by a dubious press, thereby leading unsuspecting authors to submit their manuscripts to the *Doppelgänger* journal. The recent hijacking of the *International Journal of Philosophy and Theology* provides a distinctive example of this phenomenon (ten Have and Gordijn 2017: 159–160). In situations where an author has mistakenly published an article in a dubious outlet, an informal retraction might be the only recourse for her or him, apart from undertaking a costly legal action with great uncertainty of success.

Other members of the academic community can be involved in informal corrections of the scholarly record. In 1998, American anthropologist Charles F. Urbanowicz pointed out that the 1973 book *The Last of the Tasmanians* was not novel, since it simply repackaged with a new title and under a new author of record much of the 1884 ethnological account *The Lost Tasmanian Race* by genuine author James Bonwick. Even though the 1973 book was never retracted, an informal correction of the scholarly record occurred when the National Library of Australia updated the catalogue entry to declare, “Plagiarized with slight modernization of style, from J. Bonwicks” (National Library of Australia n.d.). In discussing this scenario, David Koepsell rightly notes, “having such a reference for one’s book harms one’s own academic and scientific standing” yet when no corrections are issued “science is harmed” (2017: 34). The annotation by the National Library of Australia that disclosed the plagiarism now also appears in the description of the book in Google Books (n.d.), further exhibiting that even in absence of a formal retraction by the publisher, influential stakeholders in the academic community can provide informal corrections of the scholarly record.

8.8 Additional Disclosures of Corrections of the Scholarly Record

Even in the best-case scenario where a publisher issues a retraction that is maximal in determination, credit, and availability, and where the change in status of the article is fully represented in the specialized databases of the relevant discipline, there are further ways to disclose to the research community that a certain article is deficient. In what might be considered a distinctive type of editorial, short essays that explain the process by which a plagiarized article came to be published in a previous issue sometimes are authored by journal editors. Editorials of this genre can be found in all major disciplines, and they are customarily independent of the published retractions that also appear in the affected journals.[2] Often providing a context for the published retractions, these editorials generally recount the process that lead to the discovery of the plagiarism, identify counter-measures the journal will take to minimize the risk that other plagiarized articles will appear in future issues, and express regret that the readers of the journal were exposed to plagiarism. Sometimes these editorials warmly invite readers to send any information about questionable publications, thereby offering support to potential whistleblowers. In some cases, the editors who have issued such editorials supplement them by publishing in full the previous correspondence between the various parties, including the whistleblowers, journal editors, and the plagiarists themselves, and this correspondence can reveal details about the oftentimes lengthy background processes that culminate in a published correction (e.g., Rydzyński 2006; Brentlinger et al. 2009; cf. Autor 2011). The reasons for such editorials are likely manifold, such as the belief that "the most effective means of plagiarism abatement and enforcement is public disclosure and discussion of cases of misconduct" (Lewis et al. 2011: 493; see Kock and Davison 2003.). Editorials of this genre increase awareness of the change in status of an article, and they support the correction of the scholarly record by disclosing the correction to a wider audience.

The rare appearance of one such editorial in a philosophy journal in 2015 indicates that the indifference to academic plagiarism in some quarters of philosophy is lessening. The editor of the Springer philosophy journal *Topoi*, Fabio Paglieri, published an editorial about a case of academic plagiarism discovered in a previous issue of the journal. With a blend of condemnation and invective, Paglieri warned:

> In short, if we catch you plagiarizing, we will banish you forever from this journal. Moreover, I will personally take steps to spread this ostracism as widely as possible [...]. But there are other ways of blacklisting plagiarists, e.g. online publicity, and these will be strenuously pursued [...]. Moreover, even attempted plagiarism will be publicly denounced, once discovered: if your submission turns out to contain plagiarism, not only it will be rejected and you will become *persona non grata* to our journal; we will also do our best to

---

[2] Examples from various disciplines include: Diokno 2010 (urology); Höffken and Gabbert 2013 (oncology); Ballor 2014 (economics); Sheppard 2016 (environmental science); Martin 2007 (innovation studies); Robinson et al. 2009 (animal nutrition); Laine 2017 (internal medicine); Van Steirteghem and Williams 2011 (reproductive biology).

> alert the whole academic community of the fraud you tried to commit, so that others can either refuse to deal with you, or at least approach such dealings with caution. (2015: 5)

Paglieri underscored his commitment to the public character of retractions, arguing for the necessity that they be broadcast widely:

> [T]he retraction of a plagiarized article should never be a sordid affair to hush up, but rather a public event that attracts as much attention as possible. It is a civil rite to be consumed in the *agora*, not behind closed doors. It is not enough to cause maximum harm to the plagiarists; everybody must also know that this is what happens to plagiarists, once caught. The publicity of the punishment is crucial. (4)

Such an approach blends two distinct courses of action: correction of the scholarly record and the punishment of plagiarists. In the last decade there has been a general shift away from absolute author bans, on the view that journal editors should concern themselves with correcting the published literature and leave the matter of punishment to the offender's home institution or some formal investigating body.[3]

Perhaps inspired by the increasing reliability of pre-publication manuscript evaluation software, improved peer-review methods, and the requirement that authors make public their data in repositories, some editors in the sciences now offer second chances to previous offenders rather than issuing bans. The editor of the journal *Nature Plants*, Chris Surridge, published an editorial in 2017 that defended the journal's decision to publish new work by an author who had several articles retracted for significant data problems. Surridge explained his decision, stating, "we have jurisdiction only over what is happening in front of us: that is the particular study under consideration. We must in the first instance trust the work as it is presented to us, and when serious problems arise we must trust scientific institutions to investigate and resolve them appropriately" (2017: 1). In responding to a journalist, Surridge added, "It is not our role to investigate scientific misconduct or determine appropriate sanctions" (Kwon 2017a). Nevertheless, author bans need not be viewed as an ongoing punishment meted out by editors for bad behavior, however; they could be seen as a self-protective measure that editors might institute to insure the reliability of articles appearing within the pages of their journals.

In contrast to the permissive approach exemplified by Surridge and others, there are some who call for the criminalization of all research misconduct, including plagiarism, and they justify this view by pointing to the failure of self-regulation by researchers, grant agencies, and educational institutions. One defender of this approach states, "I propose that we begin with the existing criminal code with its various felony classes and sentences" (Hadjiargyrou 2015: 28).

In my view, the best course is to maintain a strict distinction between the role of publishing corrections of the scholarly record and the role of issuing punishments for research misconduct, and a journal editor need not fulfill the latter role. Editorials that discuss the retraction of articles on account of plagiarism can disclose the cor-

[3] For an earlier policy by Springer that mandated 5-year banning of authors who submitted plagiarized manuscripts, see Springer 2009: 10. A revised Springer policy restricts banning to "exceptional cases" (Springer 2013: 7).

rection of the scholarly record more widely to the research community. By contextualizing the published correction of articles, the editorials can furthermore be viewed as essentially instructive to the community and only incidentally punitive to the plagiarist.

8.9 Solutions

Some researchers lament that retracted articles are still cited in later literature by other researchers who are apparently unaware that publishers have changed the status of the cited articles. In the words of one research team, "retracted research frequently continues to be cited as if the record has not, in fact, been set straight" (Madlock-Brown and Eichmann 2015: 128). The ongoing citation of retracted papers, in the view of some commentators, is a serious issue because the practice "spreads misinformation throughout the scientific literature, providing a false premise for future research, and thus seriously affecting the advancement of science" (Cosentino and Veríssimo 2016: 677).[4] This practice of continued citation of retracted and corrected articles, as if no changes have occurred, has been described by other theorists as a situation where error is compounded to the greater detriment of science: "A manuscript that references another manuscript that contains errors in effect promulgates the error deeper into the literature, biasing the downstream literature" (Teixeira da Silva and Dobránszki 2017: 65). The widespread phenomenon of citing retracted articles, as if the versions of record had not been corrected, is much lamented but not sufficiently understood. In my view, the major contributing causes are the presently unresolved *Database*, *Anthology*, *Platform*, *Repository*, and *Pretend-it-Didn't-Happen* problems discussed above.

The *Database*, *Anthology*, and *Platform* problems are each relatively easy to solve. The way out of the *Database Problem* is the adoption of the practice of the biomedical sciences of updating database entries for articles to reflect any subsequent retractions or other corrections. This work can even involve crowdsourcing by database users, as the U. S. National Library of Medicine invites researchers to submit requests to correct any errors or omissions found in the *MEDLINE* database.[5] The most straightforward solution to the *Anthology Problem* would be if the copyright-holding publishers were to insert in all permissions to republish a contractual requirement necessitating that secondary publishers will reflect the change in status

[4] See also Bar-Ilan and Halevi 2017: 548 "Continued citations, or post-retraction citations, of articles that were withdrawn especially due to plagiarism, data falsification or any other unethical practices interferes with the process of eliminating such studies from the literature and research overall."

[5] One study estimates that retraction statements are linked to retracted papers in 89% of cases in the PubMed database (Schmidt 2018: 326). This less-than-perfect success rate still compares favorably to the 0% linkage rate in the two major philosophy databases discussed above. Schmidt states that in PubMed, published statements of retraction "are predominantly electronically linked to the respective RPs [retracted papers]. In other databases, this is seldom the case" (319).

of the work should any retraction, erratum, corrigendum, or expression of concern ever be issued. Finally, the solution to the *Platform Problem* is an increased diligence on the part of publishers in the prompt promulgation of a change in status of works across all platforms for which they exercise significant ownership or contractual authority. There is reason to be optimistic that the *Database*, *Anthology*, and *Platform* problems can be solved, as the solutions proposed here all fall within the scope of either the database operators or publishers in their daily activities. Furthermore, some disciplines, such as those in the biomedical sciences, fare quite well in these matters when contrasted to humanities disciplines in general (and philosophy in particular).

Overcoming the *Pretend-it-Didn't-Happen Problem* and the *Repository Problem* will require short-term and long-term interventions. If institutions allow researchers to continue to take laurels for retracted work, the incentive for the status quo will largely remain. A short-term intervention would be for the institutional homes of authors of record to treat the dissemination of uncorrected versions of corrected works to be a form of misconduct by authors of record. Similarly, institutions could treat as misconduct those acts whereby authors of record cite their own works without referencing the relevant corrections in published articles, in research grant applications, and on any online institutionally hosted *curriculum vitae*. A long-term intervention to these problems is the creation of a comprehensive database of retractions that allows researchers to submit the names of authors of record to see whether a publisher has changed the status of a published work. Funded by the MacArthur and Arnold foundations, such a database is currently under construction and it will assist in promulgating corrections of the scholarly record and mitigate the harm of the problems addressed in this chapter.[6]

Should researchers cite retracted articles by others in their own publications, or should deficient work be simply ignored in the subsequent literature? Some academic librarians argue that as a rule for maintaining information literacy, retracted articles should never be cited (Thielen 2018: 188). Such an approach is too restrictive, however. In some situations, a plagiarized article has been published in a venue with a much higher profile than the venue of original source publication, and therefore the plagiarized article has received many more citations in the subsequent literature. Case 4 discussed in Chap. 4.3.1 was an article that appeared in a high-profile international journal in medieval philosophy and theology that plagiarized from a revised dissertation published 6 years earlier by a small academic press in Finland. The article was cited frequently in relevant literature that had appeared in the nine nears between publication and retraction, but the published dissertation received comparatively few citations. In cases such as this one, the joint citation of the plagiarized article with its retraction assists in promulgating the correction of the

[6] For background on the *Retraction Watch Retraction Database*, see McCook 2015, 2018. The need for a reliable retraction database is noted by Madlock-Brown and Eichmann 2015: 135, who reference the no-longer-updated *Retraction Database* maintained online at Rutgers University. For an account of the now-defunct *Déjà vu* database, see García-Romero 2014.

scholarly record. A bibliographic entry for Case 4 could be composed as the following:

> [Name of Author of Record]. "The Origins of Probabilism in Late Scholastic Moral Thought: A Prolegomenon to Further Study," *Recherches de Théologie et Philosophie médievalés* 67 (2000): 114–157. Retracted for plagiarism in: "A Note from the Editorial Board," *Recherches de Théologie et Philosophie médievalés* 76 (2009), v–vi.

Citing both an influential article and its published retraction together can assist future students and researchers in knowing about the quality of a work. Supplementing such a reference with a quotation from the retraction statement may also be helpful in contextualizing the retraction, and doing so is especially valuable if the article has been frequently cited in the literature and the retraction is not well known. If the published retraction statement is deficient in credit by not identifying the original sources that were appropriated in the plagiarism, a conscientious researcher could provide such information with the citation of the retraction. These citation practices are not common however; very rarely in humanities disciplines is a statement of retraction also cited in print when the retracted work is cited.

A related difficulty is how a researcher should deal with published items of the scholarly record that are demonstrably deficient but their status has not been changed through retraction or through any other kind of correction by editors or publishers. If the version of record of such an article remains unchanged, how should such an article be treated in the literature? Merely avoiding references to the article is not always an option, particularly if the article is often cited in the relevant literature. For manuscript authors simply to avoid citing such an article could give the mistaken appearance that they are not familiar with the established literature on a particular subject, and this appearance could lead to the hasty editorial rejection of what are high-quality manuscripts. A footnote or a bibliographical entry is generally not a place to set forth an accusation of plagiarism or any other kind of research malpractice. Yet, some authors follow this practice (Stern 2018). A safer alternative is for manuscript authors to provide a single quotation from the original source and then exhibit that the same text appears, without attribution, in a second, later source. Doing so credits the original author and also discloses that the text appears in the later literature with a different author of record, without making any explicit accusation of wrongdoing. A text that follows this practice could look like this:

> Barbiero (2004) explains Gilbert Ryle's notion of *knowing-that* by stating that it "is characteristic of the person learning a skill through explicit instruction, recitation of rules, attention to his or her movements, etc. While such declarative knowledge may be needed for the acquisition of skills, the argument goes, it no longer becomes necessary for the practice of those skills once the novice becomes an expert." Similarly, [—] et al. (2009) write that it: "is characteristic of the person learning a skill through explicit instruction, recitation of rules, attention to his or her movements, etc. While such declarative knowledge may be needed for the acquisition of skills, the argument goes, it no longer becomes necessary for the practice those skills once the novice becomes an expert" (408–409).

A single example like this one will indicate to manuscript peer-reviewers that a researcher is familiar with the existing relevant literature and at the same time the researcher will avoid overtly or unqualifiedly commending deficient but uncorrected literature. The citation practices of researchers play an indispensable role not only in keeping the scholarly community informed about the change in status of corrected articles but also about articles that might benefit from a correction. Dealing with the aftermath of plagiarism and other research defects is not something simply to be relegated to editors and publishers, but researchers themselves have a responsibility for handling defects in the published literature through their careful citation practices.

On occasion—particularly in the empirical sciences—an article on which later articles are substantially dependent is retracted, and this scenario creates a domino effect in which secondary articles must then be corrected (McCook 2016). Researchers who learn that their work is based on the deficient publications of others will be faced with the question of either initiating self-corrections with editors and publishers or pursuing internal corrections of the scholarly record in their later publications. These decisions should be based on the degree of dependency that later work has on the prior deficient publications, and in many instances internal corrections will suffice.

8.10 The Future of the Scholarly Record

The words *publish* and *publication* are cognates of the Latin *publicare*, "to make public, make generally known" (Glare 1982: 1512). To publish one's research is to disclose widely to others what one has learned and discovered. As noted at the beginning of this book, the medieval motto *To contemplate the truth and to share with others the fruits of that contemplation* has been aptly applied to the present-day endeavor of academic publishing, which today remains the primary means of disclosing research within the world of learning. Most—if not all—researchers enjoy having their works cited by other researchers in the years that follow initial publication; such citations testify to the influence of their publications within the research community.

Despite these ideals, a significant portion of published research in all fields eventually turns out to be unreliable. For much of that deficient research, internal corrections suffice, as other researchers refute or expand upon initial findings in later publications. For some defective research, however, the more drastic intervention of external corrections are required, and editors and publishers issue retractions and various other kinds of corrections to insure the reliability of the scholarly record. In this book, I have critically analyzed the manner in which corrections by editors and publishers are used in many disciplines, and I have argued for a refinement of the ways in which published corrections of the scholarly record are disseminated in the world of learning. Both the deficient implementation of corrections by editors and publishers and the deficient promulgation of them produce an illusory pre-lapsarian scholarly universe free from the needed retractions, errata, corrigenda, and expressions of concern. Research conducted in such an illusory pre-lapsarian universe invariably creates problems in the downstream literature, as researchers unwittingly cite defective works on the assumption that they are reliable.

On the other hand, many editors and publishers understand what is at stake and are committed to their gatekeeping responsibility in the dissemination of knowledge. Many researchers find that working with these editors and publishers to restore the integrity of the scholarly record—after plagiarism and other forms of research misconduct have been discovered—is indeed a great privilege. The world of learning has historically responded well to the changing landscape of the dissemination of knowledge, so there are reasons to be hopeful.

References

[—], et al. 2009. Subjectivity out of irony. *Semiotica* 173: 397–416.

American Psychological Association. n.d. Pilot: Monitoring of unauthorized internet posting of journal articles. http://www.apa.org/pubs/journals/resources/unauthorized-internet-posting.aspx. Accessed 6 July 2018.

———. 2017. APA refocuses pilot program aimed at curtailing unauthorized publication of journal articles. http://www.apa.org/news/press/releases/2017/06/curtailing-journal-articles.aspx. Accessed 6 July 2018.

Anonymous. 2014. The previously published article [...]. *Journal of Markets & Morality*. http://www.marketsandmorality.com/index.php/mandm/article/view/391. Accessed 6 July 2018.

———. 2015. *It has been brought to our attention [...]*. Leiden: Brill. https://brill.com/view/title/159. Accessed 6 July 2018.

Autor, David H. 2011. Letter. *Journal of Economic Perspectives* 25 (3): 239–240. http://pubs.aeaweb.org/doi/pdfplus/10.1257/jep.25.3.239. Accessed 6 July 2018.

Balehegn, Mulubrhan. 2017. Increased publication in predatory journals by developing countries' institutions: What it entails? And what can be done? *International Information & Library Review* 49 (2): 97–100.

Ballor, Jordan J. 2014. Plagiarism in a digital age. *Journal of Markets and Morality* 17 (2): 349–352.

Barbiero, Daniel. 2004. Tacit knowledge. In *Dictionary of philosophy of mind*. https://sites.google.com/site/minddict/knowledge-tacit. Accessed 6 July 2018.

Bar-Ilan, Judit, and Gali Halevi. 2017. Post retraction citations in context: A case study. *Scientometrics* 113 (1): 547–565.

Bohannon, John. 2016. Who's downloading pirated papers? Everyone. Data from the controversial website Sci-Hub reveal that the whole world turns to it for journal articles. *Science* 352 (6285): 508–512.

Brentlinger, Paula E., et al. 2009. Plagiarism. Plagiarism: Authors' response. Plagiarism: Editor's note. *Transactions of the Royal Society of Tropical Medicine and Hygiene* 103 (8): 855–857.

Dalmeet Singh Chawla. 2017. Publishers take ResearchGate to court, alleging massive copyright infringement. *Science*. October 6. http://www.sciencemag.org/news/2017/10/publishers-take-researchgate-court-alleging-massive-copyright-infringement. Accessed 6 July 2018.

Coalition for Responsible Sharing. 2017–. http://www.responsiblesharing.org/coalition-statement. Accessed 6 July 2018.

Cosentino, A. Mel, and Diogo Veríssimo. 2016. Ending the citation of retracted papers. *Conservation Biology* 30 (3): 676–678.

Diokno, Ananias C. 2010. Editorial comment: Hippocratic oath and plagiarism. *International Urology and Nephrology* 42 (3): 709.

Dougherty, M. V. 2015. Plagiarism. In *Thomistica: A site for the academic study of St. Thomas Aquinas*, May 27. https://thomistica.net/news/2015/5/27/plagiarism. Accessed 6 July 2018.

EBSCO. *Research databases*. https://www.ebsco.com/products/research-databases. Accessed 6 July 2018.

Esposito, Joseph. 2016. Sci-Hub and the four horsemen of the internet. *The Scholarly Kitchen*. March 2. https://scholarlykitchen.sspnet.org/2016/03/02/sci-hub-and-the-four-horsemen-of-the-internet. Accessed 6 July 2018.

Faust, Jeremy S. 2016. Sci-Hub: A solution to the problem of paywalls, or merely a diagnosis of a broken system? *Annals of Emergency Medicine* 68 (1): 15a–17a.

García-Romero, Antonio. 2014. A bibliometric analysis of plagiarism and self-plagiarism through déjà vu. *Scientometrics* 101 (1): 381–396.

Glare, P.G.W., ed. 1982. *Oxford Latin dictionary*. Oxford: Clarendon Press.

Google Books. n.d. *Entry for Last of the Tasmanians*. Sydney: Shakespeare Head. 1973. https://books.google.com/books/about/The_Last_of_the_Tasmanians.html?id=Gc1lAAAACAAJ. Accessed 6 July 2018.

———. 2016. In *Entry for Social studies of health, illness and disease: Perspectives from the social sciences and the humanities*, ed. Peter Twohig and Vera Kalitzkus. Leiden: Brill/Rodopi., 2008. https://books.google.com/books?id=JJt4Fhqy_0QC&source=gbs_navlinks_s. Accessed 6 July 2018.

Grabill, Stephen J., ed. 2007. *Sourcebook in late-scholastic monetary theory*. Lanham: Lexington Books/Rowman and Littlefield.

Greshake, Bastian. 2017. Looking into Pandora's box: The content of Sci-Hub and its usage. *F1000Research* 6: 541:1–15. https://doi.org/10.12688/f1000research.11366.1. Accessed 6 July 2018.

Hadjiargyrou, Michael. 2015. Scientific misconduct: How best to punish those who consciously violate our profession's integrity? *Journal of Information Ethics* 24 (2): 23–30.

HAL Archives Ouvertes. https://hal.archives-ouvertes.fr. Accessed 6 July 2018.

Heathers, James. 2016. Why Sci-Hub Will Win. *Medium*, May 2. https://medium.com/@james-heathers/why-sci-hub-will-win-595b53aae9fa. Accessed 6 July 2018.

Heidegger Studies. 2011–. Errata and omissions in recent English translations of the *Gesamtausgabe*. 27 (2011): 261–262; 28 (2012): 267; 29 (2013): 293; 30 (2014): 229; 31 (2015): 221.

Himmelstein, Daniel S., Ariel R. Romero, Stephen R. McLaughlin, Bastian Greshake Tzovaras, and Casey S. Greene. 2017. Sci-Hub provides access to nearly all scholarly literature. *Peer Journal of Preprints* 5: e3100v1: 1–27. https://doi.org/10.7287/peerj.preprints.3100v1. Accessed 6 July 2018.

Höffken, K., and H. Gabbert. 2013. Plagiarism revisited. *Journal of Cancer Research and Clinical Oncology* 139 (2): 179.

Jamali, Hamid R. 2017. Copyright compliance and infringement in ResearchGate full-text journal articles. *Scientometrics* 112 (1): 241–254.

Journal of Markets & Morality. 2005. Contents. 8(1): 161–163. http://www.marketsandmorality.com/index.php/mandm/article/view/391/pdf. Accessed 6 July 2018.

Kamoun, Sophien, and Cyril Zipfel. 2016. Scientific record: Class uncorrected errors as misconduct. *Nature* 531 (7593): 173. https://doi.org/10.1038/531173e Accessed 6 July 2018.

Kock, Ned, and Robert Davison. 2003. Dealing with plagiarism in the information systems research community: A look at factors that drive plagiarism and ways to address them. *MIS Quarterly* 27 (4): 511–532.

Koepsell, David. 2017. *Scientific integrity and research ethics: An approach from the ethos of science*. Cham: Springer.

Kwon, Diana. 2017a. How journals treat papers from researchers who committed misconduct. *The Scientist*. August 18. http://www.the-scientist.com/?articles.view/articleNo/50124/title/How-Journals-Treat-Papers-from-Researchers-Who-Committed-Misconduct. Accessed 6 July 2018.

———. 2017b. Judge recommends ruling to block internet access to Sci-Hub. *The Scientist*. October 4. http://www.the-scientist.com/?articles.view/articleNo/50563/title/Judge-Recommends-Ruling-to-Block-Internet-Access-to-Sci-Hub. Accessed 6 July 2018.

Laakso, Mikael, Juho Lindman, Cenyu Shen, Linus Nyman, and Bo-Christer Björk. 2017. Research output availability on academic social networks: Implications for stakeholders in academic publishing. *Electronic Markets* 27 (2): 125–133.

Laine, Christine. 2017. Scientific misconduct hurts. *Annals of Internal Medicine* 166 (2): 148–149.

Lewis, Bruce R., Jonathan E. Duchac, and S. Douglas Beets. 2011. An academic publisher's response to plagiarism. *Journal of Business Ethics* 102: 489–506.

Luis de Molina. 2015. *A treatise on money*. Trans. Jeannine Emery. Grand Rapids: Christian's Library Press.

Madlock-Brown, Charisse R., and David Eichmann. 2015. The (lack of) impact of retraction on citation networks. *Science and Engineering Ethics* 21 (1): 127–137.

Martin, Ben R. 2007. Keeping plagiarism at bay—A salutary tale. *Research Policy* 36 (7): 905–911.

Matthews, David. 2017. Publishers seek removal of millions of papers from ResearchGate. *Times Higher Education*, October 5. https://www.timeshighereducation.com/news/publishers-seek-removal-millions-papers-researchgate. Accessed 6 July 2018.

McCook, Alison. 2015. New retraction watch partnership will create retraction database. *Retraction Watch*, November 24. http://retractionwatch.com/2015/11/24/new-partnership-will-create-retraction-database. Accessed 6 July 2018.

———. 2016. What should you do if a paper you've cited is later retracted? *Retraction Watch*, November 1. http://retractionwatch.com/2016/11/01/what-should-you-do-if-a-paper-youve-cited-is-later-retracted. Accessed 6 July 2018.

———. 2017. Following outcry, American Psychological Association "refocuses" takedown notice program. *Retraction Watch*, June 16. https://retractionwatch.com/2017/06/16/following-outcry-american-psychological-association-refocuses-takedown-notice-program. Accessed 6 July 2018.

———. 2018. Is our database missing a retraction? Tell us! *Retraction Watch*, January 18. https://retractionwatch.com/2018/01/18/database-missing-retraction-tell-us. Accessed 6 July 2018.

Meyer, Carol Anne. 2011. Distinguishing published scholarly content with Crossmark. *Learned Publishing* 24 (2): 87–93.

National Library of Australia. n.d. *Entry for Last of the Tasmanians*. Sydney: Shakespeare Head. 1973. ID: 21566816. http://trove.nla.gov.au/work/21566816. Accessed 6 July 2018.

Paglieri, Fabio. 2015. Reflections on plagiarism. *Topoi* 34 (1): 1–5.

Palus, Shannon. 2015. Two retractions cost economic historian book chapter and journal article. *Retraction Watch*, July 20. http://retractionwatch.com/2015/07/20/two-retractions-cost-economic-historian-book-chapter-and-journal-article. Accessed 6 July 2018.

PhilPapers. https://philpapers.org. Accessed 6 July 2018.

Price, Richard. 2017. Session with Richard Price founder of Academia.edu. *Quora*, February 7. https://www.quora.com/session/Richard-Price/1. Accessed 6 July 2018.

PubMed. https://www.ncbi.nlm.nih.gov/pubmed. Accessed 6 July 2018.

Reller, Tom. 2013. *A comment on takedown notices (with update)*. Elsevier, December 12. https://www.elsevier.com/connect/a-comment-on-takedown-notices. Accessed 6 July 2018.

Retraction Database. Rutgers university. http://retract.rutgers.edu. Accessed 6 July 2018.

Robinson, P.H., P. Udén, G.G. Mateos, and J. Pluske. 2009. Plagiarism, science and animal feed science and technology. *Animal Feed Science and Technology* 154: 292–293.

Rydzyński, Konrad. 2006. Message from the Editor-in-Chief: Expression of concern due to plagiarism. *International Journal of Occupational Medicine and Environmental Health* 19 (2): 149–150.

Schmidt, Marion. 2018. An analysis of the validity of retraction annotation in PubMed and the Web of Science. *Journal of the Association for Information Science and Technology* 69 (2): 318–328.

Sheppard, S.C. 2016. Handling of recent plagiarism in JER. *Journal of Environmental Radioactivity* 165: 296.

Social Science Research Network. https://www.ssrn.com. Accessed 6 July 2018.

Springer. 2009. *The Springer policy on publishing integrity*. http://www.springer.com/cda/content/document/cda_downloaddocument/SpringersPolicyOnPublishingIntegrity01_06_2009.pdf. Accessed 6 July 2018.

———. 2013. *Publishing ethics for journals: A guide for editors-in-chief, associate editors, and managing editors*. Version 1.0. http://static.springer.com/sgw/documents/1393202/application/pdf/Publication_Ethics_Guide_for_Editors_from_Springer_27052013.pdf. Accessed 6 July 2018.

Stern, Victoria. 2018. Unusual: Author uses a reference list to accuse a paper of plagiarism. *Retraction Watch*. February 5. https://retractionwatch.com/2018/02/05/unusual-author-uses-a-reference-list-to-accuse-a-paper-of-plagiarism. Accessed 6 July 2018.

Surridge, Chris. 2017. Giving research a sporting chance. *Nature Plants* 3 (17113): 1. https://doi.org/10.1038/nplants.2017.113. Accessed 6 July 2018.

Teixeira da Silva Jaime, A., and Judit Dobránszki. 2017. Compounding error: The afterlife of bad science. *Academic Questions* 30 (1): 65–72.

ten Have, Henk, and Bert Gordijn. 2017. Publication ethics: Science versus commerce. *Medicine, Health Care and Philosophy* 20 (2): 159–161.

The Philosopher's Index. http://philindex.org. Accessed 6 July 2018.

The Retraction Watch Retraction Database (Beta). http://retractiondatabase.org. Accessed 6 July 2018.

Thielen, Joanna. 2018. When scholarly publishing goes awry: Educating ourselves and our patrons about retracted articles. *Portal: Libraries and the Academy* 18 (1): 183–198.

U. S. National Library of Medicine. 2017. Errata, retractions, corrected and republished articles, duplicate publications, comments (including author replies), updates, patient summaries, and republished (reprinted) articles policy for MEDLINE. https://www.nlm.nih.gov/pubs/factsheets/errata.html. Accessed 6 July 2018.

Van Steirteghem, André, and A.C. Williams. 2011. Plagiarism, retraction and the future. *Human Reproduction* 26 (8): 1940.

Web of Science. https://www.webofknowledge.com. Accessed 6 July 2018.

Index

M. V. Dougherty, *Correcting the Scholarly Record for Research Integrity*, Research Ethics Forum 6, https://doi.org/10.1007/978-3-319-99435-2

MIX
Papier aus verantwortungsvollen Quellen
Paper from responsible sources
FSC® C105338

If you have any concerns about our products,
you can contact us on
ProductSafety@springernature.com

In case Publisher is established outside the EU,
the EU authorized representative is:
Springer Nature Customer Service Center GmbH
Europaplatz 3, 69115 Heidelberg, Germany

Printed by Libri Plureos GmbH
in Hamburg, Germany